THE OXFORD ILLUSTRATED HISTORY OF
SCIENCE

The thirteen historians who contributed to *The Oxford Illustrated History of Science* are all distinguished authorities in their field. They are:

PETER BOWLER, Queen's University Belfast

SONJA BRENTJES, Max Planck Institute for the History of Science, Berlin

JAMES EVANS, University of Puget Sound

JAN GOLINSKI, University of New Hampshire

DONALD HARPER, University of Chicago

JOHN HENRY, University of Edinburgh

STEVEN J. LIVESEY, University of Oklahoma

IWAN RHYS MORUS, Aberystwyth University

AMANDA REES, University of York

DAGMAR SCHAEFER, Max Planck Institute for the History of Science, Berlin

CHARLOTTE SLEIGH, University of Kent

ROBERT SMITH, University of Alberta

MATTHEW STANLEY, New York University

THE OXFORD ILLUSTRATED HISTORY OF

SCIENCE

Edited by
IWAN RHYS MORUS

OXFORD
UNIVERSITY PRESS

OXFORD
UNIVERSITY PRESS

Great Clarendon Street, Oxford, OX2 6DP,
United Kingdom

Oxford University Press is a department of the University of Oxford.
It furthers the University's objective of excellence in research, scholarship,
and education by publishing worldwide. Oxford is a registered trade mark of
Oxford University Press in the UK and in certain other countries

Published in the United States of America by Oxford University Press
198 Madison Avenue, New York, NY 10016, United States of America

British Library Cataloguing in Publication Data
Data available

Library of Congress Control Number: 2016955204

ISBN 978–0–19–966327–9

Printed in Italy by
L.E.G.O. S.p.A.

For Dave Rees, who was looking forward to reading this book

ACKNOWLEDGEMENTS

It was Simon Schaffer who first made me understand what an important and fascinating field the history of science was. I would like to thank him as well as my fellow students at the Department of History and Philosophy of Science in Cambridge for making my time there so invigorating and intellectually exciting. Since then I have had too many conversations with too many people in too many places about the history of science to be able to give individual thanks to everyone who has helped shape my thoughts over the years. I would, however, like to mention two in particular—Rob Iliffe and Andy Warwick. Special mention should go to my wife, Amanda Rees, without whom I would be completely lost.

I would like to thank everyone at Oxford University Press who has been involved with this project, particularly Matthew Cotton and my copy editor Tim Beck. Finally, of course, I would like to thank the chapter authors for their hard work, professionalism, and enthusiasm.

CONTENTS

x *Contents*

Introduction

IWAN RHYS MORUS

THE first book in the history and philosophy of science that I can remember reading as an undergraduate thirty years ago was Alan Chalmers's *What is This Thing Called Science?* The copy is still on my shelves somewhere. The title has stuck with me for a number of reasons. The question it asks appeals because the answers turn out to be so unexpectedly elusive and slippery. At first sight it appears obvious what science is—it's what scientists do. Very few people now would deny the critical role that science plays in underpinning contemporary life. Science, and the technological offshoots of science, are everywhere around us. Modern science does not just provide us with technological fixes, though. Its ideas and assumptions are embedded in a very fundamental way in the ways we make sense of the world around us. We turn to science to explain the material universe, and to account for our spiritual lives. We routinely use science to talk about the ways we talk to each other. But what do we really mean by science? It's a uniquely human activity, after all. At its broadest level, science sums up the ways we make sense of the world around us. It's the set of ways we interact with the world—to understand it and to change it. It is this humanity at the heart of science that makes understanding its history and its culture so important.

Despite the fact that science matters so much for our culture, we often treat it as if it were somehow beyond culture. Science is assumed to progress by its own momentum as discovery piles up on discovery. From this point of view, science often looks like a force of nature rather than of culture. It is something unique about science itself that makes it progress, and there is something inevitable about its progress. Science therefore doesn't need culture to move forwards, it simply carries on under its own steam. Culture gets in the way sometimes, but in the end science always wins. If science exists outside culture like this, then it should not really matter where it happened. This means that the history of science should not really matter either. At best, it can tell us who did what when, but that is a matter of chronicle rather than history, which tries to interpret the past as well as simply record it. Advocates of this sort of view of science often talk about the scientific method as the key to understanding its success. Some pundits now talk about peer review or double blind testing as the hallmarks of scientific objectivity, for example, despite the fact that these are cultural institutions of very recent provenance.

As a historian—obviously—I think that the history of science does matter a great deal. If we want to understand modern science we do need to understand how it developed and under what circumstances. Even the most cursory survey of the history of science should be enough to demonstrate that the scientific method itself has been understood very differently by different people at different times. A century ago, most scientists would have agreed that science progressed by accumulating observations. According to this hypothetico-deductive account, science proceeded by formulating hypotheses based on observation and testing those hypotheses through further experiment. If the hypotheses passed the test of experiment then they were confirmed as true descriptions of reality. More recently, the idea that science proceeds by falsifying rather than confirming hypotheses has gained currency. History tends to show us that in practice, it is difficult to discern any kind of consistent method in what scientists do. On the contrary, different accounts of scientific method are very much the products of particular historical circumstances. After all, no one could have argued that the modern system of peer review was at the heart of the scientific method before the cultural institutions that support that practice (like scientific journals, professional societies, and universities, for example) had appeared.

Looking at science's history suggests that the answer to the question, what is this thing called science, is that there is no single thing called science. Science is certainly not a unified and continuous body of beliefs. Neither is it captured by a single scientific method. If we want to understand what science is—the different ways knowledge has been produced both now and in the past, and how different peoples and cultures have come to hold the beliefs they have about the natural world—we simply have no choice but to look at history. To do this successfully, we need to overcome that modern condescension for the past (and for other cultures) that is often particularly prevalent in the history of science. Clearly, by the light of our current knowledge, most of what people in the past thought they knew was false. It is worth remembering, though, that by the same pessimistic induction, most of what we think we know now will turn out to be wrong by the future's standards. History of science should not be a game of rewarding winners and losers in the past, giving ticks to the likes of Galileo, or crosses to Pope Urban VIII.

In recent years, very few historians of science have tried to produce anything like a comprehensive history of science through the ages. Until fifty years or so ago such attempts were not uncommon. Historians of science then had more confidence than their current counterparts that they knew what science looked like. These kinds of histories usually began with an overview of the origins of scientific thought in Egypt and Mesopotamia. This would be followed by chapters on science in China and the Indian sub-continent before turning to ancient Greece and the beginnings of the Western scientific tradition. Chapters on Islamic science and science in the medieval West would follow. It was usually only when they arrived at the Scientific Revolution of the sixteenth and seventeenth centuries that the authors would start to provide a more detailed narrative that focused on the specific and heroic achievements of specific individuals such as Copernicus, Galileo, Kepler, and Newton. It will not have escaped the attention of those of you who have already inspected the contents page that the pattern I have just described is not entirely dissimilar to the pattern of this book. Even for historians who have been taught to regard these kinds of ambitious 'big picture' histories of science and their arcs of scientific progress with a distinctly jaundiced eye, the narrative they have established can still be difficult to avoid.

One way in which this book differs from those big picture histories from the middle of the twentieth century is obvious. This is a collaborative exercise that draws on the combined expertise of more than a dozen historians of science. I hope that gives it a breadth and a depth of analysis that was sometimes missing from previous accounts. Whilst I would not want to fall into the progressivist trap, I am also confident in saying that we now know far more about the history of science than our predecessors sixty or seventy years ago. We know more because new sources have become available over the past fifty years and because we have developed fresh ways of looking at the sources we already had. We have also started looking for science in new and different historical locations. Most importantly, though, the questions that historians of science ask have changed significantly over the past half century. Whilst historians fifty years ago were often more interested in dissecting past natural knowledge in order to identify those elements that contributed directly to modern science, historians of science now are more interested in understanding the totality of past science. We are as interested in understanding what past science meant to its original producers and their audiences as we are in understanding how it led to us.

Looking at the history of science like this, it becomes clear that beliefs and practices that we might regard as unscientific, such as religion or magic, were integral to science in the past. In most past cultures, up until and beyond the Scientific Revolution, the motivation for engaging in science at all was most often a religious one. Discovering the laws of nature was regarded as a way of directly engaging with God, or of demonstrating the existence of a Creator. Someone such as Isaac Newton would have regarded the idea that science and religion were necessarily antithetical, promoted by some contemporary science pundits, as laughable. In fact the metaphor of warfare between science and religion which has been popularized in recent books such as Richard Dawkins's *The God Delusion* is a very specific product of late nineteenth-century debates around Darwinism and would have appeared very peculiar indeed to the previous generation. In a similar way, practices that we might regard as pertaining to magic, rather than science, were an integral part of doing science in ancient China or Greece. Again, even during the Scientific Revolution, many of the proponents of the New Science (including Isaac Newton) were also keenly interested in magical practices such as alchemy or astrology, and did not really differentiate between those kinds of activities and ones that we now might recognize as scientific.

This is an illustrated history and so pays particular attention throughout to the visual cultures of the sciences. The emphasis on the importance of visual culture for understanding science that is evident in all its chapters is not just a product of the fact that this is an illustrated book though. Science throughout its history has had a complex relationship with seeing things. If I were searching for a way of trying to grasp at its most fundamental level just what kind of activity science was, I think that thinking about science in terms of seeing would be a very good place to start. For many of its practitioners and their audiences, at least, making knowledge about nature has always been concerned with trying to make visible the otherwise hidden relationships between objects and processes in the natural world. Paying attention to the visuality of science is also a good way of reminding ourselves about its material culture. Science isn't just inside people's heads. It's also in the instruments they make and manipulate and the ways they develop to make abstract relationships appear concrete. It is often embodied in the bodies of its practitioners—as in the powerful images of scientific authority conveyed by portraits of Charles Darwin, or Albert Einstein, for example.

Given the vastness of the potential field, any history of science like this one is going to be in some degree both partial and selective. The book is divided into two parts. In the first part, different chapters survey science in the ancient Mediterranean world and in the ancient East respectively. They are followed by chapters on science in the pre-modern East and in the medieval West and Middle East. This last chapter in particular is designed to emphasize the extent to which Islamic and Christian scientific traditions and practices during this period were deeply intertwined and cannot be understood in isolation. These are followed by chapters on the Scientific Revolution and Science in the Enlightenment. The chapters in this first part are broadly speaking chronological. The six chapters in the second part of the book are more thematic in character. This is partly because the volume of material that would need to be condensed in order to deal adequately with the nineteenth and twentieth centuries is so vast, but also because the increasing specialization and diversification of science during the modern era makes a thematic approach more useful and revealing. These thematic chapters will provide focused accounts of the emergence of key features of the modern scientific world-view and modern scientific institutions. Chapters will deal with the rise of experimental culture, the development of new ways of encountering nature in the field, and the emergence of new ways of thinking about the origins of life. Subsequent chapters will discuss the rise of cosmology and the development of more powerful and systematic ways of mapping the universe, and the development of a distinct culture of highly technical theoretical speculation during the nineteenth and twentieth centuries. The final chapter will look at the ways in which scientific individuals and institutions during the modern period have developed strategies and opportunities to communicate science to its growing audiences.

One important recurring theme in the early chapters is the exploration of the ways in which knowledge circulated between different cultures. James Evans explores early Egyptian, Babylonian, and Greek ideas and practices and their relationships. Greek philosophers borrowed Babylonian astrological concepts, for example, and modified them for their own purposes. He shows how new philosophies circulated around the Hellenistic and Roman worlds. Similarly, Donald Harper in his overview of science in ancient China describes the channels through which scientific knowledge was transmitted between elite philosophers and their audiences. He suggests how sciences such as cosmology developed to serve specific political interests and points to the imperial court as an important location for the circulation of knowledge. In their chapter on science in the medieval West and Middle East, Sonja Brentjes and Stephen Livesey explore how ideas and practices moved around the Mediterranean. They show how, for example, Islamic court society sought out, adapted, and built upon Greek knowledge and how this new knowledge then circulated both east and west. Dagmar Schaefer's discussion of science in the pre-modern East also emphasizes the extent to which scientific knowledge served the interests of Chinese political elites, and charts the avenues through which knowledge was circulated.

Schaefer suggests that Chinese scholars were themselves well aware of the importance of reliably disseminating knowledge. John Henry similarly starts his discussion of the Scientific Revolution by emphasizing the importance of new media in offering new ways for knowledge to circulate. Henry highlights the importance of voyages of discovery (themselves facilitated by the invention of the magnetic compass) in generating new knowledge about the natural world as well as the re-discovery of alternatives to Aristotelian philosophy. Particularly important in this account is the way the centres of scientific authority shifted away from the universities. As this happened, new ideas also developed about just how reliable knowledge should be generated. Empiricist philosophies argued for the reliability of the human senses rather than ancient authority as the proper foundation for science. As Jan Golinski argues in his chapter on science in the Enlightenment, scientific knowledge was increasingly recognized as belonging to the public sphere. Scientific experiments were performed in public and a growing print culture facilitated the circulation of new ideas and discoveries. Men of science and their activities were increasingly sponsored by the state. What science had to say about nature was also increasingly regarded as having consequences for the state of society.

All the thematic chapters in the second half of the book continue with these themes of circulation and authority. In my own chapter on the development of new experimental cultures during the nineteenth and twentieth centuries I discuss how the laboratory emerged as a new kind of space for science. During the course of two centuries laboratories were transformed from small-scale operations to massive and resource-intensive institutions. The experimental culture of the laboratory, and the authority that came with experiment, came increasingly to rely on the resources of a highly industrialized economy. Amanda Rees shows in her chapter on exploring nature that science outside the laboratory also required the mobilization of considerable resources. Geologists, naturalists, and others who carried out their science in the field had to find ways of turning the untamed spaces where they worked into sites where knowledge could be produced. Peter Bowler's survey of different attempts over the past two centuries to explain the origins of life charts the ways in which some scientists now sought to colonize new intellectual territories and compete with orthodox religion for authority. It was in the context of this new materialism that the view of science and religion as opposites first found its supporters.

Robert Smith in his chapter on mapping the universe explores the ways in which modern astronomy developed through new institutions and new technologies that transformed the ways in which the heavens were understood. Telescopes peered further into the depths of the universe, discovering and classifying new kinds of celestial objects. During this process the scale of the known universe expanded dramatically and fuelled theoretical speculation about the origins of the cosmos. Matthew Stanley's chapter charts the growth of theory alongside the rise of laboratory science from the beginning of the nineteenth century. Theorizing developed into a highly disciplined and technically challenging process that required the application of considerable mathematical expertise. By the end of the nineteenth century, the discipline of theoretical physics, distinct from experiment, had emerged. The twin theories of relativity and quantum mechanics that came to dominate twentieth-century physics spawned an industry in mathematical theory that mirrored the industrial scale of experimental big science. Charlotte Sleigh's chapter surveys the development of new strategies and new institutions for communicating science to a wider public over the past two centuries. As all the chapters have emphasized, communicating and circulating knowledge have always been central to what knowledge-makers do. By the end of the twentieth century, the business of communicating science had become big business too.

The history offered throughout this book emphasizes that science is a human activity. Science is produced by people, and since people are very much the products of culture, so is the science they produce. This is as true for twentieth-century (or twenty-first-century) theoretical physics as is was for ancient Babylonian astrology. This is a history that emphasizes the ways in which the search for understanding of the natural world was intertwined with other aspects of human culture. The quest for knowledge could be religious, or secular, utilitarian or disinterested. It was inevitably political as the order of nature was taken to provide a model for the proper order of society, or the state. The English philosopher and courtier Francis Bacon was quite right when he noted that knowledge is power. This history also emphasizes the ways in which science has always been a collective activity. The image of the lonely sage in his tower or the isolated scientist in his lab leaves the rest of the world out of the picture. It is a seductive image, but the history of science shows us that the rest of the world was still there, nevertheless, and that science could not take place without its resources, both cultural and material. Newton may have felt that his science depended on standing on the shoulders of giants, but modern science rests on the shoulders of armies of workers whose activities are usually invisible. This history of science tries to show that science is (and always has been) made by everyone, and therefore belongs to everyone.

PART I
SEEKING ORIGINS

1 Science in the Ancient Mediterranean World

JAMES EVANS

ANCIENT societies personified the gods and deified nature. This was not an irrational thing to do, for we always try to understand the unknown by analogy to things we know better. What phenomena offered a better model for the behaviour of nature and of the gods than the dynamics of a human family? Thus in the Babylonian creation story (second millennium BCE), Marduk becomes the chief god when other gods prevail on him to slay his imperious grandmother, Tiamat. He slices her body in two and sets up one part as the land, another part as the sky. In a similar way, the Greek poet Hesiod (*c.*700 BCE) described the genealogy of the gods in his *Theogony*. Ocean, rivers, Sun, and Moon are produced almost as an afterthought: the main story is Zeus's coming to power by overthrowing his father, Kronos (Saturn), just as Kronos had overthrown his own father, Uranus (Heaven). Key themes in both stories are sexual reproduction as the creative force of nature and generational conflict as a destructive force but also a mode of change. But Hesiod wrote another didactic poem, the *Works and Days*, which connects the work of the farmer and the sailor with the rhythm of the year. Hesiod uses the summer and winter solstices, the equinoxes, and the risings and settings of prominent stars to know the time of year. For example, the morning setting of the Pleiades (when the Pleiades are seen setting in the west just before the Sun comes up in the east) is a signal of late fall and the time for ploughing and sowing. Thus a theocentric view of nature could exist right alongside practical scientific knowledge.

Ancient Egypt

The Egyptians pictured the sky as a goddess, Nut, who stretched over the Earth-God, Geb. The Sun was carried each day from horizon to horizon in a boat. The Egyptians used shadow clocks to tell time by the reign of Set I (*c.*1300 BCE). The day was divided into twelve parts and the night into twelve. Filtered through the later Greek tradition, this was the source of our own twenty-four-hour division of the day and night. For telling time at night, and for telling the time of year, the Egyptians used the *decans*, thirty-six star groups ranged around the celestial sphere. Approximately once every ten days, a new decan could be seen rising in the east just before sunrise. And, in the course of a given night, the hours could be told by the rising of successive decans. Figures of Nut and the decans are sometimes found painted on the interiors of coffin lids. This is another way in which systematic astronomy could be merged with religion and ritual.

An important text of early Egyptian science is the Rhind mathematical papyrus (named after a nineteenth-century Scottish collector). This text includes a note, signed by a scribe named Ahmose, who informs us that he wrote it in the thirty-third year of the reign of king Awserre (sixteenth century BCE) but that he copied it from a work composed in the reign of king Nymatre (nineteenth century BCE). Problems illustrated in the Rhind papyrus include the arithmetic of fractions, the division of a given number of loaves of bread among ten men according to stipulated rules for unequal shares, and the solution of linear equations such as $x + (\frac{1}{5})x = 21$ (though the Egyptians did not use algebraic notation). Multiplication is treated by successive doubling. Thus to multiply 17 by 13, a scribe might write the following numbers down: 13, 2×13, 4×13, 8×13, 16×13, then add up the first and the last. Other problems involve the calculation of areas and volumes. The Egyptian rule for finding the area of a circle is to take the square of $\frac{8}{9}$ of the diameter. Since our modern formula may be expressed $Area = \frac{\pi d^2}{4}$, the Egyptian rule corresponds to using $\frac{256}{81}$ for pi. This makes pi about 3.16, compared with our 3.14—an approximation good enough for any purpose in the second millennium BCE.

Nut (sky), Geb (Earth), and Sun. Detail of a papyrus from Egypt, *c.*1000 BCE. Musée du Louvre, Egyptian Antiquities E 17401.

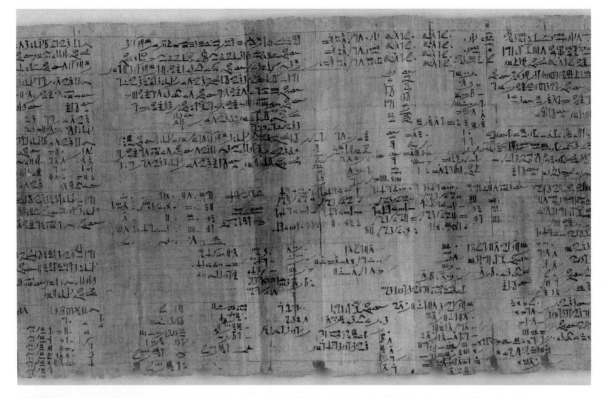

Rhind Mathematical Papyrus. British Museum 10057–8. Red ink marks the beginnings of new problems. In this portion, the scribe discusses the finding of reciprocals, the unequal distribution of goods, and the calculation of fair exchange values.

In hieroglyphic writing, the fraction $\frac{1}{2}$ had its own special symbol, as did $\frac{2}{3}$. But all other fractions were expressed by putting a small oval above the sign for a number. Modern scholars usually indicate these by an overbar. Thus, if 5 represents the Egyptian sign for 'five', the symbol $\bar{5}$ stands for 'one-fifth'. These are *unit fractions*—fractions with numerator 1. (Later on, the oval was replaced by a dot.) Aside from $\frac{2}{3}$ the Egyptians used unit fractions exclusively, so other fractions would be expanded in terms of these. The fraction we would write as $\frac{7}{8}$, an Egyptian scribe would write as $\bar{2}\ \bar{4}\ \bar{8}$ (i.e. $\frac{1}{2} + \frac{1}{4} + \frac{1}{8}$).

Babylonian Mathematics and Astronomy

Mesopotamian mathematics is well documented by texts from the Old Babylonian period (early second millennium BCE). Many of these were teaching texts and practice exercises from the scribal schools. A key feature of Babylonian mathematics is its numerical system, which uses a base-sixty notation (with some base-ten features incorporated). The medium of writing was the clay tablet, which was impressed or incised with a reed stylus. This kind of writing is called cuneiform ('wedge-shaped') because of the characteristic form of the impressions made in the clay. The numbers 1 through 9 were represented by the corresponding number of vertical wedges: ⟘2, ⟘5. Signs for 10 through 50 were made with horizontal wedges: ⟨10, ⟨40. Thus 45 would be represented by ⟨⟘. Any number between 1 and 59 could be represented in this way. The Babylonians also used a place-value convention, rather like our own. When we write 73, the 3 represents three units; the 7, seven tens. Similarly, for a Babylonian scribe, ⟨⟨⟘ ⟨⟘ could represent 24 × 60 + 45 (= 1,485). But there was no equivalent of the decimal point, so this same collection of symbols could also mean 24 × 60^2 + 45 × 60 (= 89,100)—or even 24 + $\frac{45}{60}$ (= 24.75), since fractions were written in the same way. The scribe would have to understand from context the number intended. Modern scholars use a semicolon to separate the units from the fractional part of the number, with commas separating successive sexagesimal places.

Multiplication tables and tables of the inverses of numbers (useful for doing division) are common materials. Moreover, the schools were fond of 'story problems', which could be more complicated than anything a scribe would face in real life. These include problems that we would solve by means of algebra, including the application of the quadratic formula. A small tablet from the Old Babylonian period shows a square divided by its diagonals, with an excellent approximation to $\sqrt{2}$ written on it as 1;24,51,10 (in decimal notation, 1.41421296...). The Babylonian approximation is correct to six significant figures. Another tablet from the Old Babylonian period carries a table of information about whole numbers (let us call them a, b, and c) related by $a^2 + b^2 = c^2$. The preserved portion of the tablet actually lists the values for c, b, and $\frac{c^2}{a^2}$. This demonstrates an understanding (though not a proof) of the Pythagorean theorem for right triangles a thousand years before Pythagoras. This interpretation is confirmed by a Babylonian school-text with a story problem involving a ladder leaning against a wall.

About 80 per cent of the published mathematical tablets are from the Old Babylonian period. For both earlier and later periods the record is much more spotty. But Babylonian mathematics appears to have been fairly static from the Old Babylonian period down to the middle of the first millennium. The key developments in the Persian (538–324 BCE) and Seleucid (after Alexander's conquest) periods were not in mathematics, but in astronomy. Here the established mathematical methods allowed a remarkable advance in understanding and representing astronomical phenomena.

In Mesopotamia, astronomy had a much earlier start than in the Greek world, for it possessed three advantages that were missing in Greece. First, astronomy had a social function, since it was believed that the gods sent warnings to the king through celestial signs. For example, when Jupiter reached the Pleiades, it marked a year in which the storm god would send devastation. The scribes in the temples observed regularly and sent messages to the king, citing their observations and interpreting their meaning. The ancient Greeks were no less superstitious than any other ancient people—divining the future by consultation with oracles, or interpreting the flight of birds—but they

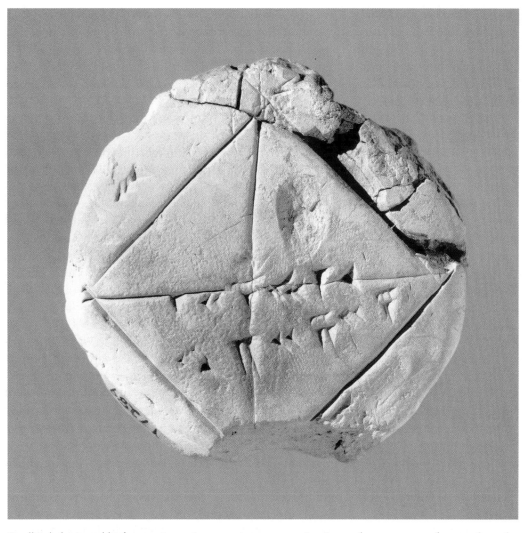

Small Babylonian tablet from *c.*1800–1600 BCE using an approximation to the square root of 2 to evaluate the diagonal of a square. Yale University Library, Yale Babylonian Collection YBC 7289.

had no tradition of divination by celestial signs. That came later, after contact with eastern wisdom. Second, in Mesopotamia there was a bureaucracy—the priests in the temples—that was charged with doing astronomy. In our culture it is common to disparage bureaucrats. But in Babylonia, astronomy flourished early on precisely because there was a bureaucracy that could be entrusted with the chore. And, finally, in Mesopotamia there was a technology for the preservation of records—the clay tablet, which is practically indestructible as long as it is kept dry, and the temple libraries where tablets could be stored for generations and recopied when they did get damaged.

The scribes' records of celestial events are today called 'astronomical diaries'. The oldest extant diaries are from the seventh century BCE, but it is likely that they began in the eighth. In a typical diary entry, the scribe would, for each part of the night, record the celestial events of interest: rising and setting times of the Moon, emergence of a planet from its period of invisibility, beginning or end of retrograde motion, and so on. Babylonian observations, made without sophisticated instruments, are not especially accurate. But having a run of several centuries' worth of data is far more important than accuracy.

The first methods of predicting planetary behaviour made use of repeating cycles. Now, in the case of the planets, events do not repeat from one year to the next. Venus travels eastward around the zodiac, but occasionally halts with respect to the background stars, reverses course for about a month, halts again, then reverts to its usual eastward motion. These retrograde motions do not occur every year. But in eight years, Venus will go through five retrograde cycles, and then the pattern does repeat very nearly. Other planets have different periods. The Babylonians found that Mars goes all the way around the zodiac twenty-five times, and makes twenty-two retrogradations, in forty-seven years. The discovery of these cycles made possible the development of a document called a 'goal-year text'. Suppose we want to predict all the planetary phenomena for 2015, which would be our goal year. We would look up what Venus did in 2007 (eight years before the goal year), what Mars did in 1968 (forty-seven years before the goal year), and so on.

By the late fourth century BCE, the scribes developed a far more sophisticated planetary theory. In these so-called mathematical planetary theories, the predictions are not made by simple repeating patterns as in the goal-year texts. Rather, sophisticated arithmetical rules are applied to each planet to deal with its non-uniform motion around the zodiac. Each event in the planet's cycle is treated as an object in its own right. Take, for example, the 'first station'—when the planet stops, just before going into retrograde motion. If we examine the first stations of Jupiter, we find that they are not spaced uniformly around the zodiac. Thus, the scribes produced a theory in which there is a fast zone and a slow zone for Jupiter. In the fast zone, the first stations are spaced uniformly, 36° apart. And in the slow zone, the first stations are spaced more closely, at 30° intervals. A scribe could produce an 'ephemeris'—a list of successive first stations of Jupiter, giving both the calendar date and the location of the event in the zodiac, over a run of many years. In more sophisticated versions of the theory, the zodiac was divided into multiple speed zones for each planet. And, in a still more sophisticated version, the speed was thought of as varying in a continuous way, in a 'linear zigzag pattern', rather than in a series of jumps. Large numbers of clay tablets preserve ephemerides from the last three centuries BCE. Other tablets, called 'procedure texts', explain the rules of computation. The procedure texts, are, however, so concise that it is unlikely that anyone could have learned how to use the theories from a procedure text alone. Rather, we should imagine face-to-face instruction in the scribal schools, with the procedure texts serving as reminders of the details.

Some tablets bear colophons, in which the owner of the tablet gave his name. He might style himself X, priest of Bel, son of Y, son of Z. (Bel is another name for Marduk.) Thus it has been possible to construct family trees for some of the scribes.

There are many versions of the theories for the Moon and planets, stretching over the whole period for which tablets are preserved. Thus, it seems that some scribes were not mere drudges but creative experimenters, who perhaps engaged in constructing new theories just out of the pleasure it gave and the opportunities for intellectual play.

The differences with Greek astronomy at roughly the same period are striking. First, the Babylonian theories are not geometrical. As far as we can tell, no effort was made to visualize mechanical or geometrical models. Rather, the speed rules are applied arithmetically. Second, there was no such thing as a Babylonian Aristotle—the Babylonians got along just fine without an elaborate philosophy of nature. And, finally, the Babylonian methods face up squarely to the non-uniformity of motion. In the Greek versions of planetary theory, non-uniformity of motion was usually masked by the need to pay homage to the doctrine of uniformity.

Greek Natural Philosophy

In the sixth century BCE, something new began in a portion of the Greek world—the thin line of settlements along the western coast of Asia Minor. Traditionally, Thales of Miletus (*c*.580 BCE) was said to have invented philosophy, along with geometry and astronomy. Of course, this only reflects the fondness of later Greek writers for attributing every discovery to one particular wise man. Nevertheless, an important philosophical tradition began with him and his successors, Anaximander and Anaximenes, at Miletus. By the fifth century, Athens became the most important place for doing philosophy, as well as for teaching and learning about it.

A key feature of the philosophical movement was the desire to remove the gods from the middle of things—to change the rules of explanation. Xenophanes of Colophon complained that the old poets (Homer and Hesiod) had ascribed to the gods shameful things, such as lying, stealing, and committing adultery, and he ridiculed his contemporaries for believing that gods have bodies and wear clothes like their own. He famously remarks that the gods of the Ethiopians have flat noses and dark skins, while the gods of the Thracians have red hair and blue eyes. And if cattle and horses had hands and could draw, horses would picture gods that look like horses, and cattle like cattle.

The philosophers sought for clues about the origins of the world in observable natural phenomena—such as the drying up of water by the Sun, or the fall of heavy bodies, or the tendency of clay to be flung outward from a spinning wheel—rather than in the whims of the gods. And they drew bold conclusions from the hints that nature offers. For example, Anaximander argued that human beings have not always been what they are now, and must have been born from a different kind of creature. His argument is simple: human infants have too long a period of dependency for them to have survived. The philosophy of the period is characterized by openness to speculative thought, coupled with insistence on close reasoning and arguments from evidence. The competitive nature of Greek society also comes through. One philosopher will mention his precursors by name, take issue with them, and argue for his own position.

A key debate raged around the competing claims of permanence and change. Heraclitus of Ephesus (*c.*500 BCE) staked out one polar position, emphasizing flux: everything flows, nothing is constant. As he famously remarked, you cannot step twice into the same river. Heraclitus believed that the world (meaning the universe) is born of fire and pays back its debt to fire—whole worlds come and go, as one world succeeds another.

The opposite position was taken by Parmenides of Elea (*c.*470 BCE), who emphasized the underlying permanence of things. For him, change is merely illusory. It is impossible for anything to come into being out of non-being, or for anything that exists to pass into non-existence. Homer and Hesiod had sung of the things that were, the things that are, and the things that will be. But for Parmenides, the essential reality is the permanence of the world: and here is all that can be said: *it is*. Carried to its ultimate, this view undercuts the possibility of any genuine change in the world. Much of later Greek philosophy can be understood as an effort to find a compromise between the positions of Heraclitus and Parmenides.

Another key argument was fought out between the monists, who held out for a single underlying principle (to account for the permanence and unity of nature) and the pluralists (who needed multiple elements to explain change). Thales is a fine example of a monist as he argued that everything is water. If one were to claim one element as fundamental, water makes sense, as every living thing requires it. But later opponents held out for fire, such as Heraclitus, or for air, such as Anaximenes. Anaximenes had some good arguments: he held that our souls are made of air. And this, in turn, would explain how one human being can emotionally and intellectually affect another. One person speaks, the voice disturbs the air, which in turn can disturb the soul of another human being. 'Just as our souls, being air, hold us together, so do breath and air encompass the world.' Anaximenes held that the Earth came into being through a process like felting, in which the air was packed together. Moreover, the Earth remains in the centre of things because it rests on air, like a flat lid, which does not cut through the air beneath it.

By the time of Empedocles (fifth century BCE), a compromise position had been reached. Empedocles opted for four elements: earth, air, water, and fire. This gave a nice solution to the problem of change, for the underlying elements provide a refuge of permanence but the combinations and dissolutions of elements can account for change. To these four material principles, Empedocles adds two immaterial ones: love and strife, or as we might say, attractive and repulsive forces. (Love and strife did not last, but the four Empedoclean elements became widely, though not universally, accepted.)

Empedocles thus also attempted a compromise in a second debate, over material versus immaterial principles. The fundamental principles of Thales and Heraclitus (water or fire) were material elements. Empedocles' introduction of the immaterial principles of love and strife along with material elements gave a flexible system with great explanatory power. But other philosophers proposed alternative immaterial principles. Anaxagoras argued that *mind* was the source of the order in the universe.

The world was initially unformed, and a rotation began, imposed by mind, which caused the heavy and the light to be separated. Yet another immaterial principle was championed by the Pythagoreans, who held that *number* is the fundamental principle of the world. All numbers were generated by the monad—the unit. Numbers were assigned qualities, even and odd, for example. Square numbers were those that could be represented as arrays of pebbles in a square pattern: their sequence is 1, 4, 16, 25. ... Triangular numbers were those for which the pebbles could be arranged in a right triangle and they make the sequence: 1, 3, 6, 10, 15. ... And then it was possible to prove theorems, for example that any square number is equal to the sum of two successive triangular numbers (16 = 6 + 10). Odd numbers were male and even numbers female. Thus the world itself might be produced from pure number. The Pythagoreans also stressed four key mathematical sciences: arithmetic, geometry, music theory, and astronomy. Arithmetic here means not elementary computation (which the Greeks called 'logistic'), but rather the study of the properties of numbers.

Plato's (*c.*429–347 BCE) philosophy was strongly influenced by the Pythagoreans. Plato—this was a nickname, for his real name was Aristocles—found his own compromise between Heraclitus and Parmenides. For Plato, the ultimate reality lies in a realm of immaterial *forms*. If we see a table, for example, we may recognize it as such because it participates in the eternal form of the table. Plato's position gains in plausibility when we consider mathematical objects. Take the case of the circle. No circle drawn on paper is ever really a circle. Thus the ratio between the circumference and diameter (the number we call pi) cannot be determined by measurement of actual circles. But this number does seem to exist independently of human beings. Another civilization, in some other galaxy, would be bound to find the same value for pi as we have. Do mathematical objects, such as right triangles and the number pi, have an existence independent of us? It might seem silly for such a commonplace object as a bed or a table to be represented by an eternal form. In the *Republic*, Plato does use the example of a bed. But he may have changed his mind, for in the *Parmenides* he is sure that undignified objects like hair, mud, and dirt do not have forms. He believes but is not quite sure whether there is a form of man, or of fire or water. But for goodness and beauty, and for mathematical things like unity, plurality, and likeness, he is certain the forms must exist.

In the *Republic*, Plato presents the allegory of the cave. People are chained inside a dark cave, with their heads constrained to face the inner wall. Behind them is a fire, and statues of people and animals are being carried back and forth in front of the fire. The prisoners in the cave see only the shadows of these statues and take them for real things. If a prisoner were able to break his chains and go up into the light of the Sun, then return to the cave, his fellow prisoners would scarcely believe him when he told them that everything they had taken as real was merely a shadow of reality. Indeed, they would kill him. (This was a reference to the fate of Plato's mentor, Socrates, condemned to death by the Athenians in 399 BCE.)

For Plato, the goal of philosophy is to bring us closer to understanding the eternal forms—to bring us into contact with the eternally true, rather than the changeable and merely apparent. In the *Republic*, Plato lays out the education necessary for the 'guardians'—the philosopher-rulers of his ideal state. Arithmetic, geometry, astronomy, and music theory (the old Pythagorean foursome) are prescribed, because they improve our minds and souls by forcing us to meditate on things that *are* and have no part in *becoming*. But these mathematical sciences are not so important in themselves; rather they prepare us for doing higher philosophy—examining goodness and justice and so on.

Plato's influence on the development of Greek science was weighty for good and ill. On the one hand, he stressed the importance of mathematics, which was crucial for such sciences as astronomy and harmonics. And he emphasized the distinction between appearance and reality—thus a science might need to seek reality below the visible surfaces of things. On the other hand, Plato disparaged the value of observation, since it relies on our faulty senses. He urged that one should do astronomy, not by looking at the sky, but by solving problems, as in geometry. It should not be surprising that the most important Greek astronomers sought their philosophical foundations, not in Plato, but in Aristotle, who was much more sympathetic to observation.

Aristotle, like his teacher, Plato, accepted Empedocles' four terrestrial elements. These elements are themselves rooted in qualities, four in number: hot, cold, wet, dry. For example, water is cold and wet, while fire is hot and dry. Of course, an element would hardly ever be seen in its pure state. We could imagine the water we see as being made mostly of elemental water. Adding fire to it can transform it into steam. Moreover, Greek philosophers sometimes distinguished between a principle and an element. Thales regarded water as primordial and ungenerated and so, for him, it was the fundamental principle. Similarly, Empedocles regarded his four elements as eternal and called them the 'roots'. But Aristotle thought the elements could change into one another, and so for Aristotle they are merely elements and not first principles.

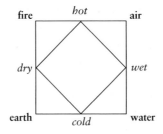

'Aristotelian qualities' of the four elements.

Closely associated with Aristotle's element theory is a theory of natural motions. The natural motion of an object is the motion it would follow if left to itself. Thus heavy things, made mostly of elemental earth, fall radially downward toward the centre of the universe. But fire moves radially upward, as anyone can see. But then it is clear that celestial things—the Sun, Moon, stars, and planets—cannot be made of the four terrestrial elements. To begin with, the heavens have remained changeless for generations. Moreover, the heavens move not radially, but circularly: each day the stars rise in the east, cross the sky, and set in the west. Aristotle therefore deduces the existence of a fifth element called the *ether*, which constitutes all celestial things. The fundamental property of ether is absolute changelessness and its natural motion is circular. So, for Aristotle, the compromise between the positions of Heraclitus and Parmenides is a geographical one: the Heraclitan realm is here below, where the four elements suffer their ceaseless changes. But the upper world, from the Moon on up to the sphere of the fixed stars, is a realm of Parmenidean changelessness.

Aristotle also offers a theory of causation based on four different sorts of causes. Consider the construction of a table. The material cause of the table is the stuff of which it is made, wood perhaps. The formal cause is the 'tableness'—having a horizontal surface at a convenient height above the floor. Aristotle's formal cause clearly owes much to Plato's forms. The efficient cause is the agency that brings change about, perhaps the craftsman who builds the table. But for Aristotle, the most important is the final cause, which is the purpose for which the change occurred—perhaps the wish of the customer to order a table to be made. In a case like this one, it is easy to identify all four causes. But in the case of natural motions (e.g. the fall of particles) some of the four causes may collapse into one. In the case of a freely falling body, there is neither an efficient cause nor a formal one. We can identify a material cause in the elemental earth and, of course, a final cause, as earth strives to realize its potential to be at the centre of the cosmos.

Aristotle's insistence on the importance of final causes—that is, of purpose—in nature is an anthropomorphic aspect of his philosophy. Sometimes this mode of explanation is called 'teleological' (from the Greek *telos*, goal or purpose). Although the whims of the gods have been removed as agents, nature itself has a purpose. Today, when in ordinary life people speak of the cause of some event, there are two things that they might mean. When we ask why a friend had to die in a traffic accident, we might be asking what mechanical failure prevented the car's brakes from working, or we might be asking what purpose was served by this tragic death. At the beginning of science, there was no way to know in advance which sort of question was more likely to lead to productive answers. The renunciation of final causes in science (in favour of efficient causes) did not come about until the seventeenth century.

But it would be easy to overstress the importance of Aristotle's theory of causation. Ancient engineers were perfectly used to thinking about efficient causes when constructing machines such as catapults. And even among philosophers, not everyone accepted teleological explanations. A notable example is provided by the atomists, Leucippus, Democritus, and Epicurus. For them, the ultimate reality is atoms rushing along in infinite void space, a view which leaves no room for purpose. The atomic philosophy, considered by its critics to be rather frightful, was viewed by Epicurus as something that could free us from our fear of the gods and of death. When we realize that at death our atoms (including the atoms of our soul, which also is material) simply return into the cosmos, we will understand that death has nothing to do with us. Lucretius, the Roman poet who popularized Epicureanism in his *De rerum natura* (On the Nature of Things), provided many convincing arguments for the existence of invisibly small bodies. When we open a vial of perfume, after a period of time, the scent can be perceived across the room. He points, too, to the gradual shrinking of the right hands of statues, as passers-by grasp them for good luck over a period of years. Lucretius attacked teleological explanation by arguing that the parts of our bodies existed before there was a purpose for them. The tongue existed before there was language, so the philosophers are wrong who claim that the tongue came into being so that we might speak. For the atomists, there are no purposes or final causes in nature.

Greek Mathematics

Greek mathematicians were proving theorems in the sixth and fifth centuries BCE, but no complete works of mathematics have survived from so early. Rather, we have mentions by later Greek mathematicians of the accomplishments of the earliest geometers. The geopolitical situation changed enormously in the fourth century BCE, when the conquests of Alexander the Great put an end to the epoch of the Greek city-state and brought a huge swath of the world under Macedonian control. Alexander's empire lasted but eight years, for he died of a fever in Babylon in 323 BCE. His generals fell to squabbling over the pieces, and when the dust settled, several Greek-speaking Macedonian dynasties had installed themselves. One important empire was established in Egypt by Ptolemy I Soter. Thus a Greek-speaking Macedonian dynasty ruled Egypt from the end of the fourth century to the middle of the first century BCE, when Egypt was annexed to the Roman Empire. Alexander had been a great founder of cities—dozens of them all over his empire, and all called Alexandria. Most of these cities never amounted to much, but Alexandria in Egypt, at the western mouth of the Nile, became one of the most important cities in the eastern Mediterranean. The first kings of the dynasty, Ptolemy I Soter and his son, Ptolemy II Philadelphus, were patrons of the arts and sciences. They founded and maintained an institution called the Museum—because it was dedicated to the Muses—and a library. The fellows of the Museum lived and worked at state expense. Many later Greek scientists had some association with Alexandria. Some lived and worked there, some visited, some corresponded with colleagues who lived at Alexandria. A second major Greek-speaking kingdom was founded by Seleucus I Nicator, encompassing much of the old Persian empire, from Syria to the borders of India. This kingdom included Mesopotamia, so it is not surprising that the period of the most intense contact between Babylonian and Greek science comes during the Seleucid period.

In Alexandria, around 300 BCE, Euclid produced his *Elements* of geometry. This work was so successful that it largely destroyed its predecessors, for mathematicians ceased to copy the older works. For this reason, it is difficult to reconstruct in detail the history of Greek mathematics before Euclid. We do have, however, some enticing remarks by later commentators. For example, in the fifth century CE, Proclus wrote a *Commentary on the First Book of Euclid's Elements*, which provides information about Euclid's predecessors. Although 'elements' had existed before Euclid, his work became canonical. The book is remarkable for its logical structure, which put a permanent stamp on the form of mathematical writing. Definitions and postulates are stated, then theorems are proved in sequence, with the later theorems dependent on those that have already been proven.

Perhaps the most creative of all the ancient mathematicians was Archimedes of Syracuse (*c.*287–212 BCE). Archimedes made a trip to Alexandria, where he met other geometers. But he lived nearly all his life in the relative isolation of Syracuse. A number of his geometrical treatises are in the form of open letters, addressed to Conan, Dositheus, or Eratosthenes in Alexandria. While earlier mathematicians—Babylonian, Egyptian, and Greek—had used convenient approximate values for the ratio of the circumference to the diameter of a circle (pi), Archimedes showed how one could bracket it by an upper and a lower limit. To do this, he circumscribed a polygon of ninety-six sides around a circle, and inscribed another polygon of ninety-six sides within the circle. Then straightforward (if lengthy) geometrical reasoning led to the conclusion that pi is less than 3 1/7, but greater than 3 10/71. The most remarkable aspect of Archimedes' mathematics was his calculation of the areas under particular mathematical curves, such as the parabola. In this one branch, mathematics did not move beyond Archimedes until the seventeenth century.

Geometrizing the Cosmos

The oldest known accounts of the heavens considered as a celestial sphere were the *Phenomena* and the *Mirror* of Eudoxus of Cnidus. 'Phenomena' (or rather 'Phainomena') is a participle of the verb 'to appear', and usually refers to celestial appearances. The second book was called the *Mirror* because it aspired to offer an image, in words, of the cosmos. Eudoxus described the positions of the constellations on the sphere, and discussed the important celestial circles: celestial equator, tropic of Cancer, tropic of Capricorn, the ecliptic, and so on. Neither Eudoxus' *Phenomena* nor his *Mirror* has survived, but, around 230 BCE, they inspired a poet, Aratus of Soli, to produce a *Phenomena* in verse. In our own day, it would be tedious to learn organic chemistry, say, from a long didactic poem. But in antiquity, when poetry was a central part of education, the poem would have been far more widely read, and more often copied, than a dry scientific work in prose.

Latin translations of Aratus by Cicero, by Germanicus Caesar, and by Avienus made the Greek sphere familiar at Rome. The *Phenomena* of Aratus powerfully influenced artistic representations of the sphere and of its constellations. Three intact celestial globes have survived from antiquity, the largest and best-known of which is the Farnese globe, a marble sphere about 65 cm in diameter, supported by a statue of Atlas. Its treatment of the constellations conforms well with Aratus. For example, on the Farnese globe, the constellation that we call Hercules is shown simply as Engonasin (a kneeling man), for the identification with the hero was made shortly after Aratus' time.

The oldest Greek text that asserts that the Earth is a sphere and offers convincing evidence is Aristotle's *On the Heavens*. Aristotle regards physical or philosophical

The Farnese celestial globe, supported by a statue of Atlas. This is a Roman copy of the first century CE or late first century BCE, based on a Greek original of the second century BCE. National Archaeological Museum of Naples.

arguments as the strongest. Thus he first deduces the sphericity and the centrality of the Earth from his theory of natural motions. The natural place for elemental earth is the centre of the cosmos. The centre-seeking jostling of individual particles of earth therefore produces the rounded shape. Moreover, if somehow the Earth could be removed from the centre of the cosmos, it would move back to the centre. But Aristotle does not disdain to add evidence from observations. He points out that during a lunar eclipse, the shadow of the Earth is seen on the Moon and the edge of the shadow always has a circular shape. Moreover, the Earth cannot be very large, since if one travels southward new stars begin to be visible. Aristotle concludes by saying that certain mathematicians have measured the circumference of the Earth and have obtained 400,000 stades. The stade was a unit of distance used in Greek surveying and geography. Stades of several different lengths were used at different times and places (there were always 600 feet to the stade but the size of the foot could vary). We may be reasonably sure, however, that Aristotle means the Attic stade of about 185 m. From this it follows that Aristotle's figure corresponds to an Earth circumference of about 74,000 km, which is a bit high (the modern value is about 40,000 km), but is of the

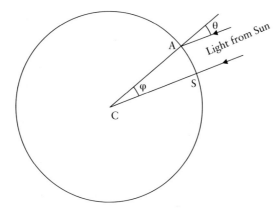

Eratosthenes' geometry for calculating the circumference of the Earth. A is Alexandria, S Syene, and C the centre of the Earth. Angles φ and θ are equal, each 1/50 of a circle.

right order of magnitude. Aristotle does not say just who these mathematicians were or how they made their measurement.

The oldest measurement of the circumference of the Earth for which we have any details comes from the next century and was made by Eratosthenes in Alexandria. Eratosthenes began with the knowledge that at Syene (modern Aswan, on the upper Nile) on the day of summer solstice, the Sun stands straight overhead at noon and there are no shadows. (Thus Syene lies on the tropic of Cancer.) But, at Alexandria, on the same day, the noon Sun stands below the vertical by one-fiftieth of a circle. (One-fiftieth of a circle is 7.2°, but Eratosthenes wrote in the third century BCE, before the Greeks had adopted the Babylonian degree). If the Sun is assumed to be so far away that all rays striking the Earth can be treated as parallel, then a simple geometrical argument leads to the conclusion that the circumference of the Earth is 50 × 5,000 = 250,000 stades. The geometry is perfectly good, but the round numbers show that this was more of an estimate than a real measurement.

Also from the third century BCE we have a remarkable treatise by Aristarchus of Samos, *On the Sizes and Distances of the Sun and Moon*. Aristarchus states his premises, including the fact that the shadow of the Earth appears to be twice the diameter of the Moon. One of the more doubtful premises is that the angle between the Sun and Moon at the time of a quarter Moon is 87° (as Aristarchus puts it, at the time of the quarter Moon the angle between the Sun and the Moon is less than a right angle by a thirtieth of a right angle.)

From seven premises, Aristarchus proves a series of propositions. He finds, for example, that the distance of the Sun is between eighteen and twenty times the distance of the Moon. To a modern reader, Aristarchus may appear to be placing limits of uncertainty on his results, to reflect the uncertainties in his original data. Nothing could be further from the truth. Aristarchus had the disadvantage of working before the development of trigonometry. Thus, he has to prove, by one demonstration using the methods of Euclid, that the Sun is at least eighteen times farther from us than is the

Moon and, by a second demonstration, that the Sun is no more than twenty times further away than the Moon.

Aristarchus' methods were improved by Hipparchus in the second century BCE and Ptolemy in the second century CE. No method available to ancient astronomers was ever able to yield an accurate value for the distance of the Sun—it is so far away that the 87° angle assumed by Aristarchus should really be 89° 51′. However, by Hipparchus' time, Greek astronomers had quite a satisfactory idea of the distance and size of the Moon.

Planetary Astronomy

The *Republic* of Plato ends with a cosmic vision. A hero named Er is killed in battle. His body lies on the field for ten days but does not decay. When Er comes back to life, he tells his companions what he saw while he was out of this world. Here Plato draws on the familiar sight of a woman spinning yarn, for Er saw the spindle of Necessity (Anangke, personified as a woman). The spindle and the yarn represent the axis of the universe, while the spindle whorl (the spinning bob to which the newly formed yarn is attached) represents the cosmos itself. But the whorl that Er saw is not like ordinary spindle whorls. Rather it consists of eight whorls nestled one inside another. Plato says they are like nested boxes one can find (but we might think of Russian dolls). The outermost whorl is the sphere of the fixed stars. Nested inside are the whorls for the five planets and the Sun and Moon. Thus each celestial body is carried around on its own spherical shell. The outer whorl turns westward (representing the daily rotation of the cosmos), but the inner ones rotate within in it, to the east, each in its own characteristic time (representing the motions of the planets around the zodiac). Riding on each whorl is a Siren who sings a single clear note. This is Plato's nod to the Pythagorean doctrine of celestial harmony. And ranged round the whole affair are the three daughters of Neccessity, the Fates who were mentioned by Hesiod in the *Theogony*. Clotho, with her right hand, helps to turn the outermost whorl. Atropos, with her left hand, helps to turn the inner spheres. And Lacheisis, alternately with either hand, touches one then the other. This seems to be a reference to the three movements of a planet—the westward daily motion, the eastward motion around the zodiac, and the oscillation responsible for the occasional retrograde motion.

This is the first appearance in literature of the 'cosmic onion'—the view of the universe as a set of concentric spherical shells. Plato here stands midway between science and myth. On the one hand, he geometrizes the cosmos, postulating a simple model to explain the complex motions of the planets, but on the other hand he has draped his image in traditional mythology.

Nevertheless, geometers took this model seriously. Eudoxus of Cnidus wrote a book *On Speeds* in which he considered a planet that rides on the innermost of a set of four nested spheres. That is, each of the planetary shells in Plato's account now

---

Opposite: Woman spinning thread. Attic vase, *c.*490–470 BCE. British Museum, registration number 1873,0820.304.

consists of four nested spheres. The outermost sphere for each planet is responsible for the daily rotation. The next sphere in is responsible for the eastward motion around the zodiac. And the innermost two spheres together produce a figure-of-eight, back-and-forth motion that accounts for retrograde motion. So, in Eudoxus, the Fates have been removed and replaced by rotating spheres. Plato and Eudoxus overlapped in Athens, so we have no way to know whether the geometer was inspired by the philosopher, or the philosopher poeticized the models of the geometer. Eudoxus' work has not survived, but we have a short account of it in Aristotle's *Metaphysics* and it was still known to Simplicius in the sixth century CE.

The nested spheres of Eudoxus were soon abandoned in planetary theory (though they dominated cosmological thought until the Renaissance). Ancient critics pointed out that in this system, although a planet is slung about on multiple spheres, since each sphere is concentric with the Earth, the planet's distance from the Earth never changes. This made it hard to understand how some planets vary in brightness in the course of their cycles. Mars, for example, is much brighter in the middle of its retrograde motion.

Around 200 BCE, Apollonius of Perga discussed the theory of epicycles and deferent circles. The new idea is that each planet travels around a circle called the epicycle, while the centre of the epicycle moves around the Earth on another circle, called the deferent. Both of these motions take place at uniform speed, in keeping with the nature of celestial things. Retrograde motion occurs when the planet is close to the Earth, on the inner part of the epicycle. For then the westward motion on the epicycle is more than enough to overcome the eastward motion on the deferent. At first, the model was intended only to be broadly explanatory, and to provide a field of play for the geometer. Apollonius' theory explains how a planet could move around the zodiac and occasionally retrograde while really executing a combination of uniform circular motions. It also nicely explains why Mars is brightest in the middle of retrograde motion. The theory (being planar) was also mathematically far simpler than Eudoxus' spherical system.

The simple version of epicycle and deferent theory discussed by Apollonius was not capable of quantitative prediction. In Apollonius' theory all the retrograde arcs of a planet would be of the same size and equally spaced around the zodiac. But a real planet shows more complexity: in one part of the zodiac, the retrograde arcs of Mars are short (about 10 long), while in the diametrically opposite part of the zodiac they are 20 long. Attempts were made to improve the theory in the second century BCE, by placing the centre of the deferent circle slightly off-centre from the Earth. (The astronomers were perfectly capable of bending the rules of Aristotle's physics when the phenomena required it.) But this by itself was not enough.

When Greeks came into close contact with Babylonian astronomy, in the Seleucid period, it must have seemed both impressive and puzzling. The Babylonians could predict planetary phenomena quantitatively. This was impossible in Greek planetary theory around 200 BCE. Greeks had been arguing about fire and water, postulating geometrical models and applying geometrical proofs, but without much of a basis in empirical data. A characteristic reaction is that of Theon of Smryna (second century CE) who tells us that both the Babylonians and the Greeks have succeeded in saving (or accounting for) the phenomena, the Babylonians arithmetically and the Greeks geometrically. But, he says, the Babylonian way is inadequate, because it is not based on a proper understanding of nature. So the new goal was to have a geometrical theory of the planets that would work quantitatively.

This proved difficult to do. And so there was for several centuries an astronomy of the low road that existed side by side with the astronomy of the high road. Greek writers steeped in geometry and philosophy wrote textbooks explaining the motions of the planets in terms of deferents and epicycles (an example is Theon of Smyrna's *Mathematical Knowledge Useful for Reading Plato*), even though these theories weren't good enough for calculation of phenomena. Meanwhile Greek astrologers, who needed quick, easy ways to obtain the zodiacal positions of planets on any particular date, adapted the arithmetic planetary theory of the Babylonians. These two ways of doing things still existed side by side in the second century CE.

A decisive step was taken by Claudius Ptolemy in Alexandria in the second century. Ptolemy's study of planetary astronomy was originally called the *13 Books of the Mathematical Composition (Syntaxis) of Klaudios Ptolemaios*, but in the Middle Ages it came to be known as the *Almagest*, the title commonly in use today. Ptolemy's work came at the very end of the creative period of Greek science. And in the case of deferent-and-epicycle theory, he built on three hundred years of work. But Ptolemy also introduced a new idea. For Ptolemy allows the centre of the planet's epicycle to travel at a *non-uniform speed* around the deferent. The non-uniformity is, however

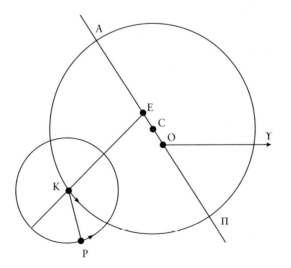

Ptolemy's planetary theory. O is the Earth, which is stationary in the centre of the cosmos. The plane of the diagram is the plane of the Solar System. The planet P moves around its epicycle, while the centre K of the epicycle moves around the deferent circle. C is the centre of the deferent circle, which is slightly off-centre from O. Finally, E is the equant point, from which the angular motion of K appears to be uniform. The theory predicts with good accuracy the direction of P as seen from O.

Three of the major fragments of the Antikythera mechanism. That on the left carries parts of the zodiac scale and Egyptian calendar scale. That in the middle contains most of the surviving gears. The large crossed wheel turned around once a year and represented the motion of the Sun around the zodiac. National Archaeological Museum, Athens.

expressed in the language of uniformity. Ptolemy imagines a third centre, distinct from the Earth and from the centre of the deferent. This third centre, which is the centre of uniform motion, came in the Middle Ages to be called the equant point. If we could stand at the equant point of Mars, we would see the centre of Mars's epicycle travelling around us at a uniform angular speed—about half a degree per day. The complete theory—epicycle and deferent, with the deferent off-centre from the Earth and a separate equant point—is very successful. For the first time, it became possible to calculate the positions of planets accurately from a geometrical theory. The Babylonians had earlier achieved numerical predictive power, but they used arithmetical rather than geometrical methods. Ptolemy's introduction of the equant point was a savage rupture of Aristotle's physics. For this, he was criticized in the Middle Ages. But Ptolemy's theory really works and is no more complicated than the motions of the planets require. Thus nearly all practical planetary computation continued to be done with Ptolemy's methods until the Renaissance.

Greek mechanicians enjoyed modelling the machine of the cosmos, using gear trains as concrete realizations of the period relations that govern astronomical cycles. (A good example of a period relation is the 'Metonic cycle', according to which 235 months = 19 solar years.) The making of such machines started by the time of Archimedes and continued into Ptolemy's day, since both authors wrote on the subject. By chance, we happen to possess a portion of one of these machines, the Antikythera mechanism, which was recovered in 1901 from a ship that sank around 60 BCE. The extant fragments of the machine contain thirty gears, and it is believed there were originally about twenty more. When it was intact, the mechanism was of about the size of a shoe box. As the user turned a knob, gear trains moved pointers that displayed the changing positions of the Sun and Moon in the zodiac, and also indicated the date in two different calendars—the Egyptian solar calendar and a Greek luni-solar calendar. Another dial indicated the months and times of day of predicted eclipses. The device probably also displayed the planetary phenomena such as retrograde motion, but this portion of the gear trains has not survived. The date of the Antikythera mechanism is controversial. The eclipses displayed on the machine would be most accurate for a run of years starting within a few decades after 205 BCE. But some scholars would prefer a construction date in the later second century, or even the early first century BCE. In either case, the Antikythera mechanism gives us a view of the mechanical cosmos well before Ptolemy's time. It also shows us how much can be learned from artefacts: although we have Greek texts that discuss gears, the devices they treat are very simple and no one would have guessed from the texts alone that a machine as complex as the Antikythera mechanism could have been built.

Geography and Cartography

Simple plans of towns and temples are known from Babylonia as early as the late third millennium BCE. From about 1500 there is a map of the city of Nippur, drawn more or less to scale. In Egypt, there was a practice of drawing maps of land plots so that an owner's tax obligation could be corrected if land were lost to the Nile. Anaximander, the philosopher of Miletus, attempted to draw a map of the whole inhabited world in the sixth century BCE.

But the big steps came in the Hellenistic period. Eratosthenes compiled a geographical treatise in three books, which is now lost. However, a good deal is known about it from the comments (often critical) of Strabo in the first two books of his own *Geography*. The typical information base consisted of travellers' accounts and sailing times. But Eratosthenes supplemented this information with more geometrical methods. For example, his measurement of the size of the Earth was linked to his programme in geography. Eratosthenes was probably the first to draw maps based on a grid of intersecting meridians and parallels. Building on the work of Dichaearchus of Messina, he chose as fundamental parallel that through the Strait of Gibraltar and the mountains north of India. His prime meridian was that through Syene and Alexandria. Along these two fundamental circles, which crossed at Rhodes, he worked out the distances between successive pairs of key places. This was the frame on which the rest of his world layout was constructed.

The culminating work of ancient cartography is the *Geography* of Claudius Ptolemy. A key problem for making a world map is finding a method of projection

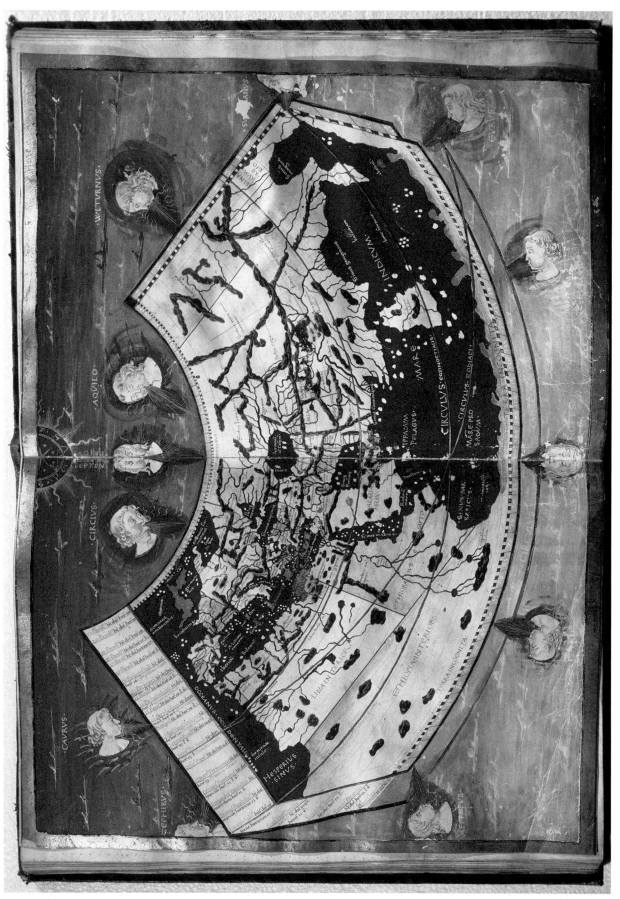

A world map in a fifteenth-century Latin manuscript of Ptolemy's *Geography*. Biblioteca Nazionale Vittorio Emanuele III (Naples), ms. V. F. 32.

so that the curved surface of the sphere can be represented in a systematic way on a flat surface. In the theoretical chapters of his book, Ptolemy describes not one, but two ways of doing this. A large part of the book is taken up by a gazetteer, giving the longitudes and latitudes of some eighty thousand places. Ptolemy measured latitudes as we do, north or south of the equator in degrees. His longitudes were measured from the meridian through the Islands of the Fortunate (the Canary Islands). It was not hard for a well-educated traveller to measure the latitude of a city, by noting the altitude of the celestial pole above the horizon, or by measuring the length of a gnomon's shadow on the day of equinox. But there were few actual latitude measurements to go by. For the great bulk of the places, the geographer had to surmise the latitude by the relation of the place to other places that were better known. Even so, most ancient latitudes are not bad. Those that were actually *measured* (by means of shadows) are usually accurate to a couple of degrees. The case is quite otherwise with the longitudes. The only possible way to get accurate longitudinal differences in antiquity was through simultaneous observations of lunar eclipses. For example, there was an eclipse of the Moon shortly before the battle of Arbela in 331 BCE, when Alexander defeated Darius III of Persia. According to the historians of Alexander's campaign, this eclipse was seen there at the fifth hour of the night. But the same eclipse happened to be observed in Carthage at the second hour. Thus, Ptolemy deduced a difference between Arbela and Carthage of three hours, or 45°. This considerably overstated the longitudinal difference (which is about 34°). But simultaneously observed eclipses were so rare that Ptolemy had no way to correct this error. As a result he considerably overestimated the width of the Eurasian continent. This, combined with Ptolemy's adoption of Posidonius' low figure of 180,000 stades for the circumference of the Earth, later served to encourage Columbus, for it made the western ocean narrower. Accurate longitude measurements did not become possible until the seventeenth century. Modern historians argue about whether Ptolemy's *Geography* actually included maps. But some of the oldest surviving manuscripts (from around 1300 CE) do, so it seems unduly sceptical to deny them.

Ancient Astrology

In early Babylonia, celestial divination was important for the king and kingdom, but astrology proper did not yet exist—the gods didn't speak in this way to ordinary people. The oldest preserved Babylonian horoscopes for ordinary people are from around 400 BCE. Evidence for the penetration of astrology into the Greek cultural sphere comes in the second century BCE, in Egypt. Ceiling zodiacs began to appear in Egyptian temples around 200 BCE. And a Greek astronomical papyrus (the 'Art of Eudoxus') contains a zodiac diagram with a rubric, 'Oracles of Hermes/oracles of Sarapis', suggesting that zodiacal prognostication had begun to be practised in the temples of Sarapis.

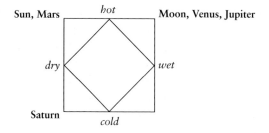

Aristotelian qualities in astrology.

Manuals of astrology were soon written in Greek and Latin. The oldest extant treatise is of the first century CE, a long Latin poem by Manilius. The most systematic of the extant treatises is the *Tetrabiblos*, written by Claudius Ptolemy in the second century CE. We have also hundreds of Greek horoscopes. Some of these are in the form of worked examples included in astrological treatises, such as the *Anthologies* of Vettius Valens. But we have many others that have come to us right out of the sands of Egypt— scraps of papyrus recovered from excavations. The oldest extant Greek horoscope is for the first century BCE, but the great bulk of them come from the first several centuries CE.

Greeks in Egypt elaborated basic astrological doctrines borrowed from the Babylonians, put their own stamp on them, and even tried to make these doctrines compatible with Greek philosophy of nature. Here are some of the basic rules. First, each planet has properties of its own. Ptolemy, the most rationalistic of Greek astrologers, explains these powers in terms of Aristotelian physical theory. Thus, each planet is either heating or cooling, and is either drying or humidifying. From these properties, Ptolemy deduces that the Moon, Venus, and Jupiter are beneficent (helpful) planets. These are the ones that are both warming and humidifying—for these are the life-giving forces. The maleficent (destructive) planets are Mars and Saturn. The Sun can be either beneficent or maleficent, which makes sense, as the Sun is essential for life but too much of it can be fatal. And, Mercury, too, can go either way. So far, it seems a sensible system. The doctrine of beneficent and maleficent planets was taken from the Babylonians, but Ptolemy has amplified it by explaining these properties in terms of the Aristotelian qualities. But when Ptolemy forges on, he divides the planets into masculine and feminine, and these associations work merely on the basis of the Greek pantheon. So Venus and the Moon are feminine, with the rest masculine, save the androgynous Mercury.

Now, each planet becomes stronger or weaker as it moves around the zodiac. Each planet has an exaltation (a zodiac sign in which its power is increased) and a depression (the diametrically opposite sign). Each planet has also a lunar house and a solar house in which its powers are increased. Moreover, the planets interact with one another. Planets tend to fight one another when they are in diametrically opposite zodiac signs. But the trine aspect (when they are approximately 120° apart) is a harmonious relation, in which planets tend to reinforce one another.

Finally, a planet's powers wax and wane in the course of a single day and night as the celestial sphere revolves. The Greek astrologers divided the sky into twelve *places*.

(These are unfortunately called 'houses' in modern astrology; recall that properly a house refers to one of the planet's zodiacal home signs.) The most important of the places is the horoscope—the wedge of sky just at the eastern horizon. A planet will be stronger when it is rising ('in the place of the horoscope'). The Greek word *horoscopos* means 'hour mark', from the custom of telling the time of night by looking to see which zodiac sign is rising. The importance of the horoscope is reflected in the fact that this place eventually gave its name to the whole chart or process by which a prognostication is made. Another place of power is the mid-heaven, where a planet is highest in the sky. By contrast, a planet tends to be weak at the descendant, when it is setting in the west.

In the *Tetrabiblos*, Ptolemy gives rules for particular investigations. Thus, for questions touching on a client's occupations, the astrologer should focus on the Sun and the mid-heaven. For an investigation of rational qualities of mind, one would focus on Mercury; for irrational qualities, on the Moon.

An astrological prognostication could be a complicated thing. The astrologer had first to work out the positions of the planets in the zodiac at the moment of birth or conception, usually from papyrus planetary tables. He then had to consider the interaction of each planet with the zodiac signs, with the other planets, and with the places. Often, these consultations took place face to face, as a modern person would consult a priest, rabbi, or psychoanalyst. The practitioner could make use of an astrologer's board, made of wood, ivory, or marble and engraved with the signs of the zodiac. On this he (or she, since Plutarch tells us that in his day Roman women also engaged in astrology) could place small stone markers, engraved with the names and figures of the planetary gods. Particular kinds of stone were deemed to have special affinity with particular planets. According to the *Alexander Romance* (a Greek adventure novel in which an Egyptian astrologer casts a horoscope for the mother of Alexander the Great), Aphrodite (Venus) is represented by a blue stone of lapis lazuli.

There was a debate in antiquity over the validity of astrology, as there was over other forms of divination. Ptolemy provides a justification for astrology in the opening chapters of the *Tetrabiblos*. He points out that there are two kinds of prediction in astronomy. First, there is the mathematical astronomy of the sort he treated in the *Almagest*, which has good claims to reliability. But then there is this less certain method of astrological prognostication, though he argues that its claims cannot be completely discounted. Ptolemy points to the obvious influences of the Sun and Moon, which no one would deny. He argues that critics of astrology are too influenced by the errors of poor practitioners or outright fakes. But one would not throw out the whole art of medicine because of a few inept practitioners. Finally, Ptolemy addresses an argument of the critics, namely that astrology is in any case useless, for (a) it is useless if it is incorrect; and (b) it also is useless if it is correct, since one cannot avoid one's fate. Ptolemy argues that it could actually be useful to know what awaits us, for we could then prepare ourselves to avoid depression or excessive elation when the event actually transpires. Moreover, the planetary influences are not forcing, but tell us of dispositions, and these can be overwhelmed by events. For example, if there is a

shipwreck and many people die all at once, it does not mean that they all had the same horoscope. Finally, he again makes an analogy to medicine. The physician can make a diagnosis, as well as a prognosis, telling the patient what the likely outcome of the illness will be if there is no intervention. But sound medical intervention may alter the outcome.

A good example of an ancient critic of astrology is Cicero, in his book *On Divination*. Cicero points to examples of twins who had different fates. He argues that the astrologers make observations of the planets at the time of birth, thus relying on sight, which is the least trustworthy of the senses. (In this criticism, Cicero misunderstands the way an astrologer would work—the planet positions were taken from theoretical tables, not from an observation made at the time of birth; after all, a client might bring in a request for an analysis of his or her child when the child was already some years old.) More plausibly, Cicero argues that the cosmos is so vast that influences from the outer planets could not reach the Earth. Finally, Cicero argues that the astrologers ignore the influence of the paternal seed. Everyone knows that children get their appearance and habits from their parents, which would not be the case if these were determined by the arrangement of the heavens at the moment of birth.

Today we would classify astrology as a pseudo-science, but this is not a helpful category for understanding ancient science. Even Aristotle had been disposed to see celestial things as causes of change on the Earth—the celestial motions stir up the fire and air, and so can have an effect on the weather. Later, with the rise of Stoicism, there was a philosophy very open to the doctrine of *sympathy*, or fellow-feeling. This word was used in music of strings that resonate together, so that the sounding of one could cause another to vibrate. The Stoics thought of the universe as enmeshed in a chain of fate, with all things subject to sympathies. Posidonius' discovery that the Moon influences the tides must have seemed a profound confirmation of this world-view. If celestial and terrestrial things really are linked together, then one might be able to glimpse aspects of the future if one could rightly read the signs. Stoicism was therefore inclined to credit most forms of divination. And when the Greeks came into contact with Babylonian astrology, Stoicism helped win it a sympathetic hearing. The fact that we can point to a few cogent critics only underscores the fact that astrology was widely accepted in the Greco-Roman world from the first century BCE onward.

Greek and Roman Medicine

Early Greek medicine involved a mixture of traditional remedies, magical charms, and pleas for divine invention. But in the late fifth century BCE a variety of learned

---

Opposite: An astrologer's board, from the village of Grand in north-eastern France, second century CE. The circle of the zodiac signs is surrounded by the circle of the decans. (Each zodiac sign is thought of as subdivided into three decans.) In the centre are decorative busts of the Sun and Moon. An astrologer would place engraved stone markers, representing the planets, at the zodiac positions appropriate for the moment of the client's birth. This would help in visualizing all the relationships that must be considered in casting a horoscope. Made of ivory and wood, this tablet probably originated in Egypt. Musée des Antiquités Nationale (St.-Germain-en-Laye), No. inv. 83675.

A small engraved gemstone of Helios, the sun god. Magical gems of this sort were probably used as markers on astrologers' boards. British Museum MG243.

medicine, influenced by the philosophical tradition, emerged around the shadowy figure of Hippocrates of Cos. Some five dozen works survive. There is no way to know which, if any, of these were written by Hippocrates himself, and there was even a dispute in ancient times about which works were more genuinely Hippocratic. Modern scholars refer to the whole set of works as the Hippocratic corpus. Most of the extant works do come from the late fifth and early fourth centuries BCE, but some are Hellenistic, and some even come from the first and second centuries CE. Needless to say, they do not reflect a unified perspective on medicine.

An early work that had a lasting influence is *On the Nature of Man*. The author says that the human body contains four humours—blood, phlegm, yellow bile, and black bile. He arrives at this point by drawing a lesson from the quibbling of philosophers over whether the universe is made of fire, or air, or water. No one of these philosophers can win an argument three times in a row, which is a sure sign that their knowledge is faulty. They agree that world is a unity, but cannot say what this unity consists in. In the same way, the physicians who say that man is a unity, and therefore claim that man is blood, or that man is bile, are also incorrect. For if man were a unity, he could never feel pain. And, if he did have illness, there could only be one cure, which is far from the truth. Moreover, how could reproduction take place out of a unity?

According to the Hippocratic author, each of the four humours waxes and wanes over the course of the year, but any person, whether sick or healthy, is made up of these four. Good health results from a proper balance among the four. Here we can see the effort to introduce a theory to make sense of the enormous complexity of human experience with health and illness. Further, the humours are attached to the qualities of hot, cold, wet, and dry. Blood is hot and wet, while black bile is cold and dry.

Other interesting Hippocratic works include *Airs, Waters, Places*, which deals with the consequences of environmental factors for health and illness, and *On Regimen* which deals with diet and exercise. In *Epidemics*, the writer gives accounts of epidemics that have stricken various cities. Often, these detail the year's weather, the time of year when the epidemic began, and the symptoms of the disease, including the particulars of a good many case histories.

Greek physicians and medical writers were famous for dividing into sects (like philosophers!) that disagreed about the conduct of medicine. The Dogmatists

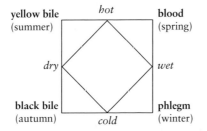

Aristotelian qualities before Aristotle, in Hippocratic medicine.

(sometimes called Rationalists) believed in the importance of seeking an underlying theory and in the value of speculative thought. The Empiricists rejected theory and insisted that only experience in treatment of patients could give a sound idea of what cures were likely to work, and that one had no need to know *why* a cure worked. These were the two chief camps, but others went in and out of fashion. With the rise of Stoic philosophy, and its emphasis on the *pneuma*—the breath, but also a kind of unifying force that pervades the cosmos—there arose a medical camp of Pneumatists. This was influenced, too, by the experiments on compressed air (pneumatics) performed by the Alexandrian mechanics.

Medical treatment involved a wide range of practices, including the use of herbs and medicines (some of which could be helpful and some more risky), stronger interventions such as blood-letting and purging, but also more conservative techniques such as the regulation of diet and activity. Medical instruments are preserved from Roman times, including a major trove found at Pompeii. A modern medical specialist can easily recognize most of the ancient instruments—scalpels, probes, tweezers, clamps, vaginal and rectal specula, retractors, and so on. But learned medicine never had a monopoly. This is clear from the huge number of *de voto* inscriptions found at ancient temples. Often, one of these may be in the form of a miniature copy of the afflicted body part, sometimes inscribed with the name of the sick person, together with an offer of thanks to the god.

In the third century BCE, largely at Alexandria, Greek anatomy and physiology flourished under Herophilus and Erasistratus. Little is known of their lives and none of their books have survived. But a considerable amount of information can be gleaned from discussions of their work by later medical writers and philosophers. A key to their anatomical researches was the dissection of human cadavers. This was rarely if ever practised in the classical period, because of legal or religious taboos. It had already become common practice to dissect animals, for the Hippocratic writers mention animal dissection and Aristotle made ample use of it. But no one had really dared to dissect human bodies for the sake of medical learning. In later antiquity, human dissection was again beyond the pale, so this brief window of time in Hellenistic Alexandria turned out to be important for the study of human anatomy. Moreover, according to Celsus, a Roman medical writer of the first century CE, Herophilus and Erasistratus not only dissected human corpses, but also cut open living men, criminals supplied from prison by the kings of Egypt.

Herophilus made the first clear description of nerves, and possibly even distinguished between sensory nerves and motor nerves (though some attribute this to his successor, Erasistratus). He discovered the ovaries in female animals, and noted their analogy to the testes in the male animal. He discovered the Fallopian tubes, but misinterpreted them, being misled by his false analogy to the *vas deferens* of the

---

Opposite: A vaginal speculum (retractor), from Pompeii, first century CE. National Archaeological Museum of Naples (Inv. no. 113264).

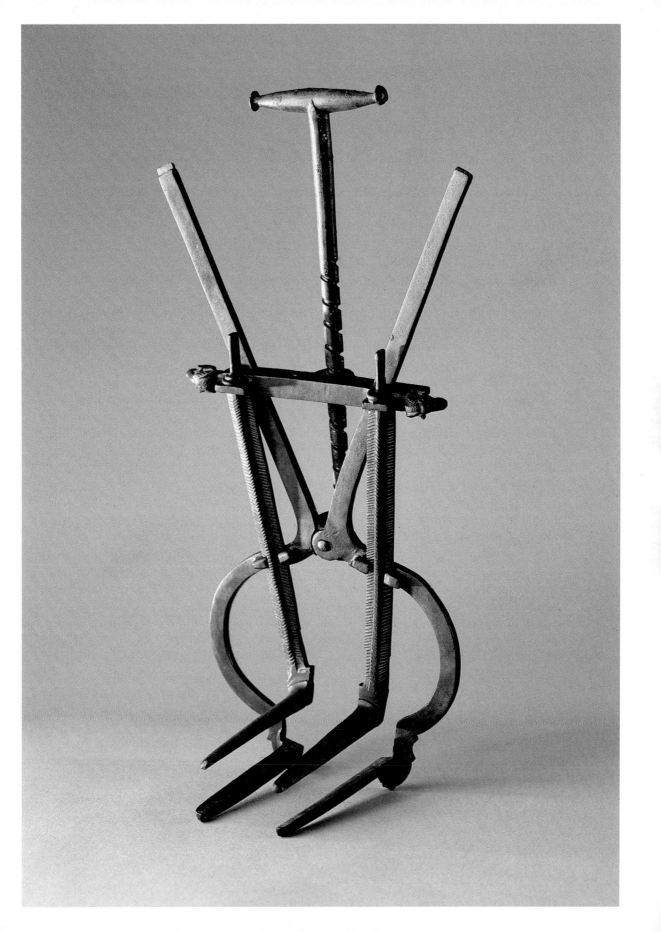

male. This is one way in which theories of female anatomy were dominated by models provided by male anatomy. This reflects, of course, the thinking of a male-dominated culture, but it also reflects the fact that the male reproductive system is more easily examined. Herophilus studied the structure of the eye, introducing terms that are still used today. In medical practice he urged the use of the pulse as a diagnostic tool. Herophilus' teacher, Praxagoras of Cos, was probably the first to make a general distinction between veins and arteries, ascribing them different functions. But Praxagoras considered that vessels coming from the right ventricle of the heart are veins, and those coming from the left ventricle are arteries. This isn't quite right since the pulmonary artery, which comes from the right ventricle, is indeed an artery: Heraphilus named it the 'artery-like vein'. Herophilus added a more secure anatomical distinction, noting that the coats of arteries are much thicker than those of veins.

Erasistratus was notable for his use of mechanical ideas and analogies. He thought of digestion as principally a mechanical process, like grinding. More impressive was his discussion of the heart as a kind of bellows. According to Erasistratus, when the heart expands it draws in blood and *pneuma* and when it contracts it pumps blood into the veins and *pneuma* into the arteries. He understood the function of the valves in the heart in preventing the backwards flow of the blood. If Heraphilus was primarily a careful anatomist, Erasistratus was more concerned with physiology and was more speculative. He believed that the veins contain blood but that the arteries contain only air-like *pneuma*. Of course when an artery is cut, blood appears, but Erasistratus explained this by a communication of blood though invisible links, called *anastomoses*, between the veins and the arteries, somewhat similar to but not identical with the capillaries discovered in the seventeenth century.

Galen, a Greek physician of the second century CE, was an extraordinarily productive and influential writer. After training as a physician, he worked for four or five years as surgeon to the gladiators at Pergamum, but spent much of his later career at Rome. Galen was strongly influenced by the Hippocratic tradition but put his own stamp on it. His preserved works cover a vast range, from introductory manuals of instruction with such titles as *Bones for Beginners* to reports of his own research, such as *On Whether Blood Is Naturally Contained in the Arteries*.

Galen disproved Erasistratus' claim that the arteries contain no blood by an experiment in which he tied off an artery above and below the place in which he intended to cut into it and then found when he cut that there was blood in it. In a sequence of experiments, he cut into the spinal cord of an animal multiple times, beginning at places low on the spinal column, then moved higher, and noted which functions were successively lost.

Influenced by a tradition that went back to Plato, Galen believed in a three-part soul. But this he attempted to correlate with the theories of human anatomy and physiology developed by Erasistratus. The rational soul has its seat in the brain, which is the source of the nerves. (This was quite a step beyond Aristotle, for whom the brain was much less important—only a sort of radiator for cooling the blood.) The passions have their seat in the heart, which is the source of the arteries. Appetites and desires have their seat in the liver, which is the source of the veins. According to Galen, the body is nourished by venous blood, which carries the products of digestion to all parts of the body. Galen disagrees with Erasistratus that digestion is merely a mechanical process—for him it involves also a kind of cooking or 'concoction'. The fire in the heart maintains the body's vital heat. The arteries carry arterial blood and life-giving *pneuma*, which is extracted from the air by the lungs. And the nerves are filled with a psychic *pneuma* which is responsible for sensation and motor functions.

In ancient physiology, there was no idea that the blood circulates all the way around the body and then returns in a complete cycle. Thus, for Galen, most of the venous blood is consumed in the tissues of the body. However, he believed that a small part of the venous blood seeps through tiny pores in the septum of the heart, from the right ventricle to the left. His argument was that the vessel bringing blood into the right side of the heart (the vena cava) is larger than the vessel taking blood out of the right side of the heart (the artery-like vein), so some of the incoming blood must somehow make it through the dense septum. These pores, which Galen admitted he could not see, do not really exist; but this error was only corrected by Vesalius in the sixteenth century.

At Rome in Galen's day there was no regular opportunity to dissect a human cadaver. So he writes that a physician ought to make use of any chance to investigate a corpse, and tells of a time when a river in flood unearthed some well-preserved skeletons that he hastened to examine. Since he did not have the opportunity to dissect human bodies, he chose animals that seem most like human beings, and favoured the Barbary ape. His forced reliance on animals caused him to make some errors in his generalizations to human anatomy. A notable example is the *rete mirabile*, which is a network of fine arteries found in the brains of some hoofed animals, and which Galen also mistakenly attributed to humans. For Galen, the *rete mirabile* is where arterial blood is refined into the psychic *pneuma*, which is then distributed to the body though the nerves.

Natural History

In Greek, *historia* means a learning by inquiry, or an account of an inquiry. And so Aristotle's *History of Animals* is a systematic study of animal life. One of his contributions was a classificatory system—a thoroughly characteristic aspect of Greek science. Aristotle divided animals first of all into the blooded and the bloodless. Animals with blood were then divided into five classes: viviparous quadrupeds (four-footed animals that are born alive), oviparous quadrupeds (four-footed animals that lay eggs, such as reptiles and amphibians), birds, fish, and cetaceans (such as dolphins). The bloodless animals were grouped as soft animals (cephalopods such as the squid and octopus), soft-shelled animals (crustaceans such as crabs and crayfish), hard-shelled animals (testaceans such as oysters and clams, but also including snails), and insects. These groups were ranged on a hierarchical scale based on vital heat. But Aristotle did not intend his classificatory system to be rigid—he used

A page from the Vienna Dioscorides showing Greek goosefoot, *c.*512 CE. Österreichische Nationalbibliothek, Vienna, Codex medicus graecus 1, Folio 273r.

different forms of it in different parts of his work—and he was never a mere classifier. He performed many dissections and had a broad knowledge of anatomy, development, and animal behaviour. For example, he dissected birds' eggs in every stage and gave a detailed account of the time development of the embryo. Some five hundred different species are mentioned in his works. For Aristotle, a causal explanation of the diverse forms of animals was the goal; but one could hardly grapple with causes until one had a reliable base of knowledge.

In later antiquity, both Greek and Roman audiences had a boundless appetite for the curiosities of nature and there arose a literature to cater to this taste. A notable example is the Latin *Natural History* of Pliny the elder (Gaius Plinius Secundus). Pliny was an inexhaustible reader and note-taker and his work is of a wholly different order than Aristotle's *History of Animals*. Pliny was a compiler, rather than an original researcher, who prided himself on the vast number of sources he had consulted. His encyclopaedic work runs to ten volumes in one modern edition. Separate books or chapters are devoted to astronomy, geology, geography, human oddities, agriculture, medicine, and gemstones among many other subjects. Pliny died in the eruption of Vesuvius in 79 CE. If his encyclopaedia is an amalgam of the reliable and the incredible, he did have access to many works that are now lost and so provides valuable historical information. Moreover, in the early Middle Ages his *Natural History* was one of the few works of ancient science that circulated in the Latin West and assumed an importance far out of proportion to its merits.

Roughly contemporary with Pliny was the Greek writer Dioscorides, who wrote a much more specialized account of plants, animal products, and minerals that can be useful to medicine. It was cited often by Galen. This work was translated into Latin by the sixth century and is often known by its Latin title, *De materia medica*. A spectacularly illustrated Greek manuscript copy was made at Constantinople around 512 CE and is now known as the 'Vienna Dioscorides'. Dioscorides described some six hundred different plants. Historians have debated whether the illustrations descend from Dioscorides' own work or were a later addition.

The Sciences in Ancient Education

We can get some idea of the place of the sciences in liberal education from a number of different sources. Strabo, the geographical writer of the first century CE, asks what sort of previous astronomy education readers of his own work are going to need. He says they need not know all the risings and settings of the constellations, but they should not be so ignorant as never to have seen a globe with its celestial circles, or be unaware that the celestial phenomena change with geographical location. But, says Strabo, this is the sort of knowledge that can be obtained from the elementary courses of mathematics. He therefore intends his *Geography* not for the uneducated, but for those who have taken the usual courses for freemen and students of philosophy. So we may infer the existence of surveys of astronomy offered to students as part of their philosophical training.

And, indeed, several nice introductions to astronomy do survive from the Hellenistic and Roman periods, and they come in different philosophical flavours. Theon of Smyrna's *Mathematical Knowedge Useful for Reading Plato* includes sections on Pythagorean arithmetic, astronomy, and music theory. There was originally also a section on geometry, but this has not survived. The astronomical portion is a competent survey of most parts of Greek astronomy, including epicycle and deferent theory. But Theon also includes frequent references to Plato's works.

A second example is the *Meteora* of Cleomedes, which also treats of standard astronomical matters and is our only source for the details of Eratosthenes' measurement of the size of the Earth (though it skips epicycles and deferents). But the *Meteora* is also steeped in Stoic physics. For example, Cleomedes discusses the *pneuma* and informs us that there can be no gaps, no empty places in the universe. But he accepts an infinite void place outside the universe. These two works, then, give us an impression of how astronomy was covered in the preparatory portions of a curriculum of Platonist or of Stoic philosophy. A third engaging text is the *Introduction to the Phenomena*, by Geminus, a writer of the first century BCE. Geminus does not show a marked philosophical affiliation. He is very much the straight-ahead astronomer, but is fond of quoting a literary figure such as Homer or Aratus in illustration of some astronomical point.

In the fourth century CE, Martianus Capella composed a work called *The Marriage of Philology and Mercury*. In Martianus' text, divided into nine books, Philology, personified as a maiden, is wed to Mercury, the god of mathematics and writing. The two opening books give an allegorical account of the betrothal and wedding. Then, in the following seven books, each of the Liberal Arts, handmaids to Philology, steps forward in turn to tell everything important to be known about her art. We receive accounts of grammar, logic, rhetoric, arithmetic, geometry, astronomy, and music. The four mathematical arts are, of course, precisely those that Plato had recommended for the school curriculum of the guardians. Of course, the level of instruction in *The Marriage of Philology and Mercury* was elementary. But in the renaissance of learning around the court of Charlemagne (late eighth to early ninth century), this book was admired as a compendium of all useful learning. With the establishment of the universities, starting around 1200, these same seven liberal arts played major roles in the curriculum. The scientific component of medieval university education therefore had deep roots in the liberal education of the ancient Greeks and Romans.

2 Science in Ancient China

DONALD HARPER

THE idea of nature and the investigation of phenomena so as to apply the knowledge gained to both satisfy human curiosity and serve human needs are basic elements of science as understood in modern society. In China, the period between the fifth century BCE and the second century CE was crucial for the formation of scientific knowledge and its application to nature as well as to human society. On the one hand we may consult the record of ancient transmitted texts to tell the early history of science in China. At the same time, since the twentieth century archaeological evidence and newly discovered manuscripts have supplemented the transmitted textual sources to give us a new understanding of ancient Chinese knowledge of the natural world. Three areas of knowledge are most relevant to the history of science: cosmogony (the study of the origins of the universe) and cosmology; astrology and the calendar; and medicine. The areas could be expanded to include mathematics, music, and geography. However, the previously named areas were central in the emergence of the model of nature that modern Western studies refer to as Chinese correlative cosmology or correlative thought. By the first century CE the paradigm of correlative thought was well established and remained the dominant paradigm in Chinese science throughout the pre-modern period.

Stated in brief, two ideas informed the mature stage of correlative thought. First, the vital stuff occurring everywhere and in everything was *qi*, literally 'vapour' in the sense of something evanescent yet material. Second, the formation of the world and its continuous operation came about as *qi* underwent processes based on cycles of *yin* and *yang*, and cycles of the *wuxing* or 'five agents' (referring to soil, wood, metal, fire, and water). The body and the body politic were linked to nature in correlative thought, which explained human physiology and was the basis for medical theory, and which also explained the stable institution of government.

From the time of its formation, correlative thought provided naturalistic explanations of phenomena, but it did not represent the emergence of a science that defined itself in opposition to religious or occult ideas and practices. Beginning with the early use of *yin* and *yang* in divination in connection with the hexagrams of the *Classic of Changes* (*Yijing*), *yin yang* never became a theoretical principle stripped of divinatory value; there also remained a relationship between the five agents and astrological or calendar-based divination systems that kept alive their occult significance at the same time as they were applied in correlative thought. Thus, the correlative thought

Silk manuscript with drawings of celestial phenomena accompanied by prognostications. Shown are drawings of comets. Mawangdui tomb 3, second century BCE. Hunan Provincial Museum.

paradigm was more than a theory of nature, and its durability was the result of a broad understanding of nature and culture that encompassed everything from phenomena in nature to politics, religion, and daily life.

For the sciences of the heavens (astrology), earth, and humankind (medicine), correlative thought was the *explanans* applied to phenomena in nature, the *explanandum*. There was a tendency for the formation of idealizing explanations, and for emphasizing commonalities that linked diverse phenomena rather than the particularities of discrete events in nature. Nevertheless, correlative thought nurtured curiosity, observation, and empirical methods in ways that produced precise knowledge of nature, at times leading to discoveries and technology before other world civilizations. For instance, the hexagonal morphology of snowflakes was first observed in China in the second century BCE, in part because the six-sided structure of the 'snow flower' contrasted with the usual five-sided structure of other flowers in nature; and paper was being made in China as early as the third century BCE.

While the correlative thought paradigm serves as a frame for the history of science in China, ancient Chinese ideas of nature are too rich to reduce to a single, monolithic outcome. Moreover, the involvement in the formation of correlative thought of different groups of people who had reason to take special interest in nature, the

Silk manuscript on planetary cycles and associated astrology. Mawangudi tomb 3, second century BCE. Hunan Provincial Museum.

diversity of ideas produced, and their applications constitute important elements of the social and cultural history of science and technology. In ancient China the leading figures in this history were not the moral and political philosophers of the classical age, from Kongzi (Confucius) in the late sixth century to Hanfeizi and Xunzi in the late third century BCE. Even the men to whom the Dao (Way) in the world and human society was a universal principle—a concept that since the third century BCE was invariably associated with the legendary figure of Laozi and the book *Classic of the Way and Virtue (Daode jing)*—did not take nature itself as their primary object of study. It was diviners, astrologers, calendar-makers, physicians, and related specialists who first observed patterns in nature and applied their knowledge for practical results. A few of their names were handed down in history, including astrologers and physicians who happened to gain prominence at the court of one of the regional rulers, but mostly they were anonymous.

The beginning point for this account of science in ancient China is the fifth century BCE, which coincides with the beginning of the period known as the Warring States

(481–221 BCE). For a number of reasons this century marks an important period of transition in ancient Chinese civilization. By the fifth century BCE, the competition for political power among the rulers of states in a region defined by the Yellow River in the north and the Yangtze in the south, with the state of Qin on the western edge and Qi on the northeast coast, stimulated speculation in many areas. By the fourth century BCE there was a flourishing manuscript culture, and written texts became one standard for defining a body of knowledge in a society in which literacy and manuscript production were valued highly. In 221 BCE Qin conquered the other states to establish its empire, replaced in 206 BCE by the Han dynasty (206 BCE–220 CE). By the second century CE the major features of forms of knowledge related to science were set and the world-view based on correlative thought was generally accepted.

As with the study of science in other ancient civilizations, we need to adjust our view of science as a modern form of theoretical and applied knowledge to accommodate the culture and society of ancient China. Divination, which includes astrology, was behind the development of systematic schemas to explain patterns in nature. Precise observation of the motions of the Sun, Moon, stars, and planets as well as of other phenomena occurring in the space between heaven and earth did not lead to a division between divination-based astrology and what we would call astronomy. In medicine, innovations in drug use owed much to the idea that drugs purged harmful agents who caused illness, including demons. Of course there were individuals who expressed views on the distinction between the human world, nature, and the spirit world. Xunzi and Hanfeizi in the third century BCE insisted on the primacy of the human world and criticized ideas that confused the boundaries. The physicians who compiled the *Yellow Emperor's Inner Classic* (*Huangdi neijing*), which dates to roughly the first century BCE, adamantly espoused 'natural' medical concepts of illness while rejecting demons and magical or religious ideas. Nevertheless, their argument was necessary in part because the correlative thought they embraced was used by others to justify occult views.

Some Historical Figures and the Formation of Scientific Knowledge

Ideas and activities begin with people. It is useful to survey the scant information about specialists who applied their knowledge of nature to serve human needs, and how their knowledge was transmitted. I claim no historical role for the men described below in the development of particular ideas. Rather, I see them as representative of the process of knowledge formation in ancient Chinese science, which proceeded alongside developments in political and moral philosophy.

Han dynasty historiography represented Zi Wei of the state of Song as one of the fifth-century BCE specialists in the 'calculations of heaven', and there was a book attributed to him entitled *Star Director Zi Wei of Song* (*Song Sixing Zi Wei*). Warring States sources confirm that Zi Wei was the astrologer at the Song court who predicted the baleful appearance of Mars in the constellation Xin (Heart) in 480 BCE; the book attributed to him did not survive beyond the Han dynasty.

In Han times the most famous Warring States astrologers were Shi Shen of Wei and Gan De of Qi (alternatively, Chu). They were reputed to have compiled star catalogues in the fourth century BCE that recorded celestial observations with measurements along the celestial equator. Their works are lost, but fragments survive in medieval Chinese sources. Modern analysis of the star data indicates that the time of observation was probably in the first century BCE, but other evidence does corroborate that they were active in the fourth century BCE.

For physicians, the *Zuo Chronicle* (*Zuo zhuan*; most likely composed in the fourth century BCE) included an event dated to the sixth century BCE when Physician He diagnosed the ruler of Jin with an ailment caused by sexual excess. His diagnosis included theoretical discussion of the 'six *qi*' in heaven that accounted for taste, colour, sound, and illness. The first two *qi* were *yin* and *yang* in the early sense of 'shade' (*yin*) and 'sunshine' (*yang*). Physician He's discussion is often cited as the earliest example of cosmological discourse in which the terms *qi* and *yin yang* occur.

Another physician, Wen Zhi, was famous for curing king Min of Qi in the early third century BCE at the cost of his own life. Eliminating the illness by causing king Min to explode with anger was the only treatment, and the ruler would not forgive the physician's offensive conduct. New evidence of Wen Zhi has come to light in the third- to second-century BCE medical manuscripts discovered in Mawangdui tomb 3, Hunan, in 1973. The text provides ten teachings on macrobiotic hygiene (including breath cultivation, exercise, dietetics, and sexual techniques). The ninth describes Wen Zhi in the late fourth century BCE instructing king Wei of Qi on sleep as the key to physical well-being and long life. Moreover, Wen Zhi tells the ruler that his medical knowledge fills three hundred manuscript scrolls. Except for the Mawangdui manuscript teaching, none of Wen Zhi's writings were transmitted.

The role of written texts was crucial in the transmission of scientific knowledge from specialists to disciples and in the spread of ideas to a wider readership among the elite. We are fortunate to have a detailed second-century BCE account by the physician Chunyu Yi of his own medical training, including his receipt of texts, which is seemingly true to life. According to Chunyu Yi, his father was not a physician. The young Chunyu loved medical recipes and received texts from his first teacher, Gongsun Guang. One of the texts that he transmitted to Chunyu was 'recipes for *yin yang* transformations', which Chunyu copied and returned, promising that he 'would not dare recklessly transmit them to others'.

When Chunyu was accepted as disciple by Yang Qing, Yang Qing chose to transmit his medical knowledge to Chunyu rather than to his sons. The texts Chunyu received included 'books of vessels of the Yellow Emperor and Bian Que' (*mai* 'vessels' are the physiological structure in the body containing blood and *qi*; Bian Que was a legendary physician). After a three-year period of training, Chunyu finally began medical practice on his own: 'I treated other people—examining ailments, judging death and life, and obtaining truly excellent results.' Chunyu's biography includes other details of his medical ideas and practice, which resembled the medicine of the roughly first-century BCE *Yellow Emperor's Inner Classic*, with important differences. Today,

Mawangdui tomb 3 medical manuscripts and another contemporary pair of medical manuscripts from Zhangjiashan tomb 247, Hubei, provide additional evidence of medical ideas and practices in the period of the third and second centuries BCE.

I save for last the figure of Zou Yan in the third century BCE. Zou Yan merits special attention because he has been characterized in some modern studies as the founder of Chinese scientific thought, which if accurate suggests a role for Zou Yan similar to the role played by a philosopher such as Aristotle in the history of ancient Greek science. However, Zou Yan's position as philosopher in the third century BCE remains unproven. His existence is virtually unnoticed in contemporary historical sources; only Hanfeizi mentioned him, condemning Zou Yan for spreading false faith in divination and astrology. This is hardly the Zou Yan described in Han historiography, where Zou Yan is described as an influential intellectual figure in Qi *c.*250 BCE; and is credited with using *yin yang* and five agent theories to speculate inductively on phenomena in nature, including cosmology and cycles of political change based on the five agents.

Zou Yan's historicity and his presence in Qi are not in doubt. Hanfeizi's condemnation notwithstanding, it is even possible that his activities marked a turning point for the acceptance of cosmological ideas in the philosophical and political mainstream. During the third century BCE the intellectual climate was changing as ideas based on *qi*, *yin yang*, and the five agents were increasingly applied to human beings and human institutions, and as human activities were modelled on the patterns in nature in a relationship of microcosm and macrocosm. Philosophy, especially political philosophy, moved in the direction of cosmology, and ideas related to nature were no longer the knowledge mainly of astrologers, diviners, and physicians. Nevertheless, during his lifetime Zou Yan probably associated more with astrologers and calendar-makers than with the philosophers.

Cosmogony

The origin of the cosmos, the creative activities that brought it into existence, its material and immaterial constituents, whether its original creation could be repeated, and the speculative uses of cosmogony for knowledge of nature as well as of the human and spirit worlds are all in evidence in ancient Chinese texts. Cosmogonic accounts or intriguing fragments of accounts occur in transmitted sources, to which we can now add new accounts from ancient manuscripts. In order to have a point of reference for the variety of evidence, let us begin with the second century and the essay 'Divine Model' ('Lingxian') by Zhang Heng (78–139). Zhang Heng was a literati-intellectual educated in the orthodox classics and simultaneously skilled in the technical arts that included astrology and the calendar. He held the position of court astrologer several times.

Zhang Heng's 'Divine Model' is a summary of contemporary knowledge of the cosmos, including the newly dominant *huntian* 'spherical heaven' theory according to which the cosmos was composed of two main parts: heaven was a rotating sphere and the earth was inside the sphere, its flat surface fixed in the lower centre. As the sphere of heaven rotated, various celestial bodies situated on it passed over the earth (the movement of the stars as if fixed to the sphere was distinguished from the motion of the Sun, Moon, and planets). Here is an excerpt of Zhang Heng's cosmogony from the 'Divine Model':

Before grand purity—in deep clarity and dark stillness, in isolated desolation and obscure silence—it was incapable of making its image. Its interior was vacuity, its exterior was nothingness. Continuing like this for a long time, this was the oceanic boundlessness. Doubtless, then, it was the root of the Way. The root of the Way was already established, and from nothingness being was generated. Grand purity first sprouted, and its sprouting was not yet manifest. Matching *qi* was identical in colour, chaos was not separated. Thus the *Treatise on the Way* [Laozi's *Classic of the Way and Virtue*] says: 'There is a thing chaotically made. It was generated before heaven and earth.' Its *qi* and body were absolutely incapable of becoming form, its motion was absolutely incapable of being measured. Continuing like this for a long time, this was the gigantic vastness. Doubtless, then, it was the trunk of the Way. The trunk was already produced, and there was a thing that became a completed body. Thereupon, primal *qi* split apart. Rigid and pliant first separated, clear and muddled occupied different positions. Heaven was formed on the outside, the earth was fixed on the inside. Heaven embodies *yang*, thus it is round and moves. The earth embodies *yin*, thus it is level and still. Moving is how it enacts and bestows, stillness is how it blends and transforms. Essence was accumulated, amassed, and bound together, and there was timeliness in producing all kinds. This was the prime of heaven. Doubtless, then, it was the fruit of the Way.

The crucial transition in the 'Divine Model' cosmogony is from *wu* 'nothingness' to *you* 'being' in the phrase 'from nothingness being was generated'. *Wu* and *you* were a contrasting pair in ancient Chinese ideas about the material conditions for existence. *You* referred to 'being' in the sense of 'existing, something existing' and *wu* 'not existing, nothingness' was the negative form of *you*. *Wu* was not understood as an abstract notion of pure negation but rather as the necessary complement to existence, hence the idea that the negation of *you* preceded *you* in time and was what generated *you* is common to many Chinese cosmogonic accounts. The precise moment when *qi* was present in the cosmogonic process is unclear in the 'Divine Model'. The word is first mentioned at the stage of 'grand purity' when *qi* was undifferentiated in a condition of chaos ('chaos' translates as *hundun*, an enveloping enclosure with the potential to be a container for the cosmos but whose interior is unknowable). Subsequently there is a prominent role in creation for 'primal *qi*', which split to produce the *yang* heaven on the outside in accordance with 'spherical heaven' theory and to produce the *yin* earth inside the sphere.

Two further aspects of Zhang Heng's cosmogony are noteworthy. First, the 'Divine Model' quotes the Warring States *Classic of the Way and Virtue* as evidence of the condition of chaos that preceded the formation of heaven and earth. The word Way (Dao) occurs in cosmogonic accounts in Warring States manuscripts discussed below. However, the manuscripts are not indebted to the concept of Way in the *Classic of the Way and Virtue*, and its importance as a source of cosmogonic ideas in the Warring

States period remains unclear. By the Han dynasty, Way was understood to be the cosmic Way in accordance with new developments in correlative thought. In the 'Divine Model', for instance, cosmogony is presented as the growth of the Way in three stages: the root of the Way, the trunk of the Way, and the fruit of the Way.

We now have several Warring States cosmogonic accounts in newly discovered manuscripts and one account in a Han manuscript that is better described as cosmological. Each account reflects different intellectual concerns; and together they attest to speculation before the consolidation of correlative thought in the first and second centuries CE.

I offer details of two of the Warring States accounts. The first cosmogony is in a manuscript composed of fourteen bound bamboo slips excavated from Guodian tomb 1, Hubei, in 1993. The approximate burial date of 300 BCE is based on comparison with other tombs in the region. The title assigned to the manuscript is taken from the first sentence of the first slip: *Grand One Generated Water (Taiyi sheng shui)*. *Grand One Generated Water* is significant in the first place as the oldest Chinese example of a water cosmogony. Modern studies of the text are divided on the issue of Taiyi (Grand One). There is fourth-century BCE evidence of Grand One as the name of a supreme deity in Warring States religion, and there is good reason to read *Grand One Generated Water* as a religious cosmogony in which Grand One initiated creation. Nevertheless, we should not overlook two possibilities: that at the time *Grand One Generated Water* was composed the name of the supreme deity was applied in a new meaning to describe a naturalistic process rather than the creation of the cosmos by a deity; or that *taiyi* ('grand one') was already an accepted term for a naturalistic first principle, and that the cosmogony was unrelated to religious ideas.

Cosmogony in *Grand One Generated Water* proceeds by stages in which the preceding stage generates the following stage. After Grand One generated water, water rejoined Grand One to generate heaven, then heaven rejoined Grand One to generate earth. Once heaven and earth existed as the first pair, the subsequent stages of creation are: heaven and earth generated *shenming* 'spirit illumination'; spirit illumination generated *yin yang*; *yin yang* generated the four seasons; the four seasons generated cold and hot; cold and hot generated wet and dry; wet and dry generated the year.

There are several distinctive features of the Grand One cosmogony. First, heaven and earth emerge from the initial interaction of Grand One and water without the stages of darkness and chaos that characterize Zhang Heng's cosmogony. Next, *yin* and *yang* are a stage of creation but they do not define heaven and earth. On this point, there is important evidence later in the text where there is another definition of heaven and earth: 'Below is soil, and one calls it the earth; above is *qi*, and one calls it heaven.' Instead of equating earth with *yin* and muddled *qi*, and heaven with *yang* and clear *qi*—and then speculating on the stage when the difference in *qi* caused them to separate—*qi* is the stuff that composes heaven while soil is another stuff that composes the earth. Looking more generally for *yin yang* and five agent ideas in the manuscript we do not find them (neither water nor soil is treated as being among the five agents). In the cosmogonic account water emerged first in the process of creation initiated by Grand One; *yin yang* are generated by spirit illumination and in their turn generate the four seasons. Later in the text the *qi* of heaven and soil of Earth are not related to *yin yang*.

The other cosmogonic account is a manuscript composed of thirteen bamboo slips that is currently in the collection of the Shanghai Museum. The title *Everness Preceding* (*Hengxian*) is written on the back side of the third slip and would have been visible when the manuscript was rolled up. *Everness Preceding* was one of many bamboo-slip manuscripts that appeared on the antiquities market in Hong Kong in 1994. Over 1,500 slips or fragments of slips, mostly still mired in mud and water from the site where they were dug up, were acquired by the Shanghai Museum. The Shanghai Museum completed conservation of the slips in 1997. Their provenance remains unclear (they might have been taken from a tomb in Hubei Province similar to Guodian tomb 1), but tests performed on samples of the bamboo and characteristics of the manuscripts attest to their authenticity as late Warring States artefacts. Like *Grand One Generated Water*, *Everness Preceding* can be assigned an approximate date of 300 BCE.

Comparing *Everness Preceding* to *Grand One Generated Water*, the first matter to note about its cosmogonic account is the statement that 'muddled *qi* generated the earth and clear *qi* generated heaven'. *Yin yang* does not occur in the manuscript, but *Everness Preceding* is at present the first cosmogonic account to describe the formation of the earth and heaven from muddled and clear *qi*. Other similarities to later transmitted sources include: nothingness preceded being; *qi* was present before the emergence of the earth and heaven; and conditions of stillness, vacuity, and dimness preceded the emergence of the earth and heaven.

Everness Preceding does not refer to *yin yang* and five agent ideas, nor is there evidence of the Way alluding to the *Classic of the Way and Virtue*. The created cosmos is referred to once as *tiandao* 'way of heaven' (this term is also used in the later part of the *Grand One Generated Water* text) and 'heaven' occurs once with a similar meaning. Mostly the text refers to the world after creation as *tianxia*, literally 'under-heaven', which was the standard term for the world, including humankind, in Warring States philosophical and political discourse. Between the two Warring States cosmogonic accounts, *Everness Preceding* makes the stronger statement of the political relevance of cosmogony and cosmology, and is less related to the calendar than *Grand One Generated Water*. The common element in both cosmogonies is the formation of the cosmos for the benefit of the human civilization that inhabits it.

The Han manuscript with a cosmological account was discovered in 1998 in Kongjiapo tomb 8, Hubei, and dates to the second half of the second century BCE. The content of the manuscript is hemerological and is related to other manuscript examples of ancient Chinese hemerology (refering to the various systems for determining favourable or unfavourable times and places for engaging in activities during the calendar year). The section entitled 'Year' ('Sui') occurs at the end of the manuscript. Among manuscript records of cosmogony and cosmology 'Year' is unique for giving prominence to the five agents and for applying five agent-based cosmology to hemerology and the calendar.

'Year' is not about creation, but about reestablishing cosmic order after the original order was destroyed. Beginning with wood in the east, each agent is installed in its proper direction with its proper symbolic colour. Once the five agents restore the cosmos, the function of the cosmos for humankind is realized:

Thereupon, this was called the regulation to fix the four directions and to harmonize *yin yang*. Feminine and masculine then were congruent. Thereupon, the sun was matched with the moon and the moon was matched with the year, each one having twelve temporal periods.

Although 'Year' is concerned with cosmology from the perspective of hemerology, we can read its emphasis on the five agents as an alternative to the priority granted to *yin yang* in subsequent Han cosmogony and cosmology. Moreover, the account encapsulates the connections between cosmogony, cosmology, the calendar, and the well-ordered world of humankind before the mature stage of correlative thought several centuries later.

Observation and the Structure of Heaven before the Han

Ideas about cosmology and the calendar grew out of observation of celestial motions and the use of observation to produce models, which in turn were applied to phenomena in general. The basic features of observation-based astrology are part of understanding the background of Han dynasty correlative thought. Warring States, Qin, and Han artefacts and manuscripts have provided abundant data to bear out Joseph Needham's characterization of Chinese astronomy as 'polar and equatorial' and 'arithmetical-algebraical': polar and equatorial because heaven was organized around the Pole Star, which radiated outward to the ring of twenty-eight *xiu* 'stellar lodge' constellations situated near the celestial equator (the Sun's path was noted but it was not the central fact informing pre-Han spatial and temporal schemes); arithmetical-algebraical because the main purpose of observation was to detect celestial regularities and to express them in the numerical categories of the calendar. The basic conception of the structure of heaven pre-dates the Warring States period. During the Warring States, astrologers and calendrical specialists gave specificity to this conception with precise observations and theoretical elaboration. Their explanations of the macrocosmic operation of heaven and earth probably contributed most to the formation of the idea that all phenomena and human activity were linked in microcosmic synchronicity; that is, their role in the formation of correlative thought was seminal.

The form and function of heaven as a turning disc situated above the earth was manifest in the motion of the constellation Beidou (Northern Dipper; corresponding to the stars of the Big Dipper in *Ursa Major*) in the north polar region and the twenty-eight stellar lodges. The bowl of the Dipper was linked to the pole while the handle pointed like the hand of a clock to positions on the ring defined by the stellar lodges.

Lacquer box lid with the Northern Dipper constellation in the centre surrounded by the twenty-eight stellar lodges. Tomb of Lord Yi of Zeng, fifth century BCE. Hubei Provincial Museum.

The oldest example of this idea and the oldest attestation of the names of all twenty-eight stellar lodges is the design on the lid of a lacquer clothes-case from the tomb of Lord Yi of Zeng in Suixian, Hubei, dated to the second half of the fifth century BCE. The ring of stellar lodges was arranged in four palaces (north, east, south, west) of seven stellar lodges each and associated with presiding spirits: the lacquer clothes-case lid depicts the dragon (east) and tiger (west).

On the lid there is a date notation by the stellar lodge Kang (Gullet): *jiayin* third day. *Jia* is from the sequence of ten *tiangan* 'celestial stems' and *yin* is from the sequence of twelve *dizhi* 'earthly branches'. In combination the ten stems and twelve branches formed a cycle of sixty binomes that were used to count days in a sexagenary cycle (the sexagenary binome was in addition to the numerical count of days within each lunar month). The branches were also used to designate the division of the celestial equator into twelve equal stations associated with the Sun. Beginning with the first branch *zi*, which marked north and the winter solstice (positioned at the bottom of the circle), the branches were enumerated clockwise passing from the north to the east (vernal equinox at *mao*), south (summer solstice at *wu*), and west (autumnal equinox at *you*). The twelve branches assigned to the stations also designated the clockwise sequence of months. This sequence was the reverse of the path of the Sun through the stations, which is counterclockwise.

The date notation on the lid attests to the relationship between the twenty-eight stellar lodges and the twelve stations, as well as to the early use of the handle of the Dipper to mark the clockwise sequence of the twelve months. Observed regularly at dusk, the handle rotated clockwise through one-twelfth of the ring of the stations in unison with the branches and months of the calendar. Among the various Warring States calendars, some set the first month of the civil year at *zi*, corresponding to the beginning of the tropical year with the winter solstice; others set the first month of the civil year at *yin* to coincide with the beginning of spring. The fixed correlation between the station branches and the stellar lodges placed Kang (Gullet) within the station marked *chen*. Depending on the calendar in use, the sexagenary cycle day '*jiayin* third day' recorded by Kang (Gullet) on the lid designated the third day in the fifth month (if the first month was *zi*) or third month (if the first month was *yin*) of the year in question. It happens that *jiayin* occurred as the third day in the fifth month (or third month) in 433 BCE, the same year attested on an inscribed bell in the tomb. Some scholars have argued that the date on the bell should not be treated as proof of the date of Lord Yi's death and burial, which may have occurred slightly later; and 402 BCE is the next year when the sexagenary binome *jiayin* once again occurred as the third day in the appropriate month.

Decorative and informational at the same time, the lacquer clothes-case lid serves as a fifth-century BCE point of reference for the standard representation of cosmic structure in the form of 'two cords' and 'four hooks' enclosed within a circle (reference to cords and hooks first occurs in second-century BCE texts). The cords are the cross formed by lines from *zi* (north) to *wu* (south) and from *mao* (east) to *you* (west). The hooks span the corners on the circumference of the circle defined by the central cross: *chou* and *yin* on either side at the northeast corner, *chen* and *si* at the southeast corner, *wei* and *shen* at the southwest corner, *xu* and *hai* at the northwest corner. The cord-hook design represented the basic structure holding the cosmos together and providing the framework for cosmic processes to take place. The ends of the cords and hooks that intersect the circumference of the circle numbered twelve, corresponding to the twelve branches.

Han Cosmology, Astrology, and the Calendar

As all who watch the sky know, the sky and the calendar are not forever fixed. The archaeological material discovered to date is notable for presenting a schematic model of heaven derived from astrological and calendrical calculations at the cost of observational precision. Historical sources of the Han period are especially valuable for evidence of the ongoing pursuit of observational precision and cosmological speculation, with corresponding influence on the calendar. While the centre of activity was often among officers at the Han court, it is noteworthy that the court did not monopolize the flow of knowledge and discovery. The calendrical reform ordered by Emperor Wu in 104 BCE, and the emergence of the *huntian* 'spherical heaven' theory sometime between the first century BCE and first century CE are of interest both for the ideas brought forward and the people involved.

The calendrical reform ordered by Emperor Wu was intended to produce a new calendar that would inaugurate a new beginning on the first day of the first year of the new reign era called Grand Inception (*taichu*). Sima Qian—who compiled the first comprehensive history, *Scribe's Records* (*Shiji*)—was court scribe/astrologer and a member of the group ordered to produce the new calendar. Court officers made measurements of the twenty-eight stellar lodges using gnomons and water-clocks to time the passage of a lodge's determinative star across the sightline established by the gnomon. This allowed them to calculate the degree-extension of each lodge in relation to the rotation of heaven. They then calculated the phases of the Moon to correct the sequence of the lunar months and determined the annual solar cycle of solstices and equinoxes. However, the court officers encountered difficulty in their calculations and requested assistance. More than twenty additional men were brought to court, including 'calendrical specialists from among the people'. When the calendar was completed, principal credit for the work went to two of the outsiders: Tang Du, identified only as a *fangshi* 'recipe gentleman' (the name typically referred to a specialist in occult methods) who measured the stellar lodges; and Luoxia Hong from Ba Commandery in the southwest (present-day Sichuan), who made the arithmetic calculation of the calendrical cycles.

There were probably clusters of specialists across the Han realm who engaged locally in activities about which we are poorly informed due to the court-centred bias of our sources. The vitality of the local exchange of ideas is evident in some of the newly discovered manuscripts. While cosmology, astrology, and the calendar increasingly served the political and ideological objectives of the Han court, and the court was an important centre for the exchange of ideas, the social milieu of the specialists was not subsumed within the milieu of the court nor did the court control the growth of knowledge. It is likely that Tang Du's knowledge and activities extended beyond his expertise in astrology and the calendar to include other skills that earned him the label of recipe gentleman in his region.

The idea of the *huntian* 'spherical heaven' had superseded the older view of heaven as a *gai* 'umbrella' situated above the earth by the time Zhang Heng composed the 'Divine Model' in the second century CE. The notion of heaven as the round umbrella held by a central pole above the square carriage chassis of earth is attested no later than the third century BCE. As a cosmological idea, the umbrella was the disc on which were placed the stars, Sun, Moon, and planets at a certain distance above the square earth. The visible movement of stars overhead was the disc of heaven turning clockwise about the pole, the cosmic axis located near the Pole Star; the Sun, Moon, and planets were thought to follow a counterclockwise path on the disc. The resemblance to the cord and hook model based on the Dipper and stellar lodges is evident. Missing from the umbrella model is a clear concept of the ecliptic, the path of the Sun through heaven, which is not attested unambiguously in Chinese sources earlier than the first century CE. Awareness of the importance of the path of the Sun along the ecliptic for more accurate measurement of the solar year coincided with additional evidence of the new spherical model of heaven.

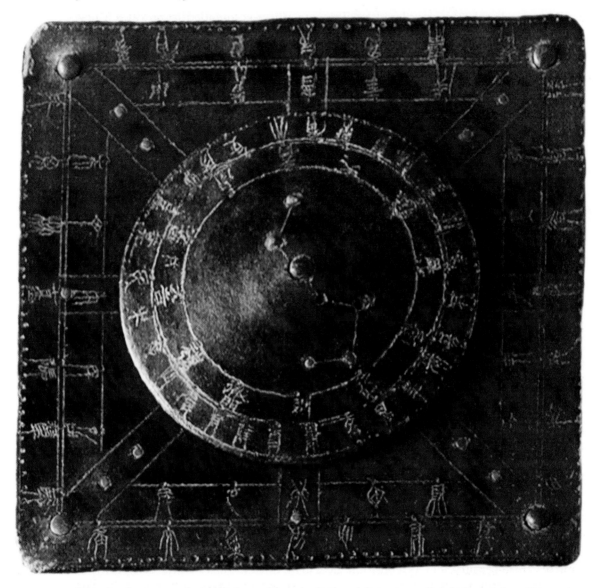

Two-piece lacquer astrological device used for calendar-based divination, called *shi* ('cosmic board') in ancient sources. The upper disc with the Northern Dipper rotates to align with positions on the square base-plate. Second to first century BCE. Gansu Provincial Museum.

One of the most important documents for the early history of the idea of spherical heaven, the use of the ecliptic, and the use of an armillary-ring instrument for observation of positions along the ecliptic (in place of the older gnomon and water-clock method) is the discourse presented to the Han court in 92 CE by Jia Kui. At issue was continued dissatisfaction with calendrical inaccuracies going back to the Grand Inception calendar of 104 BCE. Jia Kui noted that current calculations by court scribe/astrologers were inaccurate because their measurements for the Sun and Moon were based on the 'red way' (the celestial equator), whereas accurate measurements based on the 'yellow way' (the ecliptic) were already being obtained by a man named Fu An and others.

We know nothing about Fu An except for Jia Kui's reference to him, but clearly he was not one of the court scribe/astrologers whom Jia Kui criticized for their failure to embrace the idea of the ecliptic that had already produced verifiable results. In 103 CE the court finally ordered production of an 'ecliptic bronze instrument'—an armillary-ring instrument—for observational use, but the court scribe/astrologers made little use of it. The situation at court improved by the time Zhang Heng held the position of scribe/astrologer under Emperor An (r. 107–125) and Emperor Shun (r. 126–144). Zhang Heng's writings provided detailed accounts of the spherical heaven and armillary-ring instruments.

In 92 CE Jia Kui was writing from the perspective of someone who knew and accepted the spherical heaven idea and was familiar with armillary-ring instruments. His discourse mentioned that Geng Shouchang, minister of agriculture for Emperor Xuan (r. 74–49 BCE), used a 'diagram instrument' in 52 BCE to make measurements that were still based on the celestial equator, not the ecliptic. This instrument might have been as simple as a disc representing the plane of the celestial equator equipped with something to facilitate sighting. If it was more elaborate, it might have been an early form of armillary-ring instrument with a single ring for the celestial equator. The main difference between instruments of this type and the older gnomon was that the former could have been used to measure angles and distances between celestial bodies in three-dimensional space whereas the gnomon was used to observe the transits of celestial bodies overhead and to make azimuthal measurements of the degrees that separated celestial bodies in a circle.

At the end of the first century Jia Kui was still asserting the superiority of the idea of spherical heaven for its greater accuracy in measuring the motion of the Sun and Moon. By the second century, spherical heaven influenced cosmological and mathematical ideas in the *Zhou Gnomon* (*Zhou bi*), a work composed by unknown specialists active outside the Han court. If there was continued controversy at court over the better idea—spherical heaven or umbrella heaven—it was resolved by the time that Zhang Heng served as court scribe/astrologer. However, cosmology and the calendar were not isolated from other intellectual and spiritual currents at court. The essential elements of the theory of nature based on a unified concept of *qi*, *yin yang*, and the five agents were already in place by the first century CE. Many of its details presumed correlations among things and phenomena that derived from older cosmological ideas, and that the adoption of spherical heaven did not displace.

Iron ruler with gold inlay. Mancheng tomb 2, second century BCE. Hebei Provincial Museum.

Bronze clepsydra. Second century BCE. Hebei Provincial Museum.

Qi, Yin Yang, *and the Five Agents*

A summary of Han ideas in the first century CE is the best approach to discussion of older ideas about *qi*, *yin yang*, and the five agents before their consolidation in a comprehensive theory of nature. By the first century *qi* was regarded as the stuff that was everywhere in the cosmos; it accounted for underlying factors in the occurrence of phenomena and for the medium in which phenomena occurred. From the perspective of modern ideas we may argue about the materiality or immateriality of *qi* and its function. For Han people *qi* was the irreducible base on which everything else depended, from the 'air' they breathed ('*pneuma*' and 'breath' are both used to translate *qi*) to the dynamic aspect of cosmological processes (whether 'energy' is an appropriate translation is arguable). *Yin yang* and the five agents served to classify the characteristics of *qi* as it was identified in cycles of time, in configurations of space, or in the life process of individual organisms. Further, if before the Han period *yin yang* and the five agents had not yet emerged as the twin concepts of dualism and pentadic organization in the cosmos, as correlative cosmology coalesced around the ideas of *qi*, *yin yang*, and the five agents each term acquired new, broader conceptual significance.

The mature stage of Han correlative thought has been described as abstract, meaning that cycles of the universal *qi* were explained by reference to *yin yang* dualism and the pentadic organization of the five agents. If we focus on the emergence of the ruling theory for an all-embracing cosmological system and its uses in new forms of speculation, we may say that *yin yang* and the five agents signified aspects of cyclical change within the cosmos as represented by the theory. In Han astrological and calendrical theory, medical theory, or political theory, *yin yang* and the five agents functioned as signifiers to explain how *qi* operated. The resultant knowledge was then applied to the matter at hand, be it forecasting events, medical diagnosis, or government policy. The status of correlative thought as a theory of everything and the use of *yin yang* and the five agents as classifiers and signifiers tended to highlight their theoretical utility while minimizing their identification with concrete entities and their association with properties specific to those entities. In the case of the five agents—wood, fire, water, soil, metal—the emphasis on pentadic organization and cycles was probably a factor in the Han preference for the term *wuxing* rather than Zou Yan's older term *wude* 'five virtues' (understanding *de* as 'virtue' in the sense of the intrinsic power of something). 'Motion' was one standard meaning of *xing*, and would have underscored the idea of the cyclic motion of the five agents. In modern studies of *wuxing* theory the translation 'five phases' has become conventional, emphasizing the notion of passing through a sequence of cyclically recurring stages.

However, in the pre-Han understanding of the term *wuxing*, as attested in the 'Grand Model' ('Hongfan') chapter of the *Classic of Documents* (*Shujing*), *xing* denoted the distinct 'processes' that were characteristic of the five substances considered singly, not the idea of cyclically recurring stages. The more concrete understanding of the five substances that constitute the *wuxing* was maintained during the Han dynasty and remained part of correlative cosmology in later centuries. The section entitled 'Five Conquerors' ('Wusheng') in the Kongjiapo hemerological manuscript is the most explicit demonstration of the concrete application of the five substances. To have favourable conditions for travel, the person carried the substance that conquered the substance correlated with the direction of travel: carry water when traveling south in order to conquer the fire of the south, carry soil wrapped in cloth when travelling north to conquer the water of the north, and so forth. There is no ideal translation for *xing* in the term *wuxing*. The use in this chapter of the word 'agent' (in the sense of 'a substance that causes change') and the use of 'five agents' to translate *wuxing* recognize the material aspect of *wuxing* theory that is absent in the translations 'phase' and 'five phases'.

We cannot reconstruct the historical development of *qi*, *yin yang*, and five agent ideas in exact detail. The first matter to note about the situation before the Han dynasty is that *qi*, *yin yang*, and five agent ideas had not yet merged nor were they accepted as common knowledge. The ideas were still rarely mentioned in the texts of third-century BCE philosophers, or were mentioned critically, as, for instance, the condemnation of Zou Yan and other specialists in divination and astrology by Hanfeizi. However, reviewing the available evidence consistently indicates the Warring States as the period of transition from ideas that originated in the milieu of specialists (diviners, astrologers, calendar-makers, physicians, and others) to a more widespread and new view of the natural world that led to the formation of Han correlative thought.

Silk brocade bowman's armguard with astrological design and auspicious prediction in Chinese, which reads 'The Five Planets appear in the east and are favourable for the Middle Kingdom (i.e. the Han empire).' Probable date, first or second century CE. Xinjiang Archaeology Institute.

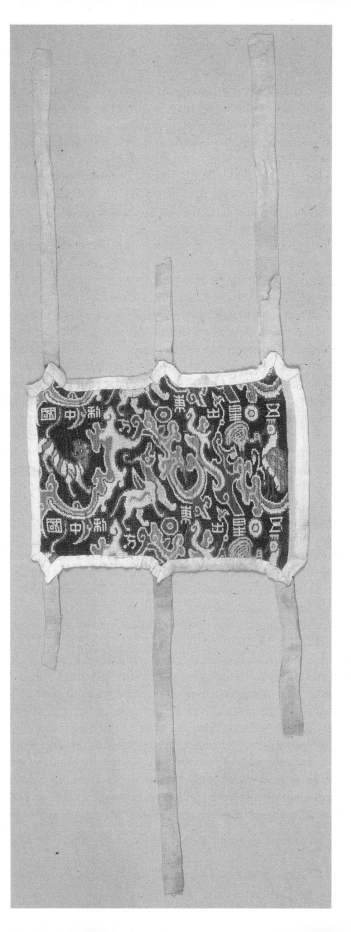

Let us begin with *qi*. The idea that *qi* and blood together were the essential components of human life is reliably documented only in fourth-century BCE sources, by which time *qi* already referred to the omnipresent stuff of the phenomenal world. It is not clear whether *qi* was initially a word for atmospheric vapours (clouds, steam, and so forth) that was generalized to encompass the source of human vitality and everything else, or whether *qi* was a term for the life-sustaining stuff received from food, drink, and air or breath, which was extended to the natural world. By the fourth century BCE the word was already a fixture in discourse on nature; barring new evidence, the question of etymological derivation is moot. While it is becoming accepted not to translate *qi* in Western languages, modern uses of the word in Chinese and in other languages often distort the ancient meaning. '*Pneuma*' and 'breath' are the most used translations, emphasizing the association of *qi* with human vitality; 'vapour' favours the association with volatile and pervasive stuff in the world at large.

The *Classic of Changes*, originally a divination book, played a key role in the formation of *yin yang* dualism. We now know that the oldest form of the mantic diagram associated with milfoil divination and the *Changes* was based on recording numbers derived from the sorting of milfoil stalks. By the fourth century BCE solid and broken lines were already replacing the numbers to form the six-line hexagrams as we know them in the *Changes*. Processes leading toward the reinterpretation of the *Changes* as a book of wisdom and cosmology were well underway by the end of the fourth century BCE, at which time the hexagram lines were already associated with *yin* (broken line) and *yang* (solid line). The *Attached Statements* (*Xici*) commentary of the third century BCE is the oldest transmitted text to claim that the *Changes* embraced the cosmos within its *yin yang* system and that the *Changes* provided humankind with knowledge of the operation of the cosmos.

Beyond validating the *yin yang* concept, the intent of the *Attached Statements* was to demonstrate that the *Changes* itself was the true manifestation of the concept; that is, the *Attached Statements* aimed to refute other *yin yang* claimants and establish the *Changes* as the foundation of cosmology. The *Attached Statements* argument belongs to the third-century BCE debate over cosmology, when *yin yang* and five agent ideas were being taken up by various specialists and became part of wider intellectual discussion.

We may glean a number of early ideas about the five agents from passages in the *Zuo Chronicle*. In addition to using the term *wuxing* in the sense of 'five processes', soil, wood, metal, fire, and water were also identified as the *wucai* 'five materials', and were understood to be resources of the earth that were exhaustible. The term *wuxing* referred to the materials from the perspective of their characteristic qualities; each 'did something' or exemplified a 'process'. According to the 'Grand Model' chapter of the *Classic of Documents*, water 'wets and descends', fire 'flames and rises', wood 'bends and straightens', metal 'conforms to change', and soil 'is sown and harvested'.

The sequence of the five processes in the 'Grand Model' chapter and their spatial arrangement in the *Zuo Chronicle* both suggest coordination with cardinal points around the centre: water and fire (first and second in the 'Grand Model') formed the north–south axis, wood and metal (third and fourth in the 'Grand Model') the east–west axis, and soil (fifth in the 'Grand Model') was at the centre. Only after *wuxing* correlations were extended to the seasons (not attested in the *Zuo Chronicle*) would the older spatial arrangement have served as the basis for the generation sequence of the *wuxing*.

A final development was needed before soil, wood, metal, fire, and water could function as rubrics for broad cosmological classification: the conception of the *wuxing* not just as material resources with their characteristic processes, but rather as the defining signifiers of the cycle of universal *qi*. There is no evidence that it occurred before the time of Zou Yan and his theory of the mutual conquest of the *wude* 'five virtues'. Transmitted sources consistently associate the term *wude* with Zou Yan, and the new term may have been adopted to enhance the conception of soil, wood, metal, fire, and water as possessing 'virtues' that extended beyond the substances and their processes, and that could be applied to other sets of five in cosmology.

Zou Yan also proposed the cycle of political change, which began with the Yellow Emperor who 'observed the ascendance of soil *qi*' and modelled his government on the symbolic correlations of soil. In each subsequent age, the new ruler recognized the ascendance of the next virtue in the conquest sequence: Yu founded the Xia dynasty when he observed wood *qi* ascendant (wood conquers soil); Tang founded the Shang dynasty when he observed metal *qi* ascendant (metal conquers wood); king Wen founded the Zhou dynasty when he observed fire *qi* ascendant (fire conquers metal). The successor to the Zhou dynasty would necessarily recognize the signs of water *qi* ascendant (water conquers fire), and in fact water became the emblem of the Qin dynasty after 221 BCE.

By the late third century BCE, *wuxing* correlative classification was expanding by imposing itself on older six *qi* correlations and by extending the range of phenomena it incorporated. In concept the *wuxing* paralleled *yin yang*: both provided a means to classify phenomena based on the characteristics of *qi*. At this point we can understand *wuxing* as 'five agents' in a cycle defined by *qi* that was applied broadly to phenomena. However, there was not yet a unified *yin yang* and five agent correlative cosmology, nor did the new ideas divorce themselves from divination and other forms of occult practice. In some areas of speculation about the natural world, *yin yang* ideas had already been incorporated while five agent ideas were only beginning to be applied.

One sign that the paradigm of correlative thought based on *qi*, *yin yang*, and the five agents had reached the mature stage by the first century CE is the definition of relevant graphs in the *Explanation of Basic Graphs and Analysis of Composite Graphs (Shuowen jiezi)*, compiled by Xu Shen in the first century (the first dictionary to analyse the composition of graphs and to define the words they represented). Xu Shen identified the graphs for wood, fire, metal, and water as the *xing* 'agents' associated with the directions east, south, west, and north respectively; and the definitions of the graphs for their correlated colours were also linked to the directions. He used the idea of primal *qi* when defining the graph for the word *di* 'earth', and the definition was itself a capsule account of cosmogony: 'When primal *qi* first separated, the light, clear, and *yang* portion made heaven; the heavy, muddled, and *yin* portion made earth.' The basic ideas of correlative thought must have been generally accepted for Xu Shen to have used them as definitions, and they influenced developments within specialized fields of knowledge as well as in society as a whole.

The *Assessment of Arguments* (*Lunheng*) of Wang Chong offers a different perspective on correlative thought in the first century. Wang Chong's world-view was grounded in contemporary correlative thought: he accepted the idea of primal *qi*; he regarded the 'Grand Model' chapter of the *Classic of Documents* as the fundamental text of five agents doctrine; and his thinking about humankind and nature was generally informed by *yin yang* ideas. However, Wang Chong was consistent in his conviction that nature operated without special regard for human activity and he was critical of what he regarded as excessive literalism in the application of *yin yang* and five agent ideas by some intellectuals and the ignorant populace.

For instance, Wang Chong criticized omenology based on correlative thought and the belief that phenomena in nature were intentional actions by which nature rewarded or punished people for their behaviour. Similarly, he ridiculed people for using substances associated with metal, wood, water, fire, and soil in *wuxing* theory in order to protect themselves from harm, which according to Wang Chong contravened the idea of the *wuxing* (no doubt he had in mind cases such as the use of substances for travel in the Kongjiapo hemerological manuscript). The *Assessment of Arguments* is valuable evidence of one literati-intellectual's views on contemporary correlative thought in society, yet Wang Chong did not engage in theoretical discussion of cosmology, astrology, medicine, or other fields of knowledge where the paradigm of correlative thought informed the ideas and activities of astrologers, physicians, and other specialists.

Medicine

The formation of a learned medicine between the fourth and first centuries BCE was accomplished by the *yi*, translated in this chapter as 'physician'. The word was applied to a range of people who practised medicine, and sometimes meant 'witch doctor', but Physician He in the *Zuo Chronicle* represents the kind of learned physician who by the fourth century BCE thought that *qi* was an essential concept for human physiology and pathology. After listing the six *qi* of heaven, Physician He linked them to six types of illness and associated illness with excess of *qi*. In third- and second-century BCE discussions of medicine there was an ever greater insistence among certain physicians that theories of *qi* were the only valid basis for medical practice, and these physicians increasingly condemned medical practices that did not conform to the new ideal.

Mai 'vessels' carrying blood and *qi* were the obvious structure for classifying *qi* in the microcosm of the human body. By the second century BCE, Chunyu Yi expressed a theoretically elaborate conception of the medical body in his account of the 'model of the vessels':

Ailment names are mostly alike and are unknowable. Thus the ancient sages created the model of the vessels for them, with which to initiate the measurement of dimension and volume, fix the compass and square, suspend the weights and balance, apply the marking-cord ink, and blend *yin* and *yang*. They differentiated the vessels in humans and named each one. Matching with heaven and the earth, (the vessels) combine in the human being to form a trinity. Thus they then differentiated the hundred ailments, distinguishing between them.

In Chunyu Yi's conception, the body was not simply fashioned in the image of the cosmos—its every function was synchronized concretely with the operation of the cosmos. Calculations necessary for determining the harmonious operation of *qi* in the *yin yang* cosmos were applied to the human body by ancient sages, with the result that they 'discovered' and named the vessels. The new physiological theory, attributed to sage invention, reconfirmed in medicine the cosmological trinity of heaven, earth, and humankind, and provided a new model for pathology based on diagnosing the condition of *qi* in the system of vessels rather than relying on arbitrary ailment names. The physician was both healer and cosmologist.

The *Yellow Emperor's Inner Classic* represents a digest of the new naturalistic medicine as of roughly the first century BCE. The original text does not survive, but we have three medieval recensions that preserve parts of the ancient classic: the *Plain Questions* (*Suwen*), *Numinous Pivot* (*Lingshu*), and *Grand Simplicity* (*Taisu*). In the *Yellow Emperor's Inner Classic*, *yin yang* and five agent theories were fully integrated with physiology and pathology. Moreover, given the definition of health as maintaining the harmonious circulation of *qi* in the body and illness as *qi* dysfunction, the ideal therapy espoused in the classic was acupuncture: needles inserted at strategic points on the body to correct the dysfunctional *qi* in the system of vessels.

Second-century BCE medical manuscripts from Mawangdui tomb 3 and Zhangjiashan tomb 247 provide new evidence of the medical knowledge that was current at the time of the formation of the *Yellow Emperor's Inner Canon*. For the sake of discussion, we may identify three main lines of development in medicine between the Warring States and the Han dynasty. First, recipe manuals such as the Mawangdui *Recipes for Fifty-Two Ailments* (*Wushier bingfang*) attest to the practical medical knowledge of ailment categories, drugs, and other forms of therapy (including fumigations, minor surgery, and magical treatments) that was inherited from earlier medicine and continued to grow in amount and sophistication in the centuries after the Han dynasty. As of the present, there is no evidence of a work of materia medica earlier than the transmitted text of the *Divine Agrarian's Classic of Materia Medica* (*Shennong bencao jing*), which can be dated to the first or second century CE. However, drug use in recently discovered recipe manual manuscripts provides important evidence for the formation of materia medica literature.

Second, we have texts from Mawangdui and Zhangjiashan that describe a system of eleven vessels in the body and associate particular ailments with the condition of the *qi* in specific vessels. The recommended treatment to correct the dysfunction was cauterization of the affected vessel. The vessels were classified into *yin* and *yang* vessels using nomenclature attested later in the *Yellow Emperor's Inner Classic* (which, however, proposed a twelve-vessel physiological model that was much more elaborate in terms of theory); five agent ideas were completely absent from manuscript accounts of the vessels. In addition, acupuncture was not mentioned. It is generally accepted that the Mawangdui and Zhangjiashan vessel theory texts represent an earlier stage in the development of the theory and practice of acupuncture that is described in the *Yellow Emperor's Inner Classic* and that cauterization preceded acupuncture as the therapeutic application of vessel theory.

Third, there are texts from Mawangdui and Zhangjiashan on the subject of macrobiotic self-cultivation—known in transmitted sources as *yangsheng* 'nurturing life'—with detailed information on dietetics, breath cultivation, exercise, and sex. These practices were intended to increase vitality and extend life. Since human life depended on the condition of the individual's *qi*, self-cultivation was essentially *qi* cultivation. On the testimony of the *Yellow Emperor's Inner Classic* it has been thought that vessel theory originally developed within the context of pathological theories, with the system of vessels serving as the basis for diagnosis of illness and treatment. The Mawangdui and Zhangjiashan texts indicate that vessel theory may have developed first in connection with macrobiotic self-cultivation and that it was then applied to pathology.

The Mawangdui and Zhangjiashan texts on the vessels and cauterization belonged to the same textual tradition; several are copies of the same text, such as the texts copied on the Zhangjiashan manuscript entitled *Vessel Book* (*Maishu*). Given the proximity in time (the tombs are separated by several decades at the most) and space (there was easy communication between southern Hubei and northern Hunan in the second century BCE), it seems likely that manuscripts with texts such as these circulated in the ancient south of China among specialists and non-specialists (neither of the men buried in the two tombs was a physician). At the same time, textual parallels with parts of the *Yellow Emperor's Inner Classic* demonstrate that the content of the manuscript texts represents older ideas; that is, the manuscript texts restore ante-cedents to the ideas of the *Yellow Emperor's Inner Classic*. In the case of the system of vessels, the manuscript texts describe eleven vessels, six *yang* vessels and five *yin* vessels, whereas the *Yellow Emperor's Inner Classic* describes twelve, six *yang* and six *yin*.

Further, in the *Yellow Emperor's Inner Classic* the *yin* and *yang* vessels were interconnected in a circulatory system that included internal organs, and the vessels themselves were conceived to be *jingmai* 'conduit vessels', a term that emphasized the circulatory system as a counterpart to cosmological cycles. Acupuncture applied to a point on one vessel affected the entire system. In contrast, the eleven vessels in the manuscript texts constituted separate channels for *qi* in the body. Cauterization applied to a vessel affected the *qi* in that vessel. One text, found both at Mawangdui and Zhangjiashan, differentiates the six *yang* vessels which contain the *qi* of heaven from the five *yin* vessels which contain the *qi* of the earth; and the *yin* vessels are the 'vessels of death' (meaning that ailments associated with *yin* vessels were likely to be fatal).

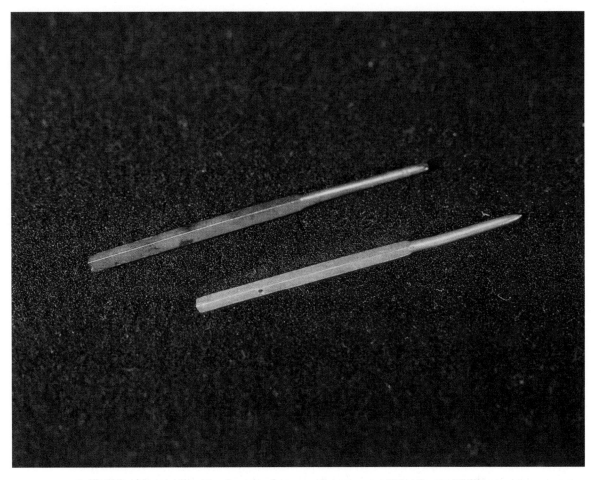

Gold acupuncture needles. Mancheng tomb 1, second century BCE. Hebei Provincial Museum.

While in the beginning vessel theory in medicine was not tied to *yin yang* and five agent ideas, its subsequent success had much to do with the growth of correlative thought. Once *yin yang* and the five agents provided the basis for understanding all phenomena, vessel theory facilitated the incorporation of correlative thought into medicine by providing an ideal schematic representation that allowed the broad application of *yin yang* and the five agents to the human body. This was the context for Chunyu Yi's 'model of the vessels'; and the *Yellow Emperor's Inner Classic* was the basis for medical theory and acupuncture in later centuries.

However, there were other physiological ideas in Han medicine. A text in the Zhangjiashan *Vessel Book* identifies six constituents of the body that each performs an essential function: bone is the pillar supporting the body, muscle is the binding, blood moistens, vessels provide channels through the body, flesh is the 'attachment' that gives the body mass, and *qi* is the 'exhalation' that vitalizes it. Each constituent has a characteristic quality of pain that is the basis for identifying the source of illness.

Vessels and *qi* are two of six components, not themselves the essential structure of the body. Yet even in the Mawangdui and Zhangjiashan manuscripts it is evident that the theoretical focus was already on *qi* and the vessels that contained it. By the time of the *Yellow Emperor's Inner Classic* any ideas about other body constituents were subsumed by vessel theory; physiological theory was vessel theory.

Vessel theory redefined health and illness by making each dependent on the condition of *qi*, which was supposed to flow evenly in the body without obstructions or harmful external intrusions (for example, by pathogenic wind). The manuscript texts and the *Yellow Emperor's Inner Classic* understood illness as a dysfunction of *qi* in the vessels that led to other physiological complications and, finally, to illness as an identifiable condition. Therapy corrected the dysfunction and restored the body to its original harmonious condition, thereby eliminating the signs of illness. Acceptance of vessel theory put into question the older conception that illness was caused by agents or pathogens, often from an external source, which sickened their victim (or patient). Demons or pathogens of a more naturalistic origin were alike in being the entities that caused certain nameable ailments; therapy cured the ailments by eliminating the entities that caused them. From the standpoint of vessel theory, such ailments were merely manifestations of the deeper dysfunction within the system of vessels and they did not have independent existence (this was the point of Chunyu Yi's skepticism about reliance on 'ailment names' that were 'unknowable'). Moreover, long-standing medical treatments for these ailments were useless because they were not designed to correct the dysfunctional *qi*.

Stated in terms of modern research in the history of medicine, we have a case of conflict between an ontological explanation of illness (which holds that ailments have an existence of their own) and a physiological explanation, vessel theory being the Chinese counterpart of humoural pathology and other types of physiological explanation in Western medicine. The Mawangdui and Zhangjiashan medical manuscripts bear witness to the coexistence of both ontological and physiological views in the third and second centuries BCE, with *Recipes for Fifty-Two Ailments* being the chief representative of the former and the vessel theory texts representing the latter. The *Vessel Book* text on the six constituents would have fitted with the ontological view of illness. Another *Vessel Book* text that classified ailments according to the part of the body where they occurred and the agent/pathogen also fitted the ontological view.

Macrobiotic hygiene is absent in the extant *Yellow Emperor's Inner Classic*. The Mawangdui and Zhangjiashan manuscripts are our first evidence in medical literature. The manuscript texts represent the baseline of self-cultivation taught by physicians such as Wen Zhi, whose instruction of king Wei in the Mawangdui text concluded with discussion of 'eating *qi*', or breath cultivation. In addition to writings on breath cultivation, the Mawangdui texts include the earliest examples of sex manuals, dietetic recipe manuals, and a manuscript with drawings of *daoyin* 'guiding and pulling' exercises for self-cultivation as well as to treat illness. The Zhangjiashan manuscript entitled *Pulling Book* (*Yinshu*) includes a text with written descriptions of therapeutic and self-cultivation exercises.

It is clear from the Mawangdui and Zhangjiashan manuscripts that vessel theory underlay ideas about self-cultivation, health, and longevity. One Mawangdui text on the 'model of the vessels' describes the body of the sage as the ideal (there is a second copy of the text in the Zhangjiashan *Vessel Book*):

Let the model of the vessels be taught to those below. The vessels are something the sage prizes. As for *qi*, it goes to the lower part (of the body) and harms the upper part; it follows warmth and departs from coolness. The sage has a cold head and warm feet. To treat ailments, take away the surplus and increase what is insufficient.

In the ideal body—the body of the sage—*qi* circulated in the vessels by moving downward; and the sage maintained warm feet to ensure the downward movement. According to the passage quoted, the 'model of the vessels' was secondarily applied to the treatment of ailments.

This view contrasts with Chunyu Yi's statement that 'sages created the model of the vessels' in order to deal with human illness. Which came first: vessel theory in pathology or vessel theory in macrobiotic hygiene? There is not a clear-cut answer. However, there is logic in supposing that vessel theory began in connection with self-cultivation practices, and over time became the basis for a new understanding of health and illness—at which point the significance of vessels and *qi* led vessel theory in a new direction in medical pathology. Once health was defined in terms of the circulation of blood and *qi* in the body, maintaining health was essentially monitoring the condition of the vessels; transposing this scheme to illness and pathology would have been a small step to take.

Conclusion

Looking at China between the fifth century BCE and second century CE for evidence of knowledge of nature and its uses, the formation of correlative thought was the result of a multiplicity of ideas and activities that over time formed a doctrine adopted by astrologers, physicians, and other specialists as well as by the elite whose world-view was influenced by correlative thought. To be sure, there were areas of knowledge and experience that fell outside the scope of correlative thought. Materia medica was nearly left out of the paradigm of correlative thought as expounded in the *Yellow Emperor's Inner Classic*, which took acupuncture as the perfect application of its medical theory, and the later history of pharmacy in China reflects this difference. Likewise, technological innovations occurred whether or not correlative thought was involved in their production.

Yet the cultural presumption that correlative thought was a theory of everything had the effect of stimulating the investigation of nature in serendipitous ways, and may account for the fact that many phenomena that were barely noticed in other ancient civilizations—snowflake crystals, comets, and atmospheric events such as parhelia, the properties of drugs, or plant domestication—were actively investigated and classified.

Opposite: Silk manuscript with drawings of therapeutic exercises. Mawangdui tomb 3, second century BCE. Hunan Provincial Museum.

3 Science in the Medieval Christian and Islamic Worlds

STEVEN J. LIVESEY AND SONJA BRENTJES

Is 'medieval science' oxymoronic? The common perception of science as a modern, inquisitive, rational, empirical enterprise may suggest that 'medieval science', if it exists at all, is superstitious, authoritarian, traditional, a remnant of an earlier and now discarded age. Yet with a little investigation, it becomes clear that it was in this past age that institutions like the madrasa and university emerged, where the nature of the heavens was discussed alongside that of the human body, where knowledge of nature itself and of numbers, areas, solids, and so on were considered, debated, and developed. Beside these teaching institutions stood hospitals and observatories, where theoretically grounded knowledge was put to work, checked, systematized, and modified. Religious, philosophical, and medical circles vigorously debated themes still of interest to us today; questions about the eternity of the world, the meaning of economic value, relationships between celestial bodies, and the points at which life begins and ends are rooted in medieval theology and philosophy.

In this chapter, we will explore the different medieval settings in which these and other themes were produced, contested, and/or discarded. We will seek not just the connections between times, regions, and communities, but also to identify their individual, particular patterns and achievements, as well as considering the unique and idiosyncratic medieval ideas that failed to become part of the Western scientific heritage. Carlo Ginzburg has noted,

The historian's task is just the opposite of what most of us were taught to believe. He must destroy our false sense of proximity to people of the past because they came from societies very different from our own. The more we discover about these people's 'mental universe', the more we should be shocked by the cultural distance that separates us from them.

Indeed, while a modern newcomer to the subject of this chapter might be surprised by the differences and distances between our intellectual ancestors and ourselves, we marvel at what connects us: the desire to know, to explore, to explain, to discuss, and to reject what we consider false or improper. Thus we will look at medieval science under successive rubrics, starting with a brief discussion of the nature of science in Latin before the twelfth century in Europe. We will then turn east to consider the role that sciences in Arabic or Persian played at different historical and geographical points, from astrology in Abbasid Baghdad to optics in Fatimid Cairo and astronomy in Mamluk Cairo and Damascus, as well as examining the part played by medicine and engineering in the contact zone between China and Iran in Mongol Tabriz. Returning west, we will look at the new notions of naturalism that emerged in the twelfth century, and preface our examination of the resulting translation movement with the earlier, prolonged, tradition of translation in Abbasid Baghdad. We will close with an analysis of the creation of new institutions in the Catholic regions of Europe, where natural philosophy and medicine also flourished, despite the nettlesome relationship of science and theology.

Early Medieval Assertoric Science

One of the most obvious features of early medieval writings on nature is its derivative character. Faced with a contracting awareness of ancient culture, many scholars in the fifth through the tenth centuries, and particularly those focused on natural philosophy in its broadest sense, attempted to preserve what they considered the central tenets of the ancient world. That conservative enterprise could take many forms. Some, like Boethius (*c*.480–524) in late antiquity, produced handbooks focused on particular disciplines, which became texts for the study of the four mathematical disciplines of arithmetic, geometry, astronomy, and music; in fact, they were frequently adaptations of translated versions of similar handbooks that had already been popular in the late ancient world. Others inserted their derivative stock within larger comprehensive works, as Cassiodorus (*c*.485–*c*.580) did when he included chapters on the disciplines of grammar, rhetoric, dialectic (elsewhere referred to as the 'trivium'), arithmetic, music, geometry, and astronomy (the 'quadrivium') within his larger exposition of Christian learning in the *Institutiones*. These sections seem not to have been replacements for the original works, but rather introductions to them, for Cassiodorus refers to his sources, presumably as an encouragement to the monks in his monastery, Vivarium, who could consult those works in the house library.

A different technique was adopted by the seventh-century Archbishop of Seville, Isidore (d. 636), whose encyclopedic *Etymologies* sought to convey 'what ought to be noted' about virtually every topic for the edification of both the clergy and Isidore's Visigothic patrons. From the text itself, it is clear that Isidore and presumably his assistants cut and pasted materials from sources at hand, sometimes misinterpreting them in the process, but also apparently unconcerned for internal consistency; thus, in Book XIII, he presents in adjacent chapters both Aristotle's causal theory of material substances and the atomist position of discrete particles in which the particles betray no causal mechanism in forming objects. All of this is organized on the principle of the etymology of names, for in contrast to Aristotle's dictum of human knowledge generated by understanding natural causes, Isidore contends that 'one's insight into anything is clearer when its etymology is known'.

Still another form of the same program involved catechetical questions and answers. A good example of this genre is an anonymous commentary on Genesis that circulated under the title *Intexuimus*, literally 'we have woven together'. It has been estimated that one-third of medieval discussions of cosmology were delivered in commentaries on Genesis (the so-called hexaemeral tradition), and the *Intexuimus* illustrates a frequently used technique. The author has woven together the positions of Isidore and the Church Fathers—especially Augustine, Jerome, and Gregory the Great— as he travels verse by verse through the first three chapters of Genesis, asking questions, then responding with orthodox positions. To the very first verse of Genesis, the author asks, 'why did God make heaven and earth?' and 'what did God do before he made heaven?' In answer, he quotes the Apocalypse (4: 11): 'you created all things, and by your will they existed and were created'. As a result, it is useless to ask why God created, why he chose a particular moment in which to act, and in fact time itself, according to the author, is a construct, for before the creation there was only God's eternity.

These four examples provided by Boethius, Cassiodorus, Isidore, and the anonymous author of the *Intexuimus* illustrate an important characteristic of science in the early medieval period, which may be referred to as its assertoric nature, proceeding by statement and assertion in contrast to the Aristotelian tradition of demonstrating effects through their causes. That is, while Aristotle had held that the fundamental characteristic of science, and by extension of humans themselves, was not just knowing the facts of the matter, but also the reasons for those facts; in each of these examples, by contrast, the focus was on the transmitted facts rather than on the reasons for those facts, and there was little concern for explaining or choosing among inconsistent theories. The assertoric nature of scientific knowledge in the period before 1050 can also be seen in the abridgement of Euclid's *Elements* (particularly in the compilations of material from Euclid and other sources known as Geometry I and Geometry II) that transmitted the propositions while removing the proofs, or in the tendency to use surveyors' handbooks like the late antique *Agrimensores veteres* or the Roman author Frontinus' writings on aqueducts as textbooks for teaching geometry.

We may make one further observation about the underlying perspective of man in nature during the period before 1050 in Latin Europe: with few exceptions, early medievals viewed themselves as vehicles for divine activity, inhabitants of a world that they neither understood nor controlled. To the extent that order was perceived, it found its place in ritual and symbolism and sacrament. Before nature, as before God, man considered himself impotent. Such a belief and trust in the divine did not exclude, however, man's desire to study God's creation and develop tools for surviving in His world. In Islamic societies, whether in Europe, Africa, or Asia, one multifaceted tradition was developing in the realm of astrology (or: studies of the stars), as we can see if we examine the conditions under which the city of Baghdad was founded.

Astrology as Science, Politics, and Guidance

In the Islamic month Jumada I of the year 145 (27 July–25 August 762), the second caliph of the newly established Abbasid dynasty, al-Mansur (r. 756–775), laid the foundations of a new city meant to become the capital of his vast empire: Baghdad.

He named it *Madinat al-Salam* ('The City of Peace'). Before the planned construction of the new capital began, the caliph had asked his most skilled scholars at court to choose with scientific means the best day for this history-changing enterprise. Our sources about this scientifically prepared city planning differ from each other. Hence, it is not clear which day was chosen for the momentous event. Neither can we be absolutely certain who the astrologers were who cast the horoscope. The task was to find the moment in time which would guarantee that no ruler would ever be killed in this new capital. Modern recalculations of the data of the horoscope determine 31 July, i.e. 4 Jumada I, as the day of foundation.

Among the people who left us reports about that day and its astrological preparation are two authors of histories and geographies, Ahmad al-Ya'qubi (d. 897/8) and Ibn al-Faqih (ninth–tenth century), along with one of the most competent scholars of the mathematical sciences in Islamic societies before 1200, Abu l-Rayhan al-Biruni (973–1048). Although Ibn al-Faqih's data agree better with the modern reconstruction, the most often cited account is that by al-Biruni. Only 27 years old, al-Biruni wrote a book about calendars, festivals, and dynasties called *The Chronology of Ancient Nations*. Here he reports that two Iranian and one Arab astrologer cast the horoscope for Baghdad's foundation, but provides as the decisive date 23 July 762, which precedes the month Jumada by four days. The three astrologers mentioned by al-Biruni were Nawbakht (eighth century), formerly a Zoroastrian and a new convert to Islam, Masha'allah b. Athari al-Farisi (d. *c.*815), a Persian Jew, and Ibrahim al-Fazari (d. *c.*777), perhaps from the Banu Fazara, an Arabic tribe in the Hijaz (today in Saudi Arabia). All three men became important representatives for the emergence of a scientific culture at the Abbasid court in Baghdad and the translation of Middle Persian and Sanskrit astronomical and astrological handbooks into Arabic.

Nawbakht came from Ahvaz in western Iran. He was the first of a dynasty of court astrologers, who served the Abbasids (750–1258) for about a century. Ibrahim al-Fazari edited the Arabic translation of a Sanskrit handbook, whose title possibly was *Mahasiddhanta* (The Great Treatise). The translation was made in 771 or 773 by an astrologer from India called Kankah, who may have come to the Abbasid court alone or as a member of an embassy from some kingdom in India. Its Arabic title appears to have been *al-Sindhind al-kabir* (The Great Sindhind; *al-kabir = maha*, *al-sindhind = siddhanta*). Ibrahim compiled his own handbooks and wrote treatises on the astrolabe, the armillary sphere, and astrology. His first handbook was based on the translated Sanskrit work and had the same title. The second handbook, perhaps commanded by Caliph al-Mansur, was computed 'according to the years of the Arabs'. Except for a few fragments, none of his works have survived the centuries.

Masha'allah came from Basra on the Shatt al-Arab, at the northern end of the Persian Gulf. When he participated in the casting of the horoscope, he was still a very young man. In the next half century, he wrote about twenty treatises, mostly about astrological topics, which made him one of the best known and most influential astrologers in Arabic and through translations also in Latin, Hebrew, and Greek. Some of his works treated purely astronomical themes, among others the composition of the universe by ten rather than eight or nine orbs and their motions. This treatise is only extant in two Latin translations, printed in 1504 and 1549 in Nuremberg. In this work, Masha'allah supported his astronomical views by references to Aristotle's *Physics*, perhaps on the basis of Syriac sources. Although he names Ptolemy and Theon of Alexandria, his planetary models are of older origin. In addition to the possible Syriac sources, Masha'allah used Byzantine writings on astrology, perhaps also in translation, and the Sasanian *Royal Tables* for casting horoscopes (the Sasanian empire was the last Iranian empire before the rise of the caliphate). On this basis he explains past historical events such as the Deluge, the rise of Christianity and Islam, or the rise of the ruling Abbasid dynasty and the birth or the beginning of the rule of important personalities of Islamic history and predicts the future of the caliphate and its rulers. The basic theoretical concepts for this kind of astrological history and 'political science' are the Sasanian belief in the particular relevance of the conjunctions of Jupiter and Saturn for terrestrial events and the Zoroastrian idea of the lifespan of the universe of twelve thousand years.

The men and their activities that played such an important role in the creation of a new capital for a newly established caliphal dynasty bring together in a nutshell central features of the rise of the new scientific culture over the next almost two centuries. Astrology, with its ancillaries of astronomy, geometry, arithmetic, and natural philosophy, was the leading science of the day. Despite later criticisms from several quarters and the measures taken in the hope of avoiding them in the future, astrology never lost its status in many Islamic societies. It was often a courtly science as well as a street practice. But it rarely entered formal education in mosques or madrasas in contrast to the other four sciences, which found a place in such institutions in several later Islamic societies. As a science it intensified its relationships with Aristotelian natural philosophy and epistemology, astronomy, geometry, and arithmetic during the ninth and tenth centuries. As an explanatory and predictive tool it retained its function as a form of political advice well into the modern period. Many horoscopes of sultans, princes, or shahs have been preserved pointing to this function. The Mughal ruler Jahangir (r. 1605–1627), for instance, asked for a horoscope of his neighbour and competitor Shah 'Abbas I (r. 1586–1629), because this would help him understand the shah's political ambitions and practices.

Optics in Fatimid Cairo

Other dynasties supported and used natural philosophy and the mathematical societies and scholars who propounded them in different ways. In the ninth century, a Shi'i movement emerged among the Berber population of today's Algeria. In a few decades, these men subdued a substantial part of North Africa and created a dynasty that claimed descent from Fatima, Muhammad's daughter and 'Ali's wife. Hence, they became known to the world as the Fatimids. Their second name, Isma'ilis, is derived from their last medieval Imam, Isma'il b. Ja'far b. Sadiq (719–?). In 909, they founded their first capital Mahdiya (today in Tunisia). In 969, they conquered Egypt and built al-Qahira ('The Victorious' = Cairo) as their new capital, from where they ruled over various regions of the southern Mediterranean and its islands until 1171. Due to their specific religious doctrines, the Fatimids created an extensive educational programme and a wide network of missionaries that reached into central Asia. Like the Abbasids, they built huge manuscript collections and other institutional support for learning. Their cosmology drew extensively on Neoplatonic theories of emanation. If the famous *Epistles of the Brethren of Purity*, written presumably in tenth-century Baghdad, indeed reflect Isma'ili thought, the Fatimids supported studies in number theory, geometry, astronomy, cosmology, geography, map-making, physics, alchemy, magic, metaphysics, psychology, and ethics, using this knowledge for arguing their views of the Imamate, i.e. the doctrine of the Isma'ili Imam as spiritual leader of the community.

Two scholars and a customs officer lend credibility to such a view, even if their connection to Isma'ili teachings are not known. The two scholars are Ibn Yunus (d. 1009) and Ibn al-Haytham (d. after 1040). Both are well-known scholars of astronomy. Ibn al-Haytham is even more famous for his revolutionary changes in optics as well as his new contributions to geometry and number theory. The customs officer is not known by name, but left us a fascinating work about astronomy, astrology, geography, trade, history, plants, and animals with marvellous illustrations, among them numerous maps.

In geometry, Ibn al-Haytham provided new ideas for calculating the volume of bodies in rotation and for squaring the circle. He discussed systematically the two main methods of problem-solving used by ancient Greek geometers—analysis and synthesis. He wrote two commentaries on Euclid's *Elements*, surveying ancient and modern critiques of Euclid's work. He often defended Euclid against his critics. In contrast, in astronomy, he wrote a work criticizing Ptolemy's models in the *Almagest* for violating basic principles of Aristotelian physics. His astronomical studies on the light of the Moon, the form of the eclipses, and the images of the Sun or the Moon seen through a hole in a window in a darkened room (camera obscura) as well as on the problems of optical illusion with regard to celestial and other bodies led him to his substantive studies of optics and his main work, the *Book on Optics* (*Kitab fi l-manazir*). In this book, which was translated into Latin in 1270 and was an important text for optical studies in the sixteenth and seventeenth century in Christian parts of Europe, Ibn al-Haytham combined Aristotelian physical and metaphysical ideas with Euclidean geometrical methods, a critical reflection on Euclidean and Ptolemaic optical theories, and atomistic ideas of Muslim rational theology for a new theory of vision and light.

Many Christian and Muslim scholars of the ninth and tenth centuries adhered to the so-called extramission theory according to which some visual matter was emitted from the eye in a cone and when an object entered this conic field it became visible. The ideas about what this visual matter was differed, but it was believed that it emanated from the brain via the optical nerve to the eye. Ibn al-Haytham rejected this theory and its concepts. His new theory, based on well-prepared observations and several experiments, explains how light shines from a self-luminous object in straight rays from every point on the object's surface in any direction. These rays extend in a translucent medium like air in rectilinear manner. They are reflected on smooth surfaces in a specific angle. If they pass through media of different transparency, light rays are refracted at the surfaces which separate these media. Analogous statements are made about illuminated opaque or partly opaque objects. Colour is different from light and a property of opaque bodies. But all observable properties of light also apply to colour. The point from which light or colour emanates needs to have a minimal size. Hence, the light ray itself needs to be of minimal size, a so-called least light, in order to apply geometry to physics. In the later parts of his seven books on optics, Ibn al-Haytham studies the reflection and refraction of light through mechanical concepts like motion, speed, density, resistance, repulsion, impact, and pressure applied to little spheres of solid material thrown at walls. He borrowed these ideas from the atomistic theories formulated in the ninth century by a group of Muslim scholars of theology known as Mu'tazilites. Later, Kepler and Descartes would pick up some of Ibn al-Haytham's mechanical comparisons.

The Sciences between China and the Islamic World in Mongol Tabriz

Between 1219 and 1221 the Mongol confederation created by Chinghiz Khan (r. 1206–1227) conquered Muslim central Asia. From 1253 to 1256 they destroyed one Isma'ili stronghold in western Iran after the other, thus ending Isma'ili power in the region. In addition to the tremendous loss of life, the Isma'ili library was burnt at the fortress of Alamut. In early 1258, they overran Baghdad, killing the Abbasid caliph and many of the city's people. Before the siege of the city, Hulagu Khan (r. 1256–1265), the new ruler of Iran, asked one of his court astrologers, Husam al-Din al-Salar (ex. 1262), to read the stars for him. The astrologer predicted that six severe calamities would befall the Mongols if they dared to attack the Abbasid capital. He was imprisoned sometime after the city's fall and later executed. Another astrologer, who had negotiated the surrender of Alamut, an agreement which the Mongols did not honour, was Nasir al-Din al-Tusi (1201–1274). He rose to become the new dynasty's vizier for religious endowments, which received funding from individuals mostly for religious purposes. Sometimes, however, such religious donations were specifically dedicated to teaching or to the building of an observatory.

One consequence of the Mongol conquest of the eastern Islamic world was the intensification of contacts between East, South, and West Asia. Astronomical and astrological knowledge came in the form of people, texts, and instruments to the court of the Mongol Yuan dynasty (1260–1368) in Khanbaliq (today Beijing), while medical, pharmaceutical, agricultural, and technical knowledge moved in the same manner to the West. Such direct contacts had existed in sporadic form since the Song dynasty (960–1279), when the participation of a Muslim astronomer in an astronomical project at the Song court is attested in Chinese sources. When Chinghiz Khan had invaded the area around Samarkand and Tashkent, one of his army astrologers contacted local Muslim experts for inquiring about their methods to predict a lunar eclipse and to calculate planetary positions. Some time later, he explained his newly acquired knowledge in a treatise. In 1244, a Muslim astrologer is said to have entered the service of the later Mongol emperor of China, Kubilai Khan (r. 1260–1294).

After Mongol rule had been established in Iran and China, the movement of people and material between the two Mongol empires increased. In Khanbaliq, the *Muslim Bureau for the Administration of the Heavens* was installed, which continued to provide observations and predictions well beyond the fall of the Yuan dynasty. At the beginning of the Ming dynasty (1368–1644), about twenty-five Muslim astrologers worked there. The bureau, including its rich library of Arabic and Persian scientific manuscripts, was transferred to the new capital of Nanjing. The first Ming emperor also ordered the translation of some of these books. The bureau was still in operation under the Manchu Qing dynasty (1644–1911), who invited the Muslim astronomers to participate in a contest with the recently arrived Jesuits.

In the thirteenth century, Chinese experts in water technology had come with Hulagu's army as far as Iraq. They participated in regulating the irrigation system between the Tigris and the Euphrates. Furthermore, Chinese, Uighur, and Indian physicians also travelled westward. Several Mongol rulers in Iran preferred the treatment that these physicians offered over that provided by local Muslim, Jewish, or Christian doctors. At the end of the century, a stable community of Chinese merchants, craftsmen, soldiers, and medical practitioners had settled in Tabriz, the second Mongol capital in Iran. There they met not only Muslims and Mongols, but Uighurs in courtly service, Genoese, Venetian, and Byzantine merchants, as well as the occasional Catholic monk and Byzantine cleric. Parts of a Frankish historical chronicle were translated at court for the universal history compiled by the physician and vizier Rashid al-Din (1247–1318). Italian seafarers mapped the Caspian Sea in this time and perhaps brought knowledge of Islamic maps back to their hometowns. In 1313, Rashid al-Din wrote a treatise on Chinese medicine called *Treasure Book of the Il-Khan on the Arts and Sciences of Cathay* (*Tanksukhname-yi Ilkhani dar funun wa-'ulum-i khata'i*). This work was the result of cooperation between Rashid (a Jewish convert to Islam) and a high-ranking Mongol nobleman named Bolad Aqa (second half of the thirteenth century), who held the position of Supervisor of Agriculture, Head of the Imperial Household Provisions, and other official posts in Khanbaliq. Both men began their respective careers responsible for the preparation of food for the palace, a position of high status and reputation, but at times also high danger, since poisoning was a favourite means to settle royal and other conflicts. Both men also cooperated in the transfer of Chinese and West Asian medical drugs, the introduction of Chinese agricultural knowledge to Iran and possibly in the creation of a new hospital at Tabriz, where Chinese medical practices like acupuncture and new forms of pulse diagnostics were applied and new Eastern drugs administered.

As in the case of astronomy and astrology, medical, pharmaceutical, and culinary knowledge did not flow only in one direction. Lombard as well as Syriac healers worked at the Mongol court in China. A *Muslim Bureau of Medicine* was established. One of the Syriac doctors was Simeon from the Church of the East, who came in the 1230s and 1240s from Upper Iraq to the Mongol court in Qaraqorum and later served Hulagu in Maragha, the first Mongol capital in Iran. These oriental Christians relied in their theory and practice primarily on Ibn Sina's great handbook *The Canon on Medicine* (*al-Qanun fi l-tibb*). It presented Galenic medicine in modified and modernized form, enriched by the drugs, practices, and interpretations of Muslim and Christian physicians introduced from the second half of the eighth century onwards. This magisterial work was taught at the same time at universities in Bologna, Paris, Oxford, or Cambridge. It also may have found its way in 1273 into the imperial library of the Mongol court in China.

One topic that was studied intensively by astrologers in Maragha, Tabriz, and elsewhere during the thirteenth century concerned the so-called difficulties (*ishkalat*) of Ptolemaic astronomy. The main goal was to resolve them by proposing alternative models for the planets. Numerous such new models were discussed, not only in Mongol Maragha and Tabriz, but also in the following centuries in Mamluk Damascus, Timurid Samarkand, or Safavid Shiraz, to name only a few places for such activities. The technicalities of these discussions are too numerous and intricate to be described here in a summary fashion. Hence, only one aspect of these new models will be briefly presented, the famous Tusi couple.

Nasir al-Din al-Tusi intended to modify Ptolemy's lunar model on two points. He wished first that the centre of the deferent (the circle which carries the epicycle, to which the Moon is attached) should remain in the centre of the universe. Second, he wanted to replace Ptolemy's mechanism by a new device, which would shorten the line connecting the centres of the deferent and the epicycle at 90° and 270° and extend it at 0° and 180°. The result of these changes would be a uniform motion of the deferent around the centre of the universe, while the new device would account for the apparent changes in visible planetary size. Tusi had developed this device in 1247 for a different model and needed to adapt it now for the particular conditions of the lunar motion. In this device, called the Tusi couple, a circle of half the diameter of a larger circle moves inside this larger circle along its circumference. When both circles move, a point on the circumference of the small circle oscillates on the diameter of the larger circle in a linear movement.

Further important issues discussed by the astrologers at the Mongol court and their successors elsewhere concern the questions of whether astronomy needed to be anchored in natural philosophy, whether the earth rotated, and what the nature and origin of comets was. All these were questions that would rise to prominence in the increasingly heated debates about the heavens during the sixteenth and seventeenth centuries in some of the Christian countries in Europe.

Many of these themes were not only discussed among the leading scholars at observatories in the Mongol capitals, but also taught at madrasas in and outside of Iran. This combination of research and education ensured the distribution of the new knowledge to Syria, central Asia, Anatolia, eastern Iran, and India, where it continued to be studied and debated into the eighteenth century.

Parallel to these discussions of astronomical theory, the mathematical texts that had been assembled since late antiquity and enriched by new texts since the ninth century were newly edited and commented upon. Nasir al-Din al-Tusi was the leading scholar in this enterprise. His new versions of Euclid's *Elements* and of the so-called *Middle Books* to be studied after Euclid's *Elements* and before Ptolemy's *Almagest* became the main teaching texts of plane, solid, and spherical geometry at madrasas and mosques across many Islamic societies from Morocco to India until the later nineteenth century.

Map-Making in Ottoman Constantinople and Istanbul

Our final example of how natural philosophy and the mathematical sciences were adopted and adapted in Islamic societies comes from the fifteenth century. In 1453, when Byzantine Constantinople had been finally conquered by the Ottoman army, the ideological orientation and demographic composition of the slowly growing Ottoman state began to shift. Large numbers of non-Muslims needed to be integrated, a task that grew with the subsequent expansion of the Ottomans into Eastern Europe and the Mediterranean. New ideologies of universal rulership were created that used Alexander of Macedonia (*Dhu l-qarnayn* = The Two-Horned) and Constantinople as potent symbols. Technological development in armament and shipbuilding, mining, and canalization were important elements to ensure Ottoman navigational power and military might. Among the sciences which the court and its different factions patronized was geography.

In a first phase, Arabic cosmographies, geographies, and travel accounts were translated into Ottoman Turkish. They set the tone and taste of the Ottoman elites and their scholars at madrasas, mosques, and the court. In the late fifteenth and early sixteenth centuries, corsairs turned Ottoman captains and admirals began to produce geographical treatises, handbooks of the coasts and islands of the Mediterranean, and portolan charts, some of which also depicted the Americas. The best-known representative of these new texts and maps is the Ottoman captain and later admiral Piri Re'is (ex. 1554). His fragmentarily preserved world maps and his two versions of the *Handbook of the Mediterranean Sea* (*Kitab-i bahriye*), dedicated to two Ottoman sultans, combined new, practical knowledge with knowledge acquired from Italian, Spanish, and Portuguese prisoners and converts from various countries, Italian books depicting Mediterranean islands, and knowledge from Arabic and Persian books. The world maps disappeared somewhere in the palace, where their fragments were discovered in the early twentieth century. The *Handbook of the Mediterranean Sea* was often copied and enriched until the nineteenth century.

Ottoman interest in maps grew over the next two centuries after Piri Re'is. Princes ordered world maps in Venice and Vienna. Historians added modernized world maps to their chronicles of the current state of affairs. European visitors sold books and maps in Istanbul. Two scholars of the seventh century, Hajji Khalifa (d. 1657) and Abu Bakr b. Bahram al-Dimashqi (d. 1691), became famous for their roles in translating Latin atlases into Ottoman Turkish. From 1654 to 1656, Hajji Khalifa, together with the French convert Mehmed Ikhlasi, translated Gerhard Mercator's *Atlas Minor* and parts of Abraham Ortelius's *Theatrum Orbis Terrarum*. Between 1675 and 1685, Abu Bakr b. Bahram al-Dimashqi translated Joan Blau's *Atlas Maior* working with a whole group of Christian and Muslim collaborators. These were major translation projects that modernized Ottoman geography and map-making above all for the regions outside the world of traditional Islamic geographical knowledge. Hajji Khalifa's new *Mirror of the World* (*Cihannüma*, version II) became an Ottoman bestseller, the first printed Ottoman geography, and a book that still drew the interest of a Qajar prince in Tehran at the beginning of the twentieth century. But not only inhabitants of Islamic societies appreciated the numerous books that resulted from Hajji Khalifa's and Abu Bakr al-Dimashqi's work. After the *Cihannüma* had been translated into Latin in the eighteenth century, map-makers and university professors in Christian countries of Europe equally used the work for new maps or discussions of geography and history in West Asia and North Africa. Before they could do so, however, the assertoric approach discussed earlier in this chapter had to be reformed. We will discuss this process, before returning to the centrality of translations and translators in the early and late periods of Islamic natural philosophy, medicine, and the mathematical sciences.

The New Naturalism and a New Society

Things in the West had begun to change about the middle of the eleventh century. As was the case in Islamic society, some of the earliest steps in this transformation centred around the recovery and study of law, first secular Roman law, and then ecclesiastical, canon law, for the study of law and both its causes and effects in the long twelfth century had much to do with the restructuring of culture in the later European Middle Ages. The study of Roman law was especially suited to a period interested in logical structures, but initially lacking sources. Law, like theology and philosophy, acquired a body of authoritative materials—in the form of the Code of Justinian, the writings of the Church Fathers, and Aristotle—but the real labour, and creativity, lay in its understanding, interpretation, development, and application. Seen from this perspective, when later medieval theologians and natural philosophers adapted old sources to create new theories, they were following techniques pioneered by earlier legal scholars who interpreted authoritatively prescribed codes of law.

Of course, the twelfth-century transformation of society was more than simply a matter of interpretation of authorities. The legal revolution of the eleventh and twelfth centuries transformed small and personal government into a vast and bureaucratic one, and in the process, through the contested spheres of power in Church and State, gave enormous impetus to the value of reason and learning. As John of Salisbury (d. 1180) noted in the middle of the twelfth century, the king must attend to the law daily, and hence an illiterate king—formerly a commonplace in European society—was now an *asinus coronatus* (that is, a 'crowned donkey').

But the study of law and the changes it produced did more than encourage scholarship; it also shifted the foundation of evidence necessary to convict or exonerate. Between 1100 and 1200 the standard of evidence that had prevailed for centuries—frequently involving the ordeal and compurgation—was subjected to scrutiny and recognized as either ineffective or fallacious. Peter the Chanter (d. 1197) discussed the matter at length in his *Verbum abbreviatum* and argued that apart from theological reasons for rejecting the ordeal, there were good empirical ones: defendants called upon to prove their innocence in trial by battle always selected champions who were skilled warriors rather than the infirm and inexperienced, and one could game the ordeal of water by practising breathing techniques. Peter even recounts the case of a man who, compelled to submit one of his sons to the water ordeal, tested each of them and selected the one who gave him the best chance of succeeding.

While the supernatural or mystical evidential rules gradually gave way to a variety of replacements, a common thread throughout the transition was the reliance on reason, observation, and the application of the rules of logic. Although their enterprises were certainly not identical, on several fundamental points jurisprudence and the emerging developments of natural philosophy and theology showed similarities: each constituted an integrated body of knowledge; in each, particular phenomena were systematically explained, usually in terms of general principles; and each relied on observation, hypothesis, and verification, although with considerable variation both temporally and geographically.

Alongside this disciplinary and methodological transformation, one can also see a profound shift in perceptions of man's place within nature. If formerly man was simply a vehicle for divine activity, an abject, powerless being before nature and the divine, the art, literature, and theology of the subsequent period emphasized the intercessory kindness of the Virgin, the pathetic sufferings of Christ, new theories of redemption that gave greater place to a covenant between God and man. Man himself acquired a new dignity; as Bernard Silvestris (*fl.* 1150) noted in his *De mundi universitate* (*Cosmographia*),

> The animals express their brute creation
> By head hung low and downward looking eyes;
> But man holds high his head in contemplation
> To show his natural kinship with the skies.
> He sees the stars obey God's legislation:
> They teach the laws by which mankind can rise.

Nor was man the only thing changing: the universe had become a more friendly, familiar, intelligible, homogeneous, and interconnected entity. Gerhoh of Reichersberg (1092/1093–1169) observed that, 'The whole structure of the universe is suitably equipped', a view that Hugh of Saint Victor (d. 1141) underscored by noting, 'The ordered disposition of things from top to bottom in the network of this universe...is so arranged that...nothing is unconnected or separable by nature or something external.' In his gloss on Plato's *Timaeus*, William of Conches (d. 1154 or 1160) sought to connect this aggregation of the world's entities with Euclid's orderly collection of propositions, when he noted, 'The world is an ordered aggregate of creatures.' If, as Bernard of Chartres (*fl.* 1119–1126) discerned in a very famous passage, the scholars of his generation 'stood on the shoulders of the giants and could see further than their great predecessors', they were only able to do so because they now had mastered the seven liberal arts and the world that was their object. More than expressions of intellectual boldness, Bernard's world-view has expanded to include the possibility that the present may surpass the past, something that was unthinkable before 1050.

This psychological transition was accompanied by a desacralization of nature. While later medieval theologians and philosophers continued to assert God's role as creator and sustainer of nature, they also affirmed that the study of nature demands a search for causes, which was a task for philosophy, but not to be expected of the Bible. The result was an emphasis on naturalism, not preternaturalism; as the twelfth-century exegete Andrew of Saint Victor (*fl. c.*1150) observed in his commentary on Ezekiel: '…in expounding Scripture, when the event described admits of no natural explanation, then and only then should we have recourse to miracles'. Now the presence of miracles was itself an indication of God's great mercy, for the many miracles of healing signalled God's concern for and love of men, effected through the agency of the saints. But Andrew is suggesting that there may be too much of a good thing: because the early medieval definition of 'miracle' was (according to its etymology) that which evoked wonder, everything not understood becomes miraculous. With the introduction of Greek and Islamic texts in the twelfth century, particularly in the Aristotelian tradition, a more precise articulation of the common course of nature was developed, and consequently Andrew's suggestion to seek natural explanations of wondrous events might be achieved. Of course, the miraculous can be banished entirely by successively widening the natural to account for such events, as T. H. Huxley noted. One of the ironies of the thirteenth century is the simultaneous tension between these two tendencies: theologians of the period creating a doctrine of transubstantiation in the Eucharist viewed with suspicion both popular accounts of miraculous bleeding hosts in the run-up to the feast of Corpus Christi as well as attempts to explain the physics of the transformation in purely philosophical terms.

The Scientific Beginnings in Abbasid Baghdad

The Greek and Arabic texts introduced to Catholic parts of Europe from the twelfth century on were the results of an earlier immensely creative and productive period in which the philosophic, medical, mathematical, and other scholarly writings of the ancient world were first translated into Arabic. Translating treatises into Arabic from other languages and times was already an important element in the work of the three astrologers who we earlier saw casting Baghdad's horoscope and the practice unfolded with increasing rapidity after their death. Cooperation across religious boundaries caused no problems. It rather was part of normal science well into the tenth century and in certain places also in much later times. Conversion to Islam among courtly scholars already had begun, often caused by direct encouragement of a caliph or another courtier. Interest in astrological practices and courtly events was not limited to students of the mathematical sciences, which included as their four main disciplines number theory, geometry, astronomy, and theoretical music, together with many other branches such as optics, mechanics, algebra, architecture, and various systems of arithmetic. Scholars of geography and history also took note.

Scientific activities had an important place at the caliphal court. This remained the case throughout the entire existence of the Abbasid dynasty, although the interests of the individual caliphs, their family members, and their courtiers differed and the kind of support they provided to scholars fluctuated. After its foundation in 762, Baghdad soon became a city which attracted merchants, soldiers, and scholars alike. Its fast growth offered opportunities for social and material advancement. But it was also a city of massive social, political, and religious conflict and unrest. Over the centuries, it saw the rise of a professional group of philosophers, astrologers, and physicians with broad scientific interests—as well, it should be mentioned, as the burning of their books. When the Mongols stormed the city in 1258 and brought the Abbasid caliphate to an end, chroniclers lamented that streams of blood flowed into the Tigris, tons of precious manuscripts drowned in its depths and the heavens were obscured by the smoky clouds of burning houses.

The formation of this scholarly class and the production of their precious manuscripts had begun with the turn of the new Abbasid dynasty (750–1258) to astrology and the translation of Middle Persian, Syriac, Greek, and Sanskrit texts on astrology, logic, dialectics, ethics, and finally medicine, metaphysics, physics, astronomy, mathematics, and other domains of knowledge. The reasons for this major shift in cultural outlook and caliphal policy are not fully clear. Several factors will have been at play.

Among the Iranian scholars and courtiers who switched their loyalty from the Umayyads to the Abbasids or came with their troops to the new capital, there were, in addition to the astrologers, important administrators, physicians, judges, language teachers, and men with an interest in issues of faith and politics. Ibn al-Muqaffa' (ex. 756) and the family of the Barmakids were among the most influential administrators who either themselves translated from Middle Persian into Arabic or patronized those men who knew Sanskrit or Greek. The texts translated in the period of their prominence between the 750s and 803 were, in addition to the above-mentioned *Mahasid-dhanta*, the famous *Kalila wa-Dimna*, Sasanian historical chronicles, Sanskrit medical and perhaps philosophical texts, and most likely Euclid's *Elements* and Ptolemy's *Almagest*, early translations of which are ascribed in some manuscripts to Barmakid patronage and in others to the support of the Abbasid Caliph Harun al-Rashid (r. 786–809). On command of Caliph al-Mahdi (r. 775–785), Timothy (780 to 823), the patriarch of the Church of the East and caliphal advisor, arranged for the translation of Aristotle's *Topics* into Arabic. He could rely in this work on other Syriac Christians and their experience with translating Aristotelian texts. Since the sixth century, members of different Syriac churches had studied Aristotelian logic either in Greek or Syriac. In the eighth century, Syriac translations of Aristotle's *Metaphysics* were produced. Moreover, Syriac clerics who also were monks like Sergius of Resh'aina (d. 536) translated medical texts into Syriac, in particular those of Galen. Many of these early Syriac translations are lost to us. Some were still available in the ninth century, when Hunayn b. Ishaq (d. 873), the most famous Arabic Christian translator from Greek into Syriac and Arabic, began to produce new translations of the Galenic oeuvre in addition to texts by Hippocrates and other ancient physicians and of philosophical texts, mostly by Aristotle.

A decisive turn away from translating Middle Persian texts into Arabic in the environment of the Abbasid court towards the translation of a very broad range of Greek works happened after 819. In 813, Tahir b. Husayn, the commander of the troops of Harun al-Rashid's eldest son 'Abdallah, had defeated 'Abdallah's younger brother and reigning caliph al-Amin (r. 809–813). Becoming Caliph al-Ma'mun (r. 813–833), 'Abdallah did not wish to rule from Baghdad, but withdrew to his stronghold Merv in northeastern Khorasan (today Turkmenistan). After six years of rule from this far-away corner of the empire, the caliph finally realized that his power had almost slipped from his hands and that a radical decision was needed. He decided to return to Baghdad and break with the Iranian orientation of his father, grandfather, and great-grandfather.

The texts from his period suggest that he and his new advisors in Baghdad cast him and his elite as philhellenic admirers and heirs of ancient Greek philosophy and the sciences. This important shift in attitude and propaganda was probably an important factor in the translation of almost the entire oeuvre of Aristotle, Galen, Ptolemy, and Euclid as well as many other Greek medical, mathematical, astronomical, astrological, and perhaps also alchemical texts during the ninth century. Influential courtiers also provided money for such activities and often were themselves scholars of the new sciences. These included Abu Yusuf Ya'qub b. Ishaq al-Kindi (d. *c.*870), as well as the three sons of a former highway robber turned into an astrologer, Muhammad, Ahmad, and al-Hasan, better known together as the Banu Musa. Notable also was a family of astrologers, men of letters and musicians, the Banu al-Munajjim. A second important social group in this process of translation were Christian doctors. They patronized translations of Greek texts into Syriac and later also into Arabic.

Parallel to these translations of philosophical, scientific, and medical texts, Greek historical chronicles, which had been translated into Syriac during the previous centuries, now began to be translated or summarized in Arabic. They provided Muslim scholars of the later ninth and tenth centuries, who actively developed a religious and political Islamic historiography, access to Christian and non-Christian historical concepts, styles, and data of previous centuries. In addition, Christian circles also began translating the Bible into Arabic and occasionally Muslim religious literature into Greek. Translating was thus an important intellectual activity with many goals in numerous different communities of Baghdad and other cities of the Abbasid Empire.

The significance and range of these translation activities can be seen in the oeuvre of three of Bagdad's many scholars. Muhammad b. Musa al-Khwarazmi was one of the courtly astrologers who had converted from Zoroastrism to Islam. He compiled a version of the astronomical handbook called *Sindhind*, modifying it through the introduction of Ptolemaic parameters and methods. While it was soon replaced by the new handbooks produced by other courtly astrologers in the 820s and afterwards, based more fully on Ptolemy's *Almagest* as well as on new observations and measurements, it became an important source for the development of astronomy on the Iberian Peninsula during the reign of Muslim as well as Christian rulers. Several of its tables are found in later astronomical handbooks in Arabic, Latin, and Hebrew. An adapted version of the *Zij al-Sindhind* and a commentary on it were translated in the twelfth and thirteenth centuries into Latin and Hebrew. Al-Khwarazmi also wrote about the astrolabe and the Jewish calendar.

In addition to his astronomical works, al-Khwarazmi wrote a book on geography with a series of maps based on one of the Arabic translations of Ptolemy's *Geography*. He added numerous new cities with their coordinates, reduced Ptolemy's excessively long west–east axis of the Mediterranean by 10° and designed new local maps, for instance of the Nile, not found in the extant Byzantine copies of Ptolemy's work. Several maps joined to geographical or historical works in later centuries illustrate the impact of al-Khwarazmi's *Book of the Image of the Earth* (*Kitab surat al-ard*) in Islamic as well as Christian societies around the Mediterranean. His depiction of the Nile is also found on several Italian and Catalan nautical charts or world maps of the fourteenth and fifteenth centuries (often referred to as 'portolan' charts).

Two other treatises with a long impact in the mathematical sciences in Islamic as well as Christian societies in Asia, Africa, the Iberian Peninsula, Italy, France, England, or German lands are al-Khwarazmi's *Algebra* (*Mukhatasar fi 'ilm al-jabr wa'l-muqabala*) and his *Indian Arithmetic* (perhaps: *al-Hisab al-hindi*). The *Algebra* is the first systematic treatment of linear and quadratic equations in Arabic with a formal terminology, algorithms, and visual demonstrations. It seems to draw on much older Mesopotamian practices, perhaps Indian concepts, and possibly Greek methods. After the chapter on algebra, al-Khwarazmi treats geometry as the determination of areas and volumes of elementary figures and problems of commercial and legal practice. Following the Greek rather than the Indian tradition, numbers are seen as positive natural numbers beginning with 2, to which 1 was added as the source or root of all numbers.

While al-Khwarazmi's *Algebra* remained for a long time one of the most influential textbooks on algebra in Islamic scholarly circles, his *Indian Arithmetic* was soon replaced by other, probably more systematic and extended treatises. As a result, this work is lost in Arabic. But since it had arrived before the twelfth century in al-Andalus (the Muslim area of the countries we now know as Spain and Portugal) it became one of the mathematical texts translated then and there into Latin. In this form it became the ancestor to all further Latin texts on Indian arithmetic and the introduction of the so-called Arabic numbers into Western mathematics, commerce, and finally everyday-life practices. Modern mathematics owes to al-Khwarazmi also its terms algebra and algorithm.

Al-Kindi, a younger Arabic contemporary of al-Khwarazmi, had the distinction to become the first Muslim philosopher and was later called the philosopher of the

Arabs. He worked with Christian translators in his quest for knowledge, who translated for him Neoplatonic and Aristotelian treatises, Nicomachus of Gerasa's (second century) *Introduction to Number Theory*, and other Greek writings. He himself is credited with hundreds of treatises. His philosophical oeuvre is strongly Neoplatonic, although the philosopher he cherished most was Aristotle. Hence, he wrote the first Arabic treatise on the harmony between the philosophies of Plato and Aristotle. Moreover, his philosophical teachings were informed by his beliefs in important Islamic tenets. He believed in God's creation of the world from nothing and the resurrection of the human soul, rejected the eternity of the universe, and classified knowledge as divine and human. His idea of God, expounded in one of his most important philosophical treatises, *On First Philosophy (Fi l-falsafa al-ula)*, is nonetheless not that of the Qur'an. It owes more to the views of the Neoplatonic philosophers Plotinus (205–270) and Proclus (d. 485) as well as to discussions among some of his contemporaries interested in matters of Islamic faith, the Mu'tazilites. Al-Kindi saw God as 'one and simple': the source of unity in all other things, who could not be adequately described in human language.

Al-Kindi's epistles on the soul and the intellect teach that the rational soul is separate from the body and an immaterial substance, while the other two parts of Plato's triple structure of the soul are situated in the human body. The intellect is in al-Kindi's view a parallel to sense perception and begins in a state of potentiality. After grasping an intellectual form, it begins to think and transforms into the active intellect. This active intellect enables humans to think about the intellectual forms at will, a state al-Kindi calls the acquired intellect. The actualization of the potential intellect does not reside in empirical experience, but in the external, non-human first intellect, a concept closely related to Aristotelian ideas in *On the Soul (De anima)*. Later Neoplatonic Aristotelian philosophers like Abu Nasr al-Farabi (d. 950) and Ibn Sina (d. 1037) deviate here (and in other important points like the eternity of the world, creation, and resurrection) from al-Kindi as well as Aristotle and take a more empiricist position.

Al-Kindi also wrote a good number of important texts on astronomy, astrology, optics, medicine, divination, the production of perfumes, and mechanics, some of which were also translated into Latin, Hebrew, European vernaculars, and perhaps Persian. He taught Muslim, Christian, and perhaps other students in matters relating to his own intellectual interests.

Thabit b. Qurra, a member of the Sabian community of star worshippers around Harran (today in eastern Anatolia), came early in his youth to Baghdad through the patronage of Muhammad b. Musa, the eldest of the Banu Musa. He made a steep ascent at court to become a court astrologer and boon companion of the caliphs. He also was a successful teacher of Muslim, Christian, and Jewish students in translation techniques, geometry, astronomy, mechanics, logic, and perhaps other sciences. He wrote epistles on Aristotelian philosophy for members of the Abbasid elite in order to popularize knowledge on the heavens and nature.

Thabit was without doubt the most gifted scholar of the mathematical sciences in ninth-century Baghdad. He translated or edited important texts by Archimedes,

Ptolemy, Euclid, Nicomachus, and other ancient scholars. In Arabic, he composed treatises on number theory (amicable numbers), algebra (the use of two theorems from Euclid's *Elements* to prove that al-Khwarazmi's algorithms for solving quadratic equations produced correct results), geometry (the calculation of areas of parabolas and hyperboles and their volumes, when they revolved around some axis), mechanics (the conditions of equilibrium of the unequal-armed balance for a finite and an infinite number of weights attached to it), and astronomy. He wrote in his Syriac mother tongue, mostly on matters concerning his own religious community, whose head he was in Baghdad, but also on a range of philosophical themes.

The work of these scholars and many others in translating and putting to work ancient philosophy in Islamic societies was replicated in the Catholic societies in Europe, as another translation movement began to make Arabic philosophy, medicine, and the mathematical as well as occult sciences—and ancient knowledge as filtered through the Islamic perspective—available to audiences on the Iberian Peninsula, France, Italy, the British Isles, and some of the German lands.

Translation and Transformation

In fact, it is possible to argue that broad structural changes in the social, political, ecclesiastical, and psychological condition of Catholic Europe were both caused and affected by the recovery of ancient texts, supplemented with commentaries and wholly original works from Arabic and other sources. While the initial stirrings of this movement can be seen in mathematical and astronomical developments in the late tenth century, the late eleventh through the last quarter of the thirteenth century marks the apex of translations. As we have seen, while there were both translations of ancient works and handbooks of scientific information in the early Middle Ages, the quality and completeness of the material left much to be desired. It was to improve the quality and fill these lacunae that Adelard of Bath (c.1080–after 1150) and Gerard of Cremona (c.1114–1187) began their translations of Euclid's *Elements* and Ptolemy's *Almagest*, respectively. Moreover, the study of logic and physics had an uneven textual tradition in the early Middle Ages. The first step in remediation was to take stock of the range of material that was lacking. Here, Arabic works like Abu Nasr al-Farabi's (c.870–950) *Catalogue of the Sciences* provided an appreciation of Aristotle's wider expertise, as a consequence of which he came to be regarded as 'The Philosopher.' Likewise, in medicine, Galen was recognized as the authority, and the early translators, like Constantine the African (fl. 1065–1085), Burgundio of Pisa (c.1110–1193), and Gerard of Cremona first compiled lists of his works and then sought the texts for translation.

If one of the goals was recovery of ancient Greek works, Latin translators drew water from two different wells. On the one hand, some, like James of Venice (d. after 1147) and Henricus Aristippus (d. 1162), took advantage of diplomatic or ecclesiastical postings in Byzantium to translate Greek language texts. Others, like Constantine the African were located in Sicily or southern Italy, where the use of Greek had never

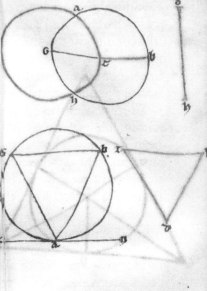

Bruges, Bibliothèque de la Ville 529, fol. 23ʳ. The Latin translation of Euclid's *Elements* known as Adelard I. In the upper-right margin, four Arabic words and their Latin translations: elmuhit/continens; elmumeth/contingens; ieohit/continet; iemet/contingit. Concerning this manuscript, see Charles S. F. Burnett, 'Some Comments on the Translating of Works from Arabic into Latin in the Mid-Twelfth Century', *Orientalische Kultur und europäisches Mittelalter*, ed. Albert Zimmermann, Ingrid Craemer-Ruedenberg, Gudrun Vuillemin-Diem. Berlin: Walter de Gruyter, 1985, pp. 161–71 at 167–8.

completely disappeared. A second source was Arabic translations of Greek works, principally in the Iberian Peninsula, and especially in the Ebro valley and the region around Toledo. Following the conquest of the city in 1085, European translators converged on Toledo in search of ancient works in Arabic translation, aided by the presence of Mozarabs who possessed Arabic texts and Jews whose linguistic facility in both Hebrew and Arabic was useful to Latin translators.

The results of this wide geographic and multilingual movement proved to be significant for Latin European knowledge of ancient science. Not surprisingly, many works were translated multiple times, sometimes because the translators were not aware of earlier translations, but also because they were dissatisfied with earlier versions. So, for example, John of Salisbury complains in the *Metalogicon* that a difficult work like the *Posterior Analytics* was not translated correctly. Because Byzantine scholars seem to have emphasized a pure classical tradition, the commentaries produced from the Byzantine Greek tradition tended to be those of the ancients, Themistius, Alexander of Aphrodisias, and John Philoponus. By contrast, Latin translators of the Arabic tradition received not only the works of the ancients, but also the more modern or even contemporary authors Ibn Sina (Avicenna) (d. 1036), Ibn Rushd (Averroes) (1126–1198), and many others.

The techniques employed in translation were almost as variable as the sources themselves. In some cases, the translation was effected by dictation, as for example the distinguished Jewish scholar Abraham b. Ezra (*c.*1090–*c.*1164/67?) undertook in producing a version of his treatise on the astrolabe with an anonymous Latin amanuensis. On other occasions, the process involved simultaneous translation from Arabic text into a vernacular, and then translation of the vernacular into Latin. Some of these intermediate vernacular translations circulated within non-Latin communities. On still other occasions, we know that Latin scholars translated directly from Arabic or Greek texts, for they included Arabic or Greek characters or transliterated terms in their texts, and sometimes they commented on the difficulty in reading diacritical marks in the manuscript. Finally, translators' approaches to rendering ancient and Arabic texts into Latin ranged over a continuum, from word-for-word translations to more stylistic or literary versions in which the translator appropriated the text into the mentality of his readers.

Motivations for the translation initiative were also diverse. With few exceptions, the translators themselves were not university masters, though many of their products became central texts within the new institutions. Those texts that did not form part of the curriculum, like works in alchemy and astrology, nevertheless circulated secretly, many times catering to the extracurricular interests of masters. As noted earlier, translators often actively sought texts to fill lacunae in the curriculum, or having heard of the existence of an ancient text, attempted to procure a copy to complete the corpus of an ancient author. On other occasions, the support of a patron was crucial; Frederick II (1194–1250) and king Manfred (*c.*1232–1266) in Italy or Alphonso X (1221–1284) in Castile sponsored translators in their territories, and ecclesiastical patrons did the same by providing positions that required no residential service. There

sistunt 2 delectatio etiam in bono 2 quod. Ex quibus patet quod pena eterna potest esse
in omnibus dampnatis equalis in intensione 2 extensione quia durat infinito tempore 2 tamen
indifferenter 2 inequales figurationes magis affliget unam animam quam aliam ut patet

quod a. 2 similiter potest dici quod eadem ratione certa persona possit magis delectare
2° unum spiritum quam alium 2 quod. 2° similiter potest accidere quod a sit por b per

intensiones inequales nam si a sit in 2° magis intensa quam b. 2 in 2°
minus extensa tamen propter dictam sunt sibi equales 2 tamen fugibili est sustinere a
quam b tollerabili non est in pena remissa per 2 dies quam intensa per 1 die
ut sit a quadratum 2 b quadratum ortogonium 2 in duplo longius quadrato sed in 2°
est bassius 2 patet quod sunt equales 2 quod 2 eodem modo dicendum est de leticia 2 aliis
actibus anime suo modo.
Omnis qualitas uniformiter difformis est tanta quanta est qualitas eiusdem subiecti uniformis
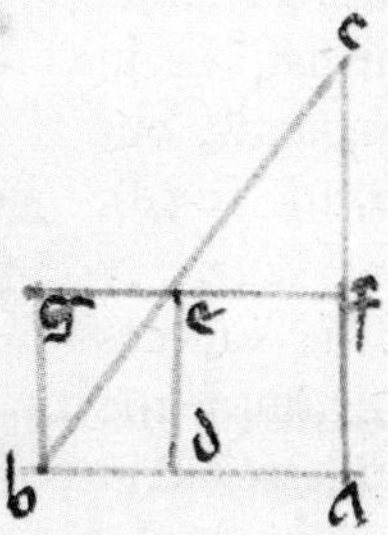
uniformis secundum gradum medium ipsius subiecti ut si qualitas est linearis tunc accipiendus
gradus puncti medii ipsius subiecti si autem qualitas sit superficialis accipiendus secundum
gradum linee medie ipsius subiecti 2 si est corporalis secundum ipsius secundum gradum superficiei medie
ipsius subiecti. Sed gratia exempli sit scilicet 3 angulus ortogonius a b c cuius subiectum
seu basis sit a b 2 gradus medius sit d e 2 per f gradus medie linee d e
igitur quadrangulus ortogonius a f g b ostendens uniformem qualitatem est equalis
triangulo a b c ostendenti qualitatem eandem uniformiter difformem ut ex geometria
2 per 26 primi patet. Item similiter dicendum est de quadrangulo a c d b ostendente qualitatem
uniformiter difformem terminatam utrobique ad gradum quadrangulo rectangulo a f g b
ostendente qualitatem uniformem secundum gradum medium 2 quod.
Ex hiis patet quod non accipiendus gradus medie subiecti proportionaliter i. quod sicut se habet a c ad
d e gradus puncti medii sic se habet d e ad b non quod aut sic se habet
a c ad h k gradus puncti h medii linee a b sic se habet h k ad b d seu
non sic sit sed accipiendus gradus qui correspondet intensive medio puncto d aut h
linee a b 2 quod. tamen iste gradus est proportionalis gradui seu intensiori correspondenti puncto
medio linee d b aut h b eadem proportione qua sibi proportionaliter intensione seu gradus a c
2 sic secundum partes proportionales subiecti a b corespondent in infinitum taliter eadem
proportione in gradibus proportionibus quod est inter a c 2 d e ut h k i. Et similiter
potest arguendum de qualibet superficiali 2 corporali 2 quod de velocitate motus 2 qualitate
successiva eodem modo dicendum est sic de istis dumtaxat loco puncti medii subiecti
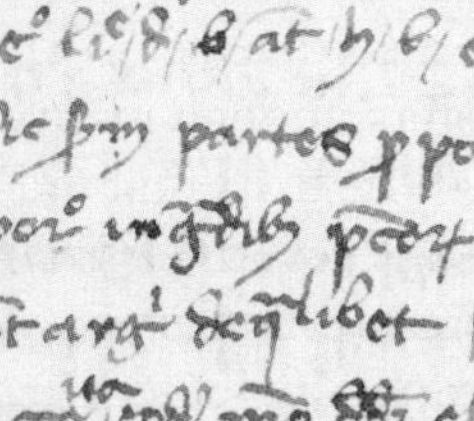

is some indication that at least in Toledo, translators selected literature to translate not through a blind groping for texts, nor because of a desire to improve scientific knowledge, but rather because of an interest in becoming more like their Andalusian hosts. Whatever the motivation or the material support, the evidence suggests that European translators were engaged in an active rather than passive effort to translate and appropriate ancient and Arabic materials.

A simultaneous, yet quite different transmission occurred in Jewish culture. Beginning about 1140 and continuing through the end of the twelfth century, Jewish scholars in Provence, the Iberian Peninsula, and Italy—but not northern France or German-speaking regions—undertook a broad and rich translation of philosophy and science into Hebrew. Most interestingly, in its early stages this activity often occurred independently of the surrounding Latin culture, except in medicine, where there was considerable importation of Latin literature into Hebrew, and in Italy, where Christian-Jewish exchanges often took place. In philosophy and science, the twelfth- and thirteenth-century translations into Hebrew passed directly from Arabic sources, particularly in Provence and the Iberian Peninsula. There may have been multiple reasons for this, including the importance of linguistic ties and a flourishing Judeo-Arabic cultural relationship, but also a simultaneous estrangement between Jewish and Latin language and culture, in which Christian philosophy was considered inferior by many Jews. On the other hand, the emergence of Jewish philosophical and scientific culture was a response—at least in Provence—to dominant Christian rationalist polemics against Judaism.

A New Home for Science: The University

Prior to the twelfth century, educational centres in Western Europe were located chiefly in monastic houses (for the purpose of educating the next generation of monks) and cathedral schools (whose purpose was the education of the non-monastic clergy). Frequently these schools had both a very limited curriculum and a small faculty, usually just a single master who taught all disciplines. But over the course of the twelfth century, partly in response to the influx of new texts, and partly in recognition of the demands of changes in European society, government, and the Church, new educational institutions—the universities—evolved, each reflecting the characteristics and requirements of the local conditions from which they sprang. In spite of these local distinctions, as a whole the universities possessed four characteristics that distinguished them from earlier educational institutions: they were highly utilitarian enterprises that equipped students for specific positions following their university training; they were corporations invested with the rights and responsibilities

---

Opposite: Firenze, Biblioteca Nazionale Centrale, Conv. Soppr. J. IX. 26, fol. 32ᵛ. A fifteenth-century manuscript of Nicole Oresme's *De configurationibus qualitatum et motuum*, with marginal diagrams illustrating the mean-speed theorem. On this text, see Marshall Clagett, *Nicole Oresme and the Medieval Geometry of Qualities and Motions; A Treatise on the Uniformity and Difformity of Intensities Known As Tractatus de configurationibus qualitatum et motuum*. Madison: University of Wisconsin Press, 1968, pp. 408–11.

...audiã gñlæ doni...	...ignñte...
...dendenaucluciones...	sčrit uico
	pamuni tt...
iǒ Trinitaticu	oǧte cutholicæ upratc frem...
iiǒ Decornu	ERA
iiiǒ Diuleaticu	
iiiiǒ Ariahmeaticu	Sardiur
vǒ Guricu	Topucijus
viǒ eo metoricu	Zmiuruedur
viiǒ Asteonomiu	Curbuculur
(39) uinque enim sūt uirtualur	Suphitur
ad supientiæ.	Jurpir
Adsiduitatur legendi.	Liquruur
& a lncossoqundimuctsær.	Aquær
& acēemnendo diuitiæ.	Amatraur
& au notēn scripatutum.	Errolicur
& ecōatēnatumundi.	Oniainur
	Berill

Ce livre fut de luy posté ll
Qui de tout veoir fait diligence
(Soy cognoissante home mortel)
Pour chasser de soy l'oubliance

Riche ne fut, dont povre
Le feist estre si vigi...
...e en tout bon scavoir
Escoute come en bien

Estant a richesse venu
Lors la laissa pour son
Quant il voulut est
De soy et doulx...

Postea fuit Francisci Raphelengii, et dono Postelli.

of fictitious legal entities; they created curricula that ensured uniform education among their alumni; and they certified the competence of their students by awarding degrees. Although universities have continued to evolve since the twelfth century, modern universities also bear these characteristics, and the university itself is one of the great legacies of the Middle Ages, both in Europe and throughout the world.

In recognition of both the quantity and complexity of the influx of new texts, scholars in the twelfth century realized that the old 'jack-of-all-trades' model of monastic and cathedral schools would not suffice within these larger institutions. As a result, pedagogy became more specialized, and the institutions themselves defined faculties whose responsibility was limited to teaching particular segments of the material. Each faculty became the intellectual home to literature retained from the early Middle Ages, supplemented by the enormous trove of texts that entered Europe in the twelfth and thirteenth centuries. The foundational unit for most of the earliest universities was an arts faculty whose curriculum focused on philosophical texts, of which Aristotelian natural philosophy and to a smaller extent mathematical sciences constituted a significant portion. Because students who were not members of religious orders matriculated in the arts faculty (and those who were received a similar introduction in the schools of their own orders), and because arts training was a prerequisite for advancement to the higher faculties of law, medicine, and theology, natural philosophy and science permeated the entire university, regardless of the student's ultimate specialization!

The comprehensiveness of Aristotle's work—comprising logic, the study of natural bodies, cosmology, epistemology, natural history, ethics, metaphysics, and other disciplines—made it a ready-to-hand arts curriculum for the nascent universities. The requirements for degrees specified that students must have heard lectures on (among other things) Aristotle's logical works, *Physics*, *De caelo*, *De anima*, *Ethica*, *De generatione et corruptione*, and *Metaphysics*, in some cases multiple times. While masters at some universities (like Oxford before 1431) were free to choose the material on which they would lecture, at others (like Vienna) the curricular materials were distributed among the teaching faculty at the beginning of each academic year, thereby ensuring that students had the opportunity to fulfil the requirements for degrees. In the mathematical disciplines, the curriculum followed a slightly different model. Although the *Elements* of Euclid had been available since the early twelfth century, geometrical instruction was limited by specifying the number of books to be taught (usually the first six) or the length of the term (often three weeks of lectures). In astronomy, the preferred textbook was a medieval composition, the *Sphaera* of Sacrobosco (*fl.* 1221–1250), which provided the rudiments of planetary motions,

Opposite: Leiden, Rijksuniversiteit, Bibliotheek Cod. Or. 231, a page from the *Glossarium latino-arabicum*. The Latin entries are alphabetized with definitions or explanations in Arabic, indicating its use as a tool for Mozarabs to study Latin. The work drew on an earlier Latin–Latin glossary as well as lexical material from a Latin manuscript of Isidore of Seville's *Etymologiae* and Arabic versions of other works. On this text, see P. S. van Koningsveld, *The Latin–Arabic Glossary of the Leiden University Library: A Contribution to the Study of Mozarabic Manuscripts and Literature*. Leiden: New Rhine Publishers, 1977.

Vni cuiqz ⁊ q̄re sup dc̄te p̄mū domiciliū firmū. Alim ū sequens firmū. Icū afirmo remotū dīc̄. Vnūq̄z hoꝛ p̄o noīe distinguunt. hau domicilia oīm iꝉioꝛ significatura sē. dicend q̄. xii. domicilioꝛ signoꝰ.

O domiciliū itaqz ortū occupans: horoscopicū dr̄ bui effect tr̄ ipsi hoīs corp̄. uirta. omisqz eī inceptiones. Domiciliū sec̄m sup possessione p̄isi. Icū sup germanos ꝯsanguineos. vxorisqz p̄pinquos s; ⁊ sup lectione itineꝰ non usqz locale. Ortū sup parentes videlicet patre ᵃematre eoꝛ ᾱ cessores. Quiaqz ᷑re parte occupaꞇ sit. thesau ros ⁊ra abscondita. Quintū sup libidine sua ⁊filios. Sextū sup egritudine. seruos qz ⁊bestias suas. Septima uxoꝛ nuptiaꝛ. Octauū amoris mores. Ilonū itineris ⁊peꝛnois. Decimū regni ⁊atus fortune. fame racinna tis officioꝛ ⁊laboꝛ. Vndecimū spei. fortune dicunꞇ fame. Sochiū. Duodecimū passionū iīmicoꝛ carceꝛ do loꝛ penuriaꝛ. Gaude au mercurii indomicilio. horoscopico. luna au itercio. Ven̄ ī q̄to. Mars insexto. Sol ī nono. Iupit in xi. Satūꝰ in xii.

IHouli ᵹ planetes se q̄ ꝑexactū diuūrū. Vnusq̄z enī ⁊erecto signoꝛ sublimeꞇ. Cū itaqz sublimaꞇ uiꞇ n̄ lucidus minor nobis uidr̄ unt̄ p̄mouet etē eū ꝫap̄s sue sublimatiōis. mīn q̄ xc. gduū exantore ⁊ p̄teriore parte conteneꞇ. Cū au ut̄qz. xc. gdꝰ fueriꞇ. iuro iq̄ititas ⁊moꝛ examinata erunꞇ. Cū ū hunc loci transcendiꞇ. iluū iq̄ititas ⁊moꝛ augebunꞇ. ¶Sunt au horę quedā inquibz cresciꞇ nūs. Quedam inqbz decrescit. Alie ⁊iquibz nec crescit nec decrescit. Cū eni fuit argiu mīn q̄ ᵉlxxx. graduū: cresc̄ Cū maḡ decliꞇ. Sin au lxxx nec h̄ uiꞇ illd̄. ¶Sciri ⁊ conuenit qz ⁊ qn ꝑpuꞇacio augeꞇur. ē qn minuiꞇ. ē qn neqz augeꞇur nec minuiꞇur. auged̄ ⁊gueriens examino medialitati sup poniꞇ. Minuiꞇur ū ᵈū eid̄ id̄ subtrahiꞇ. Atqz qn nec addiꞇur nec minuiꞇur ⁊ecto obliuiois unua sotis inueniꞇ. Sed ⁊illud q̄ō planete superiores medialē motū suū superauerꞇ: moꞇ eoꝛ auct̄ dr̄. Cū ū amedialitate subtrahiꞇ. moꞇ duū miur̄. dice nec addiꞇ nec subtrahiꞇ moꞇ examinar̄ appellaꞇ. Venus ū ⁊mercurii moꞇ primoru sol denoꞇꞇ. Si eni medialē sol motū superauerint eoꝛ moꞇ auct̄. Siu medialitate miū p̄mouiꞇ. diminuiꞇ. Sin au nec addiꞇur nec subtrahuerint medialꞇ dr̄. ¶Sz illud q̄ themeli ⁊genubi q̄niū aduia sol dux̄. A capite it. draconis sui usqz adcaudā themeli a cauda uero usqz adcap̄ genubi p̄hibenꞇ. ¶Illd̄ au p̄eundū ⁊ē qz singlē stelle sue potentie ᾱ uerꞇ ⁊ dui uiis deꞇerminatos. Sol itaqz. xv. gdꝰ ᾱ se ⁊motū ref̄ se sue potentie uiꞇ. Luna. xii. au ⁊motū retro Satūꝰ in iupr̄. xc. ᾱ ⁊motū ref̄. Mars au. viii. au ⁊motū rec̄. Ven demiqz ⁊mercurii. septē au ⁊motū retro diced̄ q̄ qᵃ adsle p̄etates recipiunꞇ. inmaiore au ᵹ ᵗsᵃgoga diffuss. hic ū necessaria ūꞇ. Qᵉqz q̄ xū plane tariū supiory Satūꝰ dico ⁊ iupr̄ siue ⁊ Mars a conuentu discesserinꞇ: respectu eī dexo diceꞇ donec ille adopp̄m gdū p̄uenerit. Cū ū sol opp̄m ⁊inseriꞇ ⁊planetra siniꞇ ei diceꞇur. Ven au ⁊mercuri asole adouen ᵘˢqᶻ tem discedens donec adeū redeanꞇ. sinistri dicunꞇ. Abeo uero recedentes q̄ usqz adp̄dictū p̄mū loci redeant dext̄ appellanꞇ. Luna qz asole disc̄dens donec adopp̄m ueniat sinistra Siu oppositū ⁊inseriꞇ dext̄ diceꞇ. ¶Satūꝰ it̄ iupr̄ ⁊mars asole p̄etates. viii. suscipiunt. Prima qn solem eod̄ loco conue niunt. Scda. qn sol asatno. xv. gdib; ⁊moue totid̄ atuarte. xviii. recesserit. Tcia qn uiꞇ sole ⁊ q̄mlibet illoꝛ tiū xc. siliꞇ gradus. Qta qn instone p̄ma existit. ref̄ gdū inciprꞇ. Qnta cū sol ⁊noppositio eī fuit. Sexta cū instatione scda fuit p̄cede incipientes. Septima: cū asole ite. xc. gdib; distiuiꞇ. Octa: cū xc. gdib; aut pautioꝛb; asole remoti fuerint. Iamqz abeo occulⁱanꞇur. ¶Ven au ⁊mercuri viii. asole p̄etates uiiꞇ.

7 ⁊ p̄ma qn sole conueniunꞇ. Scda: cur ipsi asole septē gdꝰ oriente usus recesserⁱꞇ. Tercia: cū ila ectone p̄ma uiā cursu retrogradū meuꞇ. Qta. asole. vii. gdib; uiā cū exactᶜ adeunꞇ. Quinta cū uⁱ le ᶜᵉᶜᵉᵖᵗⁱᵒⁿᵉ conuenⁱunꞇ. Sexta cū̄ aslᵉ opposito distruꞇ gdib; vii. uiꞇ p̄ma noctis apparenꞇ Septima: cⁱstasⁱoē ᵃᵇᵉᵘᵈᵒ f̄ cursū dⁱriguⁿꞇ. Octaua: cū̄ asole. vii. gdib; remoti fuerⁱꞇ. ¶Elkamur q̄ p̄etates asole. viii. P̄ma cū̄ coniugaꞇ. Scda cā abeo. xii. gd̄ recesserⁱꞇ. Tercia cā abeo. xc. gdib; distriꞇ. apparetqz dimidia. Qta cā abop posito sol. xii. gdib; disteꞇ. Quinta ceⁱ opponaꞇur. Sexta cⁱ sup oppositū xc. gdꝰ addat. Septimū: q̄ens instat eī xc. gdib; remota apparetqz dimidia. Octaua cū̄ asole. xii. gdib; disteꞇ ⁊ ꞇ

e inceps ū de planetariū habitudinib; diceꞇur ⁊ Staū. xxv. q̄rū nota sunꞇ hęc. p̄etenⁱa acceslⁱ

sdicimus istorum ett pla sub nutveni. 7 de hospiciis humili silentio uelut in cognita. 7 non cecca. primū sim. ne dice
ret ani horurbar. xii signa. xx. 7 vii. apponeim hospicia. Auia sic certa suo prespsini seyrem planetarū duodeci signa cū
recepracli 7 ascensionis earum 7 descensionis ee hospicia 7 ergastula. sic deoq hīs. xii. qb; prsint pistare aptare recepracli.
cū enim alic suppuracionem secis astt. merico bruis hospiciis. tercia parte ecia hospitari. Secūn uo duab; ecū hospitari.
7quo iduab; qui scias mecari. Tercium sexto 7 septimo. tercia parte quini disce placare. 7 sic cecis. xii. signis h rxvii. hos
picia partire. DE BELLATORIBVS. IIII.

   iuis q scire docte 7 indubitant de duob; militib; si faciant bellum an ñ. accipe planeta uni illor.
   7 eaue sisit inimica illi hore q consuluerit. 7 si e inimica sue bellum. ti sit ualde perturbatum. ce ti
et amica bellum inde non fiet. DE ADAVRIENDA PECVNIA. IIII.

   VRSVS mih cem ualde e magnus euenied. qd non excipit ex mentis philosophie adtas. ne noscu amicat
familiarib; sapiente. nec utilissimum cecto ingenioy q sup plibaui. S. dagrendo munere. Si uis scire plse an
adihi tibi puentum sit h regla qt dicet abaio tuo non latet. In pmis qdem astrum inspice tuum. deide secio qd
e ydoihracum duncis aut adqrendis ant p dedis. pq; nonū de euentib; prspicacu suarū 7 undecimū prspsin. ad
q agenda. 7 si e 7illis tue amici planete. scias pcerto auqs prticin abamico ac etere. Si uo inimici iduciat e dabit
   vo enim qspiam uult dinosce si inuilla aut malum aut bonum uenturū sit. De bono aut inisto tuliat
in pmis qde sic de hoie nom uille coputet. 7 planetā inueniat. 7 siūm onomatis uille amear. 7 matrem
uille poteniorem semina scedat. 7 coputet. 7 calculacionem q inde exeuerit cū sup dicta sūma iungat. pq;
xviii. diuidat quiens cuq; ualeat residui au nūm. p xii. signa dispge. ui finire nūt. Ille erit signum eius
7 sic euenit homini ita euenier erunt. De cels spera uoctire. vii. signa. nec non rx vii. recepraclis
   t au plani tuqnt pdicta. 7 cunctoy certi possis nosce signa ette anob subiecta figa q intra se cludens
  ptinet xii. signa. prxviii. induain hospicia. Ipsoq au noia noscas more aluaten saraceni magistri.
nemeit descpta. Quay pmū e alnair. ii. albotan. iii. arboria. iiii. adoron. v. almusan. vi. arhaia. vii. adn
m viii. anathra. ix. atharf. x. aluha. xi. alchoraten. xii. alcarpha. xiii. alacha. xiiii. alcaneth. xv. alzahar
vi. adiuuenn. xvii. alehalil. xviii. alchelb. xix. aleuetri. xx. anochrum. xxi. albelda. xxii. cahaddedua. xxiii.
cahaddola. xxiiii. Sahaelaphaut. xxv. Sahahaleabia. xxvi. palehalud. xxvii. palehalacht. xxviii. aliaoram.
sic stiga uac ostenduntur.
a. i. b. ii. c. iii. d. iiii. e. v. f. xiiii. g. x. h. xii. i. h. xvi. l. ii. m. xii. n. xviii. o. vi. p. ii. q.
x. r. iii. s. xx. t. viii. u. xx. x. iii. y. iii. z. vii.

Ries ht mansiones. Alnair. 7 Aldoraia. 7 tcia parte de Albotai | Alnair. stelle iii. forma
aurus duas partes albotaim. Aldebra. et duas partes Almiscen | stelle ii. Aldora stelle vii.
enim terciam parte almiscen. Altaia. 7 Altharan. | Aldeb. xvii. Almisce. ii. Altai. vi.
ancer ht. Alnatza. 7 Altarf. 7 tcia parte Alcebach | Aldr. ii. Alnair. i. Altharf. ii.
eo duas partes alcebatha. Alcothatei. 7 duas partes Alsarfa | Alcebar. xi. Alcotai. iiii. Msarf. i.
irgo tercia parte alsarfa. 7 Alaua. 7 Alsemech. | Alaua. v. Alsemech. v. Algofora. iii.
ibra ht Algafora. 7 Alzebenech. 7 tcia parte Alachid | Alzebenec. Alachid. vii. Alcaiu. iii.
corpi. duas partes alachid. Alcalu. 7 duas partes Alebra | Alebra. vi. Alnai. viii. Albelda. vii.
agittarius tcia de alebra. 7 Alnaam. 7 Albelda | Scaldebo. iii. Scaldbol. Scalcod. ii.
apricoanus ht. Scaldebo. Scalbola. 7 tcia parte Scadcod | Scaldadab. xii. Garfalau. ii.
quarius duas partes Scadcod. 7 Scadadabia. 7 duas partes Garfalau | Fargaliter. ii.
isces tcia parte de garfalau. Alfargaliter Aluaten. | Guaten. xc.

Alnair. i. cornu.	Alnatza .i. rastru	Algafora .i. fuscinula	Scaldebo. i. faut occasio
Aldoraia .i. ceruix.	Altharf .i. porta	Alzebenec .i. pensa	Scalbola. facit magnū
Albotaim .i. uent.	Alcebatha .i. frons	Alachid .i. ungla	Scalcod. facit bona
Aldebran .i. cor	Alcotatei .i. pett	Alcalu .i. corp'	Scaldadabia. facit dubiu
Almiscen .i. cauda	Alsarfa .i. manus	Alebra .i. arcus	Garfalau. cape
Altaia .i. capud lune	Alaua .i. tibia	Alnai .i. corda	Fartgalifer. pisces
Aldara .i. brachiū	Alsemech .i. pes	Albelda .i. sagitta	Aluaten

90

supplemented by another medieval text, the *Theorica planetarum*, which introduced the student to Ptolemy's geocentric models, and both were used in conjunction with the *Canones*, or rules, that governed astronomic tables. While the Latin translation of Ptolemy's *Almagest* continued to circulate, it was reserved for the private study of a small minority. In optics, the texts of choice were also medieval compositions, like John Pecham's (d. 1292) *Perspectiva communis* and Witelo's (c.1237–c.1280) *Perspectiva*.

The same tendency to assemble a canon of texts prevailed in the higher faculties. In medicine, the collection of texts known as the *Ars medicine* [or *Articella*]—comprised of the *Isagoge* of Johannitius, Hippocratic *Aphorisms* and *Prognostics*, *Urines* of Theophilus, *Pulses* of Philaretus, and Galen's *Tegni* (which was added in the second half of the twelfth century)—defined the curriculum until the mid-thirteenth century. In the 1250s, during a period of educational revision, a new collection of texts— Galen's commentaries on the Hippocratic *Aphorisms*, *Prognostics*, and *De regimine acutorum* and Haly Ridwan's commentary on *Tegni*—was assembled and changed medical teaching, narrowing the focus of the curriculum, expanding the Galenic presence in it, and shifting attention from primary texts to commentaries on those texts.

Similar collections of texts prevailed in the other two university faculties, law and theology, where Roman law codes and canon law compilations like Gratian's *Decretum* (written c.1140) and the Bible and Peter Lombard's *Sentences* (written 1142–1158), respectively, defined the curriculum. The designation of core texts lies at the heart of the medieval scholastic world-view, and signals the underlying assumption that knowledge derives from books and authoritative authors. Hence to understand the medieval view of nature, one must also understand the way scholastics unpacked the texts that formed the curriculum of these early universities.

So great was the onslaught of new texts in the twelfth and early thirteenth centuries that medieval scholars provided mechanisms for expeditiously explaining the ideas of the authors. Recognizing that the problem involved not just the quantity of the material but also the unfamiliarity of its organization, many pedagogues like Hugh of Saint Victor prepared systematic guides to reading and later teaching the text. Known as 'establishing the *ordinatio*', this technique entailed the creation of rubrics at the head and subsections of chapters, use of colour or symbols to visually distinguish one part of the text from another, or even divide previously unbroken texts into smaller, more easily digested sections, as for example, was done in the commentaries of Averroes on Aristotle's texts. The next step, known as *compilatio*, sought to distil the volume to more manageable quantities. With their rearrangement and amalgamation of authoritative texts, works like Vincent of Beauvais' (c.1199–c.1265) *Speculum maius*, Bartholomew Anglicus' (fl. 1220–1235) *De proprietatibus rerum*, or Brunetto Latini's (c.1220–c.1294) *Li livres dou trésor* compressed the huge collection

Overleaf: London, British Library, Sloane 2030, fol. 84ᵛ (Abū Maʿshar, *Ysagoga minor*) and 94ʳ (*Alchandreana*), containing Arabic–Latin correspondences and suggesting a vocalization procedure in the translation process. On this issue, see Burnett, 'Some Comments on the Translating of Works from Arabic into Latin in the Mid-Twelfth Century', pp. 166–7; on the manuscript, see David Juste, *Les Alchandreana primitifs: étude sur les plus anciens traités astrologiques latins d'origine arabe, Xᵉ siècle*. Leiden: Brill, 2007, pp. 331–2.

of new information into enormously popular reference materials that could be drawn upon in a variety of fields.

These compilations were seldom the subject of university curricula, but the mechanisms of *ordinatio* and *compilatio* were very much used in the pedagogical techniques of the schools. Medieval lectures on authoritative texts followed a prescribed format that derived from the *ordinatio* and *compilatio* traditions: first a reading that served as an overview of the section to be treated that day; second 'establishment of the text', that is, the correction of textual errors; third, the designation of hierarchical divisions in the text; fourth, resolution of linguistic or terminological difficulties; and finally, analysis of the important questions within the text. Such formal techniques survive in the multi-layered glossed manuscripts of works in the curriculum, in which linguistic corrections can be seen between lines, and divisions of texts are noted by gibbet-like symbols. The other pedagogical university technique, the disputation, was built upon the lecture tradition, for questions that eventually found their way into the disputation often derived from those articulated in this final aspect of the lecture. The disputation was central to university education: students were obligated to attend the master's performances, and masters assessed the knowledge and competence of students by observing disputations within the private reserve of the classroom before similar but public debates certified this more broadly. Aside from the pedagogical value of disputations, they were the venue where assimilated previous knowledge could be expanded and extended.

To see how content and technique worked together to shape medieval science, we must understand that universities and their scholastic techniques arose in the twelfth century at a time when the study of logic and language occupied a central place in the interests of intellectuals, and that this same interest preoccupied many of their successors in the fourteenth century; seen from this longer perspective, the thirteenth-century concentration on the texts and natural philosophy of Aristotle and related authors may be an interlude between two periods of linguistic and logical concerns. In the twelfth century, a significant part of the investigation focused on what was called 'speculative grammar', that is, the formal structures of language in general—what we might see as fundamental to linguistics—as distinct from the rules governing particular languages. In the fourteenth century, the emphasis was frequently on signification and supposition, that is, the way mental signs, spoken words, and natural things are related, and how terms within propositions can stand for different things, depending on the linguistic constructions and intentions of terms. But while the assimilation of Aristotelian natural philosophy may have been an interlude within this longer, more durable interest in logic and language, the Aristotelian content furnished significant material for the fourteenth-century discussions. When William of Ockham (d. 1347), for example, dismissed the possibility of the medium serving as the continued cause of projectile motion, he did so by tendering a linguistic argument, rather than invoking empirical evidence. And in the next generation, when the various Oxford Calculators (including William Heytesbury (d. 1372/1373) and Richard Swineshead (*fl.* 1340–1350)) devised theories of substantial change—referred to as intention and remission of qualities—they also derived them within logical texts

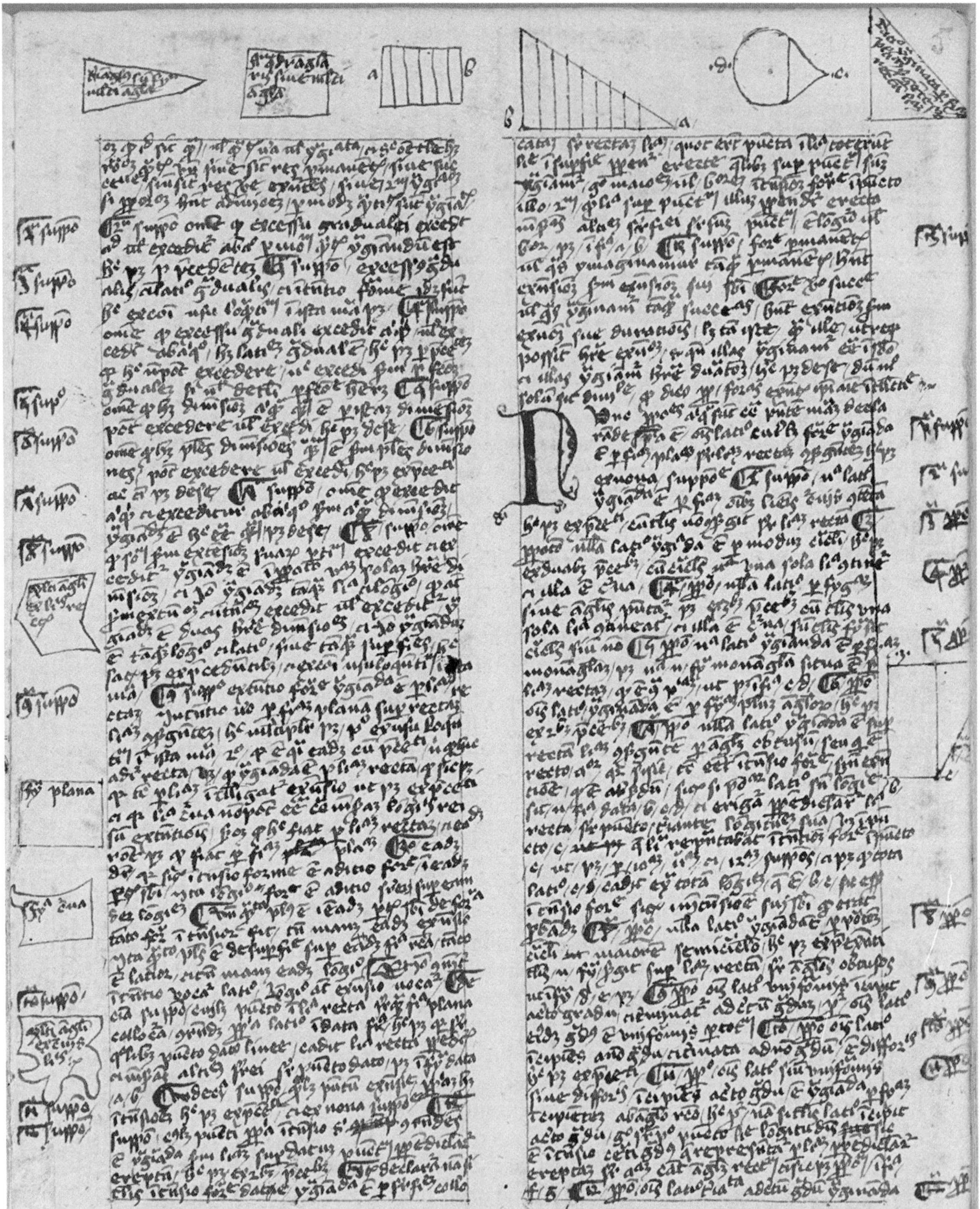

focused on linguistic issues, but with specific Aristotelian natural examples and consequences. So instrumental is this nexus of logic and linguistics and Aristotelian natural philosophy, delivered through university pedagogical exercises, that the demise of medieval science in the fifteenth century may be the result of changing tastes and conventions in education, as much as in the Renaissance humanist preference for ethics, history, and literature over natural philosophy and logic.

Medieval Science and Religion

As noted at the beginning of the chapter, one of the reasons for the apparent incongruity of medieval science is the modern conception of the relationship between religion and science. For if science appears to be the antithesis of religion, as modern reflections on a long line of contested encounters—Galileo and the Church, or evolution and creationism, to cite but two of the most prominent—would suggest, how can a period so defined by its religious nature be considered to have a scientific element? But a broader investigation of the issue, focused on the bidirectional assessment of the effect of science on religion and religion on science, reveals a much more complex story. We can see this clearly when we consider how astronomy and Muslim religious duties co-evolved since early Abbasid times in Baghdad, achieving a new institutionalized status some five hundred years later in Mamluk Cairo and Damascus. One factor behind this dialogue between religion and the mathematical sciences is the following Qur'anic prescription:

Turn your face towards the Sacred Mosque; wherever you may be, turn your face towards it.
(Qur'an 2.144)

Praying in the direction of the Ka'ba in Mecca is one of the five pillars of Islam. Hence, from the eighth century onwards, scholars developed methods for determining this prayer direction or *qibla*. They developed accurate and approximate solutions using complex and complicated as well as simple and straightforward methods. Many Arabic, Persian, and Turkish manuscripts as well as those in other languages used by Muslims show that not only experts dealt with this problem, but ordinary people without mathematical or astronomical knowledge also wanted to learn about these directions. They inquired about the practices of Muhammad and his companions, asked for the opinions of legal scholars and transmitters of tradition (*hadith*), and copied or bought simple circular plans of prayer directions either sketched on paper and bound into a manuscript or engraved on an instrument like an astrolabe or a special instrument, the so-called *qibla* pointer. After about the twelfth century, a compass needle began to appear on such a pointer.

Opposite: New York, Columbia University Libraries, Plimpton 171, fol. 5ʳ. From a late fourteenth-century manuscript of *De latitudinibus formarum*, often attributed to Nicole Oresme but actually by Jacobus de Sancto Martino, showing uniform, uniformly difform, and difformly difform configurations of qualities. See Clagett, *Nicole Oresme and the Medieval Geometry*, pp. 85–96 and T. M. Smith, 'A Critical Text and Commentary Upon *De latitudinibus formarum*', PhD dissertation, University of Wisconsin, Madison, 1954.

En nom de dieu
ci commence le liure
que aristote appelle du
ciel et du monde le
quel de comman
dement de tressouuerain et tresexcel
lent prince charle quint de cest nom
par la grace de dieu roy de france
desirant z amant toutes nobles sci
ences Je nichole oresme doyen de le
glise de rouen propose translater et ex
poser en francois ❡ Et est cest liure
ainsi intitule car il traite du ciel z des
elemens du monde en prenant cest
nom monde pour les iiij elemens
contenuz dedens le ciel et soubz le ciel
qui auiennent et corrumpent en cest
liure cest nom est pris pour toute
la masse du ciel et des iiij elemens en
semble et est cest mot pris ailleurs
en plusieurs autres signefiacions

qui ne sont pas pres de cest propos ❡ Et
en cest liure sont pluseurs liures prin
cipaulz ❡ Ou premier il determine de tout le
corps du monde selon soy et de ses pro
prietez ❡ Ou second en especial du
ciel ❡ Ou tiers des elemens selon leur
mutacion ❡ Ou quart des elemens se
lon sa position ❡ Et contient le
premier liure xxvij chapitres
❡ Ou premier chapitre il moustre
que le monde est parfait selon quanti
te ou magnitude
❡ Ou second ch il moustre come
des corps du monde sunt iij simples
mouuemens locaulz
❡ Ou iij il applique aucuns des
mouuemens locaulz a aucuns des co
rps du monde
❡ Ou iiij il moustre par dij raisons
que fors les iiij elemens il con
uient mettre un autre corps

Less widespread among the ordinary Muslim public were tables for prayer times. Once the rules governing the number and definition of these prayer times had been settled in different Muslim communities, scholars began compiling tables for different localities, which showed the beginning and end of each of the five standard periods. A third major issue of religious importance was the determination of the beginning of the month, in particular Ramadhan, the month of fasting. Here too, the astrologers of the Abbasid court in Baghdad early on began to compile tables for the new Moon, using Indian rules. While scholarly solutions for the first two issues were relatively successful and appreciated, despite continued debates and differences of opinion, the calculation of the new Moon fared less well. Confirming the appearance of the new Moon by an eyewitness was and is an important criterion for the call to fast, as can been seen still today with the different beginnings of Ramadhan in different countries of the Islamic world.

In the late thirteenth century, these religious duties and their mathematical and astronomical interpretations were brought together in a new astronomical discipline called the 'science of timekeeping' (*'ilm al-miqat*). Scholars who specialized in this new discipline were called 'timekeepers' (*muwaqqit*). This process seems to have begun in the two main cities of the Mamluk realm—Cairo and Damascus. It rearranged spherical geometry, trigonometric functions, astronomical problems, and the construction of scientific instruments into one disciplinary whole. It was taught at madrasas and mosques. At the latest from the fourteenth century onwards it spread across the Mamluk territories to Yemen, Mecca, and Medina and a short time later to North Africa, al-Andalus, and the court of the Ottoman Sultan Murad I (r. 1360–1389).

Mosques, madrasas, and other buildings were meant to be oriented towards the Ka'ba in Mecca, after Muhammad had changed the prayer direction from Jerusalem towards the city of his birth. Nonetheless, many extant buildings across the Islamic world show very different orientations, because often non-mathematical means like the rise or setting of a particular star or constellation, the solstices, wind directions, or the traditions of Muhammad or one of his companions were used. But even when mathematically determined prayer directions were applied, as was for instance the case in Mamluk Cairo, the orientation of the buildings still deviates from the modern *qibla*. The accuracy of the mathematically defined value of the *qibla* depends on the accuracy of the geographical coordinates of any given location. The latitude dates in medieval tables are usually correct within a few minutes. The true problem consists in the values for the longitude difference towards Mecca. Here, the deviation could reach several degrees. In the case of Cairo, this value was 3° too small. While latitudes could be easily measured or calculated from other astronomical data, longitudes depended on the parallel observation and measurement of lunar eclipses in two different localities and the accuracy of time-measuring instruments.

Opposite: Paris, BnF français 565, fol. 23^r. In this fifteenth-century manuscript from the collection of Jean, duc de Berry, Nicole Oresme presents his *Traité de la sphère et Livre du ciel et du monde* to Charles V, flanked by a cosmological diagram of the world, incorporating a T-O globe. Concerning this image, see Pascal Arnaud, '*Plurima Orbis Imago*. Lectures conventionnelles des cartes au Moyen Age', *Médiévales* 18(1990) 33–51 at 50.

me felix pnapium tribuendo z feliciul
me sui gubernando Et Jubeat ypsece
cy fiat Vale ad fine optim ducendo.

Racio huiusmodi comentacidis seu
collectois non fuit libror defect sz pcli
britals et pfectus. Non em quilz omis
libros habere potest. et si hret tediu esset
sertere, z fora onia In mente retinere
Curia lectio delectat certa pdest. Et In
astructoribus semper occurrunt melio
rameta saenae em per additamenta
fiut. Puer em sumus In collo
citantes qr vide possimus gsed criatas
et aliqntulum plus. Est certo In
astructoibus et assumacomb Unitas
z pfect'. Vex qr ut ait plato crthim'
ea que scribuntur breui qn expediat
diuita suo et obscura. Ea vero que
sonitiue bizentes fusasuit Ure est
liber gu reprehencoem effumat.

Et propter ho ni ad solaciun

Prius
quis extero deo
omipotenti vitam
ppetuam aiar et
samtatem corporis
medianti aioxbos
matipos per gram
quasobtul oi cir in
ex virtutibus samtate gseruantib;
et ptergentibus a langore Santi
Intellitu arte mediae z Intreiu sautl
Suimis et aiosis Intellittenab Valv
operum ad amentans et assumens
In pmis It autgrediens quadam
amentacoem seu coleacoem artis curur
tie dito mus deo Vnio; Vero g omis
tribuit esse sine g nullu rite fundat
exordium. Ad eu deuotissme trahr
rendo totis uirib; cordis mei suplicando
vt In hoc opere et in euictis aliis imitat
michi auxilium de scto z de spon tueatur

What about the situation in medieval Catholic societies? Given the centrality of the Church in the Middle Ages, it is hardly surprising that religion influenced the content and course of science during the period. In part, this reflects the ambivalence of the Church toward knowledge: it was, after all, the tree of the knowledge of good and evil (*lignum scientiae boni et mali*) that occasioned the fall of humanity, and the apostle Paul warned of the dangers of self-aggrandizing knowledge in contrast to the humble nature of love. Yet he and others also commended the word of knowledge (*sermo scientiae*) granted by the Spirit. The early Fathers of the Church likewise took both sides of the issue, with some (like Tertullian) arguing that Christians should put away secular knowledge because it led to error or unbelief, while others (like Origen) urged Christians to take what is useful and use it to augment and perfect their spiritual natures. And as early as Augustine, it was recognized that by establishing a countervailing institution that siphoned talent away toward ecclesiastical careers, the Church reduced the pool of those interested in investigating nature. More actively and aggressively, the recurrent condemnations of natural philosophical positions, like those promulgated in 1277 at Paris, certainly exercised a determinative influence on the course of science in the Middle Ages.

Yet at the same time, medieval religion also contributed materially and formally to the course of scientific inquiry. In the early Middle Ages, the monastic emphasis on timekeeping and calendar that regulated the liturgical year and the daily services gave significant emphasis to astronomy and other related mathematical activities. As noted previously, the rise of humanistic naturalism in the twelfth century that precipitated the period of intellectual activity in the high Middle Ages was driven by the complex web of religion, art, law, and science during the period. And while the fourteenth-century interest in quantifying natural changes reflected contemporary preoccupations with logic and language, this movement also depended heavily on the doctrinal issues inherent in form and quality, because it was commonly observed that divine grace in the soul of the believer is a form, and as such could experience spiritual growth (or diminution). The scholastic authors describing the 'intension' and 'remission' of qualities in nature (that is, the way qualities like heat, light, and speed increased or decreased over time and space) were in fact drawing on the very same language and terminology that their theological colleagues had used in discussions of grace and other spiritual qualities.

But while medieval science surely was shaped by contemporary religious doctrine and practice, it is also true that religion felt the effects of contemporary science. The twelfth-century introduction of ancient and Islamic texts transformed the nature of many aspects of medieval culture, including politics, law, theology, and science.

Opposite: Paris, BnF lat. 6966, fol. 4ʳ. Fifteenth-century manuscript probably produced for Duke Charles d'Orléans, depicting a physician (Guy de Chauliac?) teaching students, flanked by Avicenna, Galen, and Hippocrates. Concerning this manuscript, see Marie-José Imbault-Huart, Lise Dubief, and Bernard Merlette. *La médecine au Moyen Age: à travers les manuscrits de la Bibliothèque nationale*. Paris: Editions de la Porte verte, 1983, pp. 38–9.

No longer was it sufficient to *assert* principles or propositions, be they about nature or God; it was now incumbent upon practitioners of science and theology to present a coherent, consistent, structurally robust explanation of those principles and propositions. After a brief period of experimentation with a wide variety of models, including a mathematical or Euclidean doctrinal structure, theologians settled on the language and techniques of Aristotle, and for much of the thirteenth century the issue of the scientific status of theology was a point of departure in the theological literature. Among many scholastics, including Thomas Aquinas, John Duns Scotus, and William of Ockham, the discussion extended even to questions of whether human theology and the knowledge of the saints in heaven were related as subalternated sciences, in the same way that astronomy and optics or music were related to geometry and arithmetic, because the superior discipline provided causes for the propositions in the subordinate science. Moreover, the central doctrinal development of the thirteenth century—the transubstantiation of the elements in the Eucharist—was inconceivable without the appropriation of Aristotle's theory of matter, form, substance, qualities, and their changes.

In the end, we return to the point of departure: we have shown that 'medieval science' is a meaningful expression. It was an enterprise that was bounded by the parameters of the culture in which it was embedded, yet shaping through its own practices several features of that culture. We have seen this repeatedly in the different exemplars we have explored in Islamic societies from the eighth to the seventeenth century. Our study of the emergence of new institutions, theories, and learned practices in Catholic societies of Europe, which was firmly based in the linguistic, philosophic, mathematical, and medical work of scholars from ancient and Islamic societies, has confirmed this point. For this reason, we conclude that medieval science in Islamic as well as Catholic societies was not so different from science both before and since the Middle Ages. Religion continues to have a voice in scientific issues, though its role has shifted significantly in the modern world; in the place of the Church, new, different, but equally influential and consequential forces affect the course of science, including political and corporate entities and funding agencies. While medieval and modern competitors to science inhibited science on occasion, modern investigations of the sociology of science have taught us that these constraints also could open unexpected pathways that influenced the character of science and society both contemporaneously and for the future. Modern commentators on science would do well to learn about their medieval predecessors and their enterprises.

Opposite: Paris, BnF n.a. lat. 99, fol. 250ᵛ–251ʳ. A page from the 1338 Catalogue of the Sorbonne, containing entries for Aristotle's logical works. Concerning the library of the Sorbonne and its catalogues, see Richard H. Rouse, 'The early library of the Sorbonne', *Scriptorium* 21(1967) 42–71, 227–51.

4 Science in the Pre-Modern East

DAGMAR SCHAEFER

IN the year 1088, the minister of justice (*xingbu shangshu*) Su Song (1020–1101) presented the emperor with a small-scale wooden model (*muyang*) of a three-level pagoda-style tower. The model, which Su had produced with the help of court artisans, had a bronze armillary sphere (*hunyi*) on the top level, a celestial globe (*hunxiang*) on the middle, and opening doors and timekeeping mannequins on the lower. Despite its rather toy-like appearance the model was, in fact, built for a serious purpose: to apply for government funds to construct an astronomical clock tower. Scaled to size, the combined components of the tower would enable Su Song to observe and scrutinize the 'symptoms' (*hou*) and 'scriptures of heaven' (*tianwen*) in the midst of the Song dynastic capital Bianjing (modern Kaifeng) 'from dusk until dawn and announce the degrees to which the stars occur: such has never existed before'.

Dynastic houses in the historical region of what is now China ruled according to a heavenly mandate (*tianming*), which was manifest in the dynasty's ability to foresee and interpret usual and unusual events in heaven, on Earth, and in the human world, to identify units of time, and order the world accordingly. The observation of planetary movements, the movement of the stars, and weather phenomena were all, therefore, integral to court life and central state bureaucracy. Conversely, court and state bureaucracy had an impact on the clock's design, the working process and the historical role of Su Song and the artisans. After its completion, Su Song compiled an elaborate report on the clock's manifold components which dynastic historiographers included in the official 'History of the Song' (*Song shi*). While the 'New Essential Rules for Observing Constellations' (*Xin yixiang fayao*) was thus saved for posterity, the model was discarded and the craftsman's contribution faded from the historical view. Jurchen armies dismantled the clock tower when raiding Bianjing in the year 1127. The parts were transported to the Jin dynasty's (1115–1234) capital near modern Beijing to celebrate their triumph over the faltering Song state. Local experts, however, failed to rebuild the complicated mechanism and thus the manifestation of Su Song's knowledge was lost.

The Song era highlights the intimate relation between bureaucracy and scientific and technological efforts which historians have identified as characteristic of the pre-modern Sinophone world. Indeed, Song society and state differed markedly from its predecessors, because its rulers employed educated men, the 'literati', in its service. In contrast, the aristocracy had led Tang society (618–907) and military regiments had held power during the Five Dynasties period (907–960). By granting political power to those who had passed regional and central civil service exams, the Song imperial house thus laid the foundations for a new society in which elite status was based on the ability to read and write; and in which the right to rule was grounded firmly on the mastery of a classical canon of action and thought. Men and women, inside and outside the court, developed an orthodoxy of culture (*wen*) and knowledge (*zhi*) that succeeding generations of Yuan, Ming, and Qing scholars strove to imitate or innovate and, at times, even to surpass.

Historical shifts in knowledge cultures in the Sinophone world since the Song era assigned material qualities to paper and woodblock print. While both had been around for some time, the confluence of sociopolitical changes, commercialization, and further advances in production techniques resulted in their increased use. Emperors and court elites systematically invested in literary works, commissioning and collecting volumes on classical themes of statecraft, ritual, and morality as well as on practical topics such as architecture, hydraulics, agriculture, mathematical issues, music, botany, pharmaceuticals, and medicine. State attempts to gather or disseminate information and to control knowledge flows were challenged by individuals who published philosophical, philological, and encyclopaedic works leveraging genres such as 'private writings' (*biji*), 'treatises and listings' (*pulu*), or 'local gazetteers' (*difang zhi*), that complemented or countered the orthodox or central state view.

Paper and print did not, however, prevail without question. Increased access to texts spurred people's anxieties and thus led to a substantial review of the role of scripts, objects, or human interaction in the documentation and circulation of information. Questions were asked: should information be given to everyone? How reliable was a print or a manuscript text? How should people understand texts with regard to lineages of thought? Did they replace or constitute them? Methods of arguing and codifying knowledge diversified, as the pluralism of bureaucratic methods depicted in Su Song's case shows: a model was a visual manifestation of the ideas behind a project while a textual report accounted for the funds and the project's ultimate success. Knowledge was gained and action construed within complex structures of textual, visual, and material means, as well as oral exchange.

In the literate world epigraphical, archaeological, and text-critical approaches proliferated and the literati increasingly debated which kind of knowledge could be attained through texts. Philosophical positions from this era reflect a deep concern about the role of 'things' (*wu*), 'phenomena' (*xiang*), and 'affairs' (*wu*) discussed in terms of the 'mastery of principles' (*qiong li*), the 'investigation of things' (*gewu*), the 'origin of things' (*wuyuan*), the 'creation of affairs and inception of things' (*chengwu kaiwu*), or the relation between 'knowledge and action' (*zhi xing*). Inquiring into the nature of 'heaven, earth and man' (*tian di ren*), 'all under heaven' (*tianxia*), or the 'ten thousand things' (*wanwu*), intellectual debates resorted to ubiquitous themes such as the five

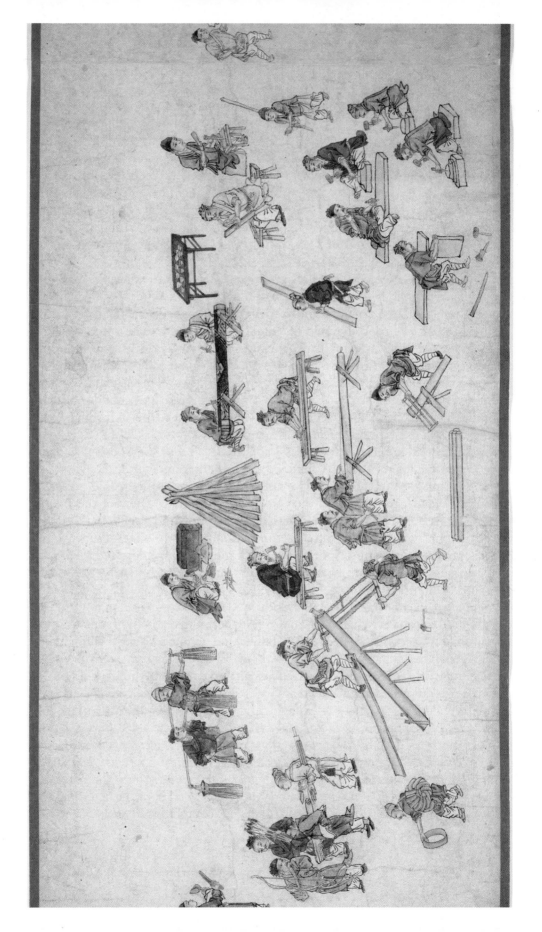

phases (*wuxing*), the *yin-yang* forces, and *dao*, *li*, and *qi*. Intellectuals and practitioners of Buddhist, Confucianist, or Daoist faiths, foreigners from the tributary states or regions as far away as western India, and believers of European-Christian or Islamic-Persian ideals, all contributed to these debates. From the Song to the Ming a multi-ethnic, multi-religious, and multi-cultural society differentiated or melded theories, practices, styles, and notions of varied schools of thought, sometimes arguing with the old and sometimes introducing novel ideas.

Not only the view of things, but also the things themselves changed: agricultural techniques and ceramic and textile production were incessantly discussed and developed. Bureaucratic record-keeping, increasing in complexity throughout the centuries, reveals changing state approaches to the practical arts and technological set-ups. In the early Song, when porcelain production began, the state had minimal involvement in the process of manufacture, imposing taxes, recruiting local experts, or enforcing market control. The Yuan state imposed quotas on crafts, and made tasks inheritable, and the Ming rulers systematically engaged with manufacture in porcelain and silk, determining raw materials and production modes, as well as designs and patterns of use. Most dynastic states actively regulated areas like shipbuilding, mining, and salt production. While farming still moulded scholars' and officials' minds, artisans of all kinds, including diviners, healers, and artists, populated the increasingly urbanized scenes of Song, Yuan, and Ming cultural and daily life.

Bureaucracy, Scholarship, and Expertise

Classical Chinese state theory portrayed farmers (*nong*) and scholars (*shi*) as pillars of society, since they nurtured society's bodies and minds. In comparison, artisans (*gong*) and merchants (*shang*), the remaining two of the four people (*simin*), were of minor relevance. Song and Ming statecraft theory particularly venerated agriculture, discussing its ritual and technical intricacies in detail in various genres. They used a far more regulatory tone when discussing crafts and trades.

In statecraft and intellectual thought from this era, practical hands often play walk-on roles. Artisans who took on official roles therefore increasingly concealed their professional origins, manoeuvring elite and, thus, scholarly rhetoric into their family genealogy and works. The Liao, Jin, and Yuan states set a slightly different tenor with their focus on the practical and martial arts, and less interest in texts. By the end of the Ming era in the seventeenth century, artisans, merchants, and entrepreneurs systematically provided their children with a literary education, in accordance with contemporary social ideals and economic concerns. Hence, historically, the notion of practically engaged literati-officials (rather than practitioners with official roles) has prevailed.

'Scholars' (*shi*), in contrast, are omnipresent in Chinese texts, but their definition is often blurred, so that the word 'scholar' indicates a social paragon imbued with varied

---

Opposite: Street scenes in times of peace.

and shifting historical meanings. At its most general level, the term 'scholar' in the period from the tenth to seventeenth centuries refers to a person who was literate in a classical canon of thought. Song scholars constituted the ruling elite, establishing the idea of a moral scholarly obligation to order society and thought. The Yuan made use of these scholarly bureaucrats, whilst downgrading their social reputation and political role. By the end of the Ming era, scholars were a heterogeneous fusionist group of literary-trained officials, wealthy merchants, the landowning gentry, prominent artists, and intellectuals—some of whom might have also had an artisanal background.

Whatever the discursive tone of any particular dynastic period, institutional structures were comprehensive, including all social and functional groups and tasks. The state needed craftsmen to build palaces, maintain waterways and roads, and produce and process food, tributary goods, and desirable luxuries. In fields such as porcelain and silk manufacture, shipbuilding, weaponry, salt production, mining, minting, and numerous other tasks, production was occasionally institutionalized, with the state controlling manufacture, use, and consumption through the lease of implements, recruitment of labour, and through taxes and other fees.

Officials, who maintained a leading political and social role, had daily contact with practical know-how, and were therefore able to manage the ritual, economic, social, and political implications of manual work and craftsmen's expertise. A tension between the scholars' intellectual stance and their public role informs notions of practice and theory, knowledge, skills, and expertise in the literature from the Song to the Ming dynasty.

The third-century BC text on statecraft and rituals 'Rites of Zhou' (*Zhou li*), which remained a general reference book for state debates, highlighted celestial officialdom (*tianguan*) in terms of the six arts (*liuyi*): ritual (*li*), music (*yue*), archery (*she*), charioteering (*yu*), calligraphy (*shu*), and arithmetic (*shu*). From the tenth century onwards, when civil service exams (*keju*) became the key to officialdom and socio-political status, scholars questioned the role of education and talent per se, weighing institutional knowledge against moral learning and the individual sophistication of the scholarly mind.

Among the many areas of self-conscious fields of learning in China some, such as mathematics (*shuxue, suanfa*), mathematical harmony (*lülü*), mathematical astronomy (*lifa, tianshu*), astrology (*tianwen*), medicine (*yi*), and medical materials (*bencao*—as part of medicine) were occasionally integrated into state governance. Wang Anshi's (1021–1086) 'new policy' (*xinfa*) called specifically for education in fields such as mathematics, medicine, pharmacy, and astronomy. His ideas impacted Song, Jin, Yuan, and Ming bureaucracy and ideals of expertise, and many of these fields were implemented into the education system and included in the civil service examinations. Other subjects, including outer or inner alchemy (*nei, waidan, fu lian*), siting or geomancy (*dili, kanyu, fengshui*), and the study of things (*wuli, wulei, xianglei, gewu*) constituted fields mainly by references to a defined textual corpus. Many of these were subject to more individual scholarly pursuits in the realm of philosophy and philology.

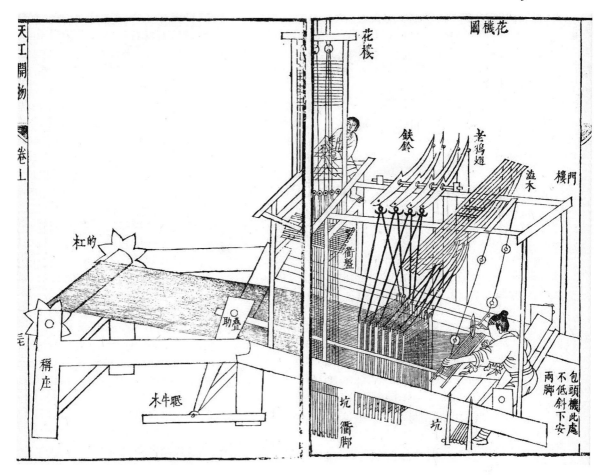

A silk loom.

Mythical emperors of antiquity represented practical tasks that were important for state and scholarship. Shennong represented the value of agriculture or medical treatments, while the Great Yu symbolized hydraulic engineering. Often a charismatic founding text such as the 'Materia Medica of Shennong' (*Shennong bencao*) was identified. Craftsmen also contributed to this textual heritage, such as the carpentry manual 'Canon of Lu Ban' (*Lu Ban jing*). Although the origin of this text is unclear and may date from the late Ming era, in its time the text served to substantiate the importance of carpentry knowledge and skills, and craftsmen used the book to initiate experts in the rituals and techniques of the trade.

Song Dynasty

Irrespective of notions such as the six arts, the reality from the Song to the Qing was that scholars were morally trained and literarily adept polymaths who understood and, as officials, supervised practical tasks, but rarely performed them. Song politicians such as vice chancellor (*pingzhang shi*) Fan Zhongyan (989–1052) and high chancellor Sima Guang (1019–1086) both became historical role models for Ming activist political scholars and those managing practical tasks because they found pride in doing more than supervise. Fan, for instance, was concerned with tributary relations as much as with managing large-scale hydraulic projects or famine control. Fan himself also defined a scholar's duty in rather practical terms. He asserted that scholars should advise the emperor, supervise military campaigns, serve their localities, or manage the education system. Fan drafted some ideas on scholarly expertise that Wang Anshi would later use in his reform policies. However, both men's notion of knowledge was, like most of their peers, mainly concerned with maintaining the identity of a scholar, not a practitioner.

Notions of expertise or the role of practice and theory also varied according to individuals' personal interests. For instance, the poet Su Shi (1037–1101), a contemporary of Su Song, worked in medicine and agriculture and was interested in material transformations (*liandan shu*). Irrespective of their individual interest, all of these men grounded scholarly education in the tenth- and eleventh-century scholarship of the learning of the changes (*yixue*). Accordingly Fan also promoted a study of the classical canon of the 'Book of Changes' (*Yijing*), in particular chapter *Xici*.

Philosophical and moral texts such as the 'Analects' (*Lunyu*), the 'Great Learning' (*Daxue*), and the 'Doctrine of the Mean' (*Zhongyong*) came to prominence in the twelfth century through Zhu Xi (1130–1200), who believed that the world was determined by a dualism between *li* and *qi*. Deeply interested in a comprehensive interpretation of the Confucian classics, Zhu Xi built on classic Chinese literature to initiate new ideas.

Zhu Xi was a successful official in his own lifetime and his syncretistic ideas became increasingly popular with contemporary elites. Zhang Zai represents a different strand of adept scholars who served in minor posts during the Song era. In contrast to Zhu, Zhang claimed that everything was part of a larger flow of *qi*, and hence the order of the universe, society, the state and the individual could not be separated. Zhang also disapproved of the state's institutionalized approach to learning, but curiously his oppositional moral metaphysics actually strengthened the social system of his time as it opened new avenues for those scholars who had focused their careers on the civil service only to realize that the state had no suitable positions for them: the role of oppositional intellectual retreating into private life became an eligible and honourable alternative to a civil servant's career.

Zhang blamed previous scholars who: 'knew men and did not know heaven. They search for virtuous men (*xianren*) and not for sages (*shengren*)'. He urged his colleagues to meet their responsibilities through moral leadership, by shaping everyday practices and values and studying the universe. Eventually, Zhang claimed, scholar-officials would be responsible for the behaviour of everyone—the emperor, the people, and themselves. Zhang's theory gave unemployed and dismissed scholars a purpose beyond state service.

Zhang himself stayed clear of Realpolitik and taught the learning of 'change' (*yi*) for a living. He was called back to the court in the year 1069, but he soon argued with the reformist prime minister Wang Anshi and returned home. Zhang's opposition gained him the praise of retired high chancellor Sima Guang who later dethroned Wang Anshi, but who also disagreed with Zhang Zai about the pragmatic realities of politics.

Yuan Dynasty

The Yuan era brought a new atmosphere to debates about scholarly skills and talents, as shown in the small booklet 'Schoolbook Guide to the Practices of Officialdom' (*Lixue zhi'nan*) which was compiled in 1301 by Xu Yuanrui, a local leader from Suzhou county. The Mongol armies of Khubilai Khan (1215–1294), had conquered Suzhou county in 1276 on their way to the capital Hangzhou to overthrow the Song regime three years later. Once in power, Khubilai utilized the Han administration, particularly practitioners of medicine, prophecy, astrology, and physiognomy. Confronted with the structural shift of dynastic change, the scholarly trained Xu compiled his 'Guide' to describe the skills and tasks of judicial clerks at the local level, explaining how ideas of practice and theory could be adjusted to meet contemporary needs.

Ostensibly, Xu's guide recommended using a combination of Confucian scholarly pursuits and personal morality, as well as listing rules and regulations which were vital for the legal role of a village clerk. Xu was aware of the inherently personal nature of a clerk's work, so he identified two relevant qualifications for the role—conduct (*xingzhi*) and abilities (*caineng*)—in his first chapter on educational schemes and migrational strategies. He considered that 'conduct' comprised virtues such as integrity and loyalty; and 'abilities' included the management of various professions (*xingqian shuxian*), a talent for debate (*yuyan bianli*), calculation skills (*suanfa jingming*), neat calligraphy (*zihua duanzheng*), familiarity with regulations and law (*tongxi tiaofa*), and a good grasp of the Confucian canon (*xiaojie rushu*).

In fact, Xu's booklet was a reaction against the leading Mongolian prerogatives, purposefully interpreting philosophical and philological study as a practical matter; and defining literary skills and the theoretical discourse of morality (which the previous Song dynasty had made a prerequisite for power) as technological tools.

Xu thus represents a group of scholars in the Yuan who remained committed to the heritage of the Song. While Mongolian rulers of that time venerated scholarship, they did not favour classical Song-literati learning. The Mongolian elite remained relatively cautious towards the Chinese elite and common people, who outnumbered them and were on their own territory. Rather than coercion, as has often been suggested, Mongolian rulers actually used a subtle combination of intimidation and non-intervention in daily life and, while skimming off market profits and agricultural goods, utilized scholarly talents to endorse their rule.

Ming Dynasty

During the early years of the Ming period, Mongolian institutions and ideals continued to play a role, but later generations emphasized the return of Ming scholarship to Han and Song thought. Once the Ming state made Cheng-Zhu learning a major aspect of the civil service exams in the fourteenth century, Song learning and debates returned in full strength.

Zhu Xi's theory of *li* gained prominence as part of Ming state doctrine, situating *qi* in opposition to state policy—a position its enthusiasts from the Song era had often taken. Zhang Zai's interpretation of *qi* became a subordinate part of Zhu Xi's theorem of *li*, but it is important because it influenced the theories of many Ming scholars interested in fields of scientific and technological thought such as the study of nature, affairs, and things.

Ming understanding of practical statecraft included practical, technical skills and historical studies—epitomized by a broad group of Ming thinkers whom the historians Ge Rongjin and Zhang Qizhi have called a collective school of 'practical learning' (*shixue*) or 'scholars of *qi*'. Although Ming philosophers like Wang Tingxiang (1474–1544), Wu Tinghan (1491–1559), Han Bangqi (1479–1556), Wang Fuzhi (1619–1692), and Fang Yizhi (1611–1671) all debated practical issues and discussed phenomena such as sound, electricity, and metrology to explain *qi*, it is highly unlikely they would have concurred with Ge Rongji's anachronistic classification because, in their view, true understanding came through an individual's innate (*liang*) Mencian knowledge of 'the good', gained from their own individual approach, rather than through corporeal learning.

Wang Yangming (1472–1529) suggested such knowledge of 'the good' needed to be realized by actions in the real world. Innate knowledge could never be separated from knowledge gained through seeing and hearing (*jianwen*). For Liu Zongzhou (1578–1645) knowledge was just as related to material things as it was to moral or ethical considerations. Jesuit missionaries had made their first forays into Chinese knowledge during Liu's lifetime, but it was people such as Xu Guangqi (1562–1633), a minister at the court, who would help the Jesuits promote Western *scientia*. Xu devised the phrase 'obtaining knowledge by making a thorough inquiry into the principle of things' (*gewu qiongli*) to address studies of nature which fed into a physical and material understanding of the world. This included moral, technical, and sociopolitical issues related to geographical, botanical, medical, mathematical, engineering, and philosophical knowledge as well as knowledge from the West—*scientia*.

Towards the end of the Ming and the beginning of the Qing dynasty, philosophers who corroborated *qi* as forming the orthodox *parem esse* became closely associated with political opposition. Thinkers such as Wang Tingxiang, Luo Qinshun, Liu Zongzhou, Tang Hezheng (1538–1619), and Song Yingxing (1582–1666?) further determined that the nature of morality and principle was the very essence of the *qi*-quality (*qi zhi*). Seventeenth-century learning, emerging at the clash of Ming and

Qing cultures, referenced Song traditions of learning bereft of the practical, concrete, and instrumental aspects that eighteenth-century scholars of the Qing reign would come to emphasize in evidential scholarship (*kaozheng*), as the Qing declared the Cheng-Zhu school of Song learning to be the new orthodox belief system.

Codifying Bodies of Knowledge

Heaven

When Su Song began work on his astronomical clock tower, the Song dynasty had reigned for almost a century. The ravages of war had been overcome and new urban centres were filling with the ambient noise of economic and cultural prosperity. A new elite was devising institutional structures of governance to manage society and the state, including aspects such as calendar calculation and astronomical-metrological studies.

From the beginning of their dynastic rule, the Song excelled at astronomical record-keeping, continuously collecting new data by observing and monitoring the stars. Ever since the Yuanyou reign (1034–1037), the Song state had checked and measured star constellations once a decade, particularly meticulously in the first two instances. Su Song had himself participated in this scrutiny during the Yuanfeng (1078–1085) era and had produced a star map. Over its 160 years, the Northern Song era initiated nine calendar reforms. After the conquest by the Jurchen Jin in 1127, the Southern Song state followed suit with ten reforms within 152 years. Most of these calendar reforms targeted the observation of the skies by increasing the accuracy and sophistication of their astronomical instruments.

Astronomers also obtained knowledge from books, and produced more textual accounts. As early as the 960s, the Cabinet for Astronomy (*tianwen yuan*) already owned 1,561 'chapters' (*juan*) for its employees. The second Song emperor, Taizong (r. 976–997), had a large library of astronomical works, totalling 2,561 chapters. The fourth Song emperor, Renzong (r. 1022–1063), compiled a brief account of Su Song's armillary sphere, which was circulated within the palace but not any further. In 983, chancellor (*pingzhang shi*) Li Fang (925–996) included part of the Cabinet's records in the encyclopaedic 'Imperial Observations of the Taiping Era' (*Taiping yulan*, 984). A century later in 1084, chancellor Sima Guang copied these into his 'Comprehensive Mirror of Governance' (*Zizhi tongjian*), adding records of the period between the Eastern Jin and the Later Zhou era about comets and eclipses (*hunxing rishi*).

One innovation of Song approaches to astronomy is that textual debates were carried on beyond the court. This openness was partially caused by the vulnerable political status of scholar-officials, who often found themselves dismissed as rapidly as they were appointed. Shen Kuo (1031–1095), one of the most prominent polymath-astronomer-mathematicians of Chinese history and Su Song's contemporary, for instance, had gained astronomical expertise as director of astronomy (*Sitian jian*). After his retirement, Shen Gua compiled a private anthology 'Brush Talks from a

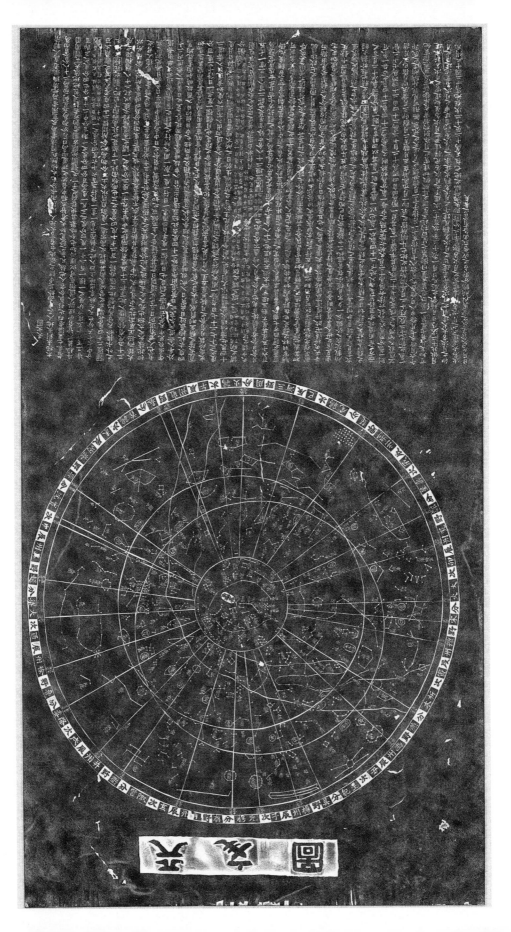

Dream Brook' (*Mengxi bitan*) in which he described astronomy and geometry (*jihe*), discussed water clocks (*louke*), bronze armillary spheres, and gnomons. Thus, scholars such as Shen Kuo also increasingly embraced their role as political and intellectual leaders, notwithstanding any dynastic recognition or state-sponsored appointment, sharing their knowledge and opinions in books. Hence, Shen Kuo continued to advise the state from outside to check the records of various departments in charge of astronomy as a way 'to guard against mistakes' (*fang xuwei*).

When Su Song began his project, heavenly data were thus very familiar to people both inside and outside the court, and the emperor and his court were under pressure to increase astronomical knowledge. During a diplomatic mission in 1077, Su Song had realized that the Khitan Liao dynasty (907–1125) which held power in Asia's northwest, could calculate the calendar with greater accuracy than the Song. The Liao's superior knowledge of the heavens posed a serious threat to the rulers of the Song, who continuously struggled to legitimate their rule by unifying the territories, through military and political means.

The year 1088 was star-crossed. The empress dowager Gao (1032–1093) protected the interests of the imperial clan, representing her 12-year-old son and emperor of the Song dynasty Zhezong (1076–1100) by patronizing the conservative chancellor Sima Guang who stood against the reformist group led by Wang Anshi. In their roles as officials, Sima and Wang argued about the right way to rule and, as scholars, they fought lively intellectual debates over their own roles, and the right way to acquire knowledge.

Su Song kept his distance from such courtly factional struggles, always refusing to align with one group or another, but sources indicate that he was prepared to build strategic alliances to support his vision of creating an astronomical clock tower, using the scale model throughout the 1080s to argue his case. In 1082, the Hanlin academic Wang Anli (1034–1095), brother of the famous reformer Wang Anshi, approached the astronomical official Ouyang Fa (1040–1089), stating that:

the instruments of the *Zhidao* (995–1007) and *Huangyou* (1049–1054) era are all deficient and without any galley proof (*ju*). Today the wooden samples produced of an armillary sphere and a water clock were prepared for presentation to the emperor.

In due course, Ouyang Fa provided more information about adjustments needed in the new instrument.

Models had always been used as working tools in astronomy. Sampling was a way to forestall the huge risk and investment needed for a bronze armillary sphere which, as a representation of the universe, provided a physical form to demonstrate dynastic claims of true knowledge and righteous rulership. The Hanlin scholar Xu Jiang (1037–1111) referred to Su Song's project: 'once the previously approved wooden model of the water driven armillary sphere is produced and it has been found exact as regards the *houtian* [display], produce another bronzed version'. What was new in the Song era was that models became part of an established institutional process of negotiation.

Opposite: Song dynasty star map.

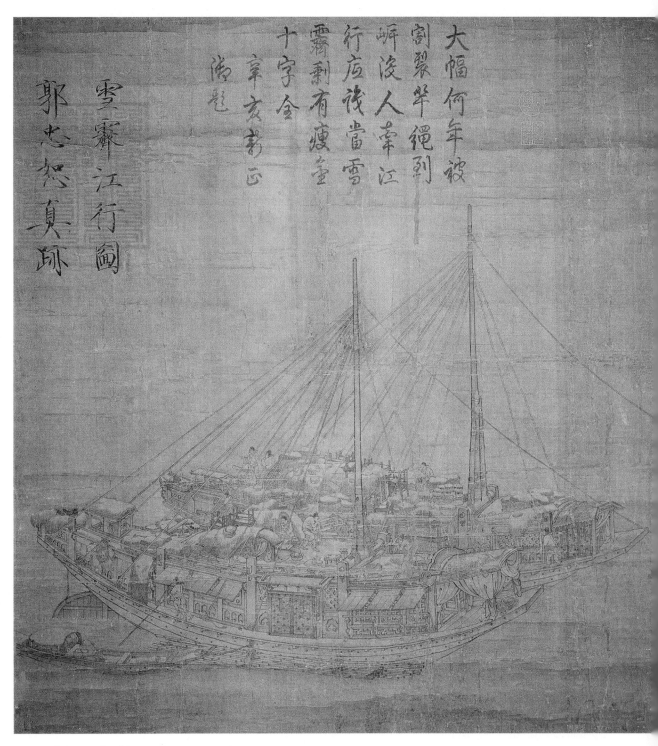

大幅何年被

割裂牟繩剄

岈淩人牽江

行庭淺當雪

霽剝有瘦重

十字全

辛亥影正

滿霓

雲霽江行圖

郭忠恕真蹟

Painting of a ship.

Within this process, a model was used to prove a project's viability. As the state archival records of the 'Continuation of the Draft Report on Finances and Statecraft' (*Xu zizhi tongjian changbian*) (originally compiled by Sima Guang) illustrate, this process required chancellors to *see* and testify to this feasibility: 'the Imperial Hall of Yanhe called in all chancellors (*fuchen*) for inspection of the newly produced model'. Once they were convinced that the model functioned, court painters began to produce plans (*tu*) for constructing the tower and the armillary sphere. Four months later, the chancellors were asked to 'assemble at the Chonghe Hall to inspect the construction plans of the armillary sphere'. The armillary sphere was then cast in bronze over the next two months, and construction of the tower began.

Bureaucratic records focus on the politics of the project, so do not reveal technical details such as how to work from a scale model or what effects different materials could have on the mechanics or design. The sources show clearly, however, that Song astronomers employed the whole gamut of communication methods in their discourse and analysis: texts, models, sample materials, sculptures, sketches, diagrammatic images, and star maps. Knowledge was dispersed through society as Su Song's armillary sphere and clock tower and similar projects were collaborative works between diverse government institutions. Han Gonglian (n.d.), a minor official in the Ministry of Personnel (*libu*) supported Su Song's proposal. Yuan Zhenggong, divisional director of the Ministry of Works (*gongbu yuan wailing*) and a former pupil of Su Song, contributed to creating an armillary sphere model and casting it in bronze in the seventh year of the Shaoxing (1094–1096) reign. Shan E (1031–1110), a palace steward, was put in charge of producing the instruments.

One after another, officials of the Northern Song applied for four different armillary spheres (*hunyi*), spending more than 20.000 *jin*—a substantial part of their annual revenue—on astronomical concerns. The trend of textualization also continued into the Southern Song. Scholars such as the poet and minister of rites (*libu shangshu*) You Mao (1127–1197) were noted as owning ninety-five *juan* astronomical treatises. By the mid-twelfth century the state corpus of astronomical knowledge had also grown substantially. Zheng Qiao (1104–1162) identified 369 different kinds (*zhong*) of records on heavenly patterns (*tianwen*) in storage at the Grand Secretariat. Zhu Xi included astronomical issues in his historiographic 'Outlines and Details of the Comprehensive Mirror' (*Tongjian gangmu*), and Ma Duanlin (1254–1324) equally manifested the place of astronomy in his historical representation of the Song state: 'Comprehensive Examination of Literature' (*Wenxian tong kao*).

As an eyewitness of the dynastic transition from Song to Yuan, Ma Duanlin compiled his administrative guide because he 'was afraid that the dynastic system would fall apart'. Like preceding historiographers, he gathered data about omens such as comets, meteors, and other unusual heavenly events, but he focused most of his attention on the institutional apparatus of the Song which he knew well, being the second of six sons of one of the last grand chancellors (*canzhi zhengshi, chengxiang*), Ma Tingluan (1222–1289). Ma Duanlin's descriptions show that Song bureaucracy concerned with the heavens was set up to find guidance on pressing contemporary concerns, and that divination and prognostication were central to the Southern Song state. Students seeking entrance to the Astronomical Bureau were accordingly required to demonstrate knowledge of three methods of divination to pass the provincial exams. Most forms of prediction involved mathematics (*shuxue*), rather than astronomical knowledge (*tianwen*). Astrology (*tianshu*) did require astronomical knowledge and thus was usually confined to the imperial court. The emperor needed astrological information to preserve his reign and his dynasty, and to avert calamities for his people by divining heaven's messages and performing the appropriate sacrifices and rituals in response.

To some extent this focus of historiography was driven by ideological motifs, and by Song state attempts to substantiate its rule against the Jin, who were not only a strong military force ruling the north of China, but also an equal competitor in the field of astronomical understanding. Indeed, in some ways the Jin were even more knowledgeable, modelling astronomical education on the Tang dynastic approach rather than that of the Northern Song. The Jin thus established three branches (*ke*) of astronomical learning: (1) a school for heavenly patterns (*tianwen*) with ninety students, (2) the calculation of calendars (*lishu*) with fifty-five students, and (3) studies of the water clock clepsydra (*louke*), which usually taught forty students. Students entered the Astronomical Bureau in accordance with this ratio.

The Jin educational system of astronomical survey hence relied on a combination of specialists in calendar production, timekeeping mechanisms, and the theoretical and empirical observation of heavenly patterns, with a quantitative emphasis on the latter. Many students had Han origins. Jurchen commoners often filled the lower ranks. In principle, any commoner aged between 15 and 30, however, could attempt the astronomy entrance exam, which took place every third year. The exam included arithmetic for calendar calculations such as the Xuanming calendar (*xuanming li*), an understanding of a ritual canon of the 'Marriage Book' (*Hunshu*), as well as the 'New Book of the *Li* of the Earth' (*Dili xinshu*), which combined burial and marriage rites and regulations. They studied the 'three forms' (*sanshi*) of fortune telling and calculations: the hidden stems (*dunjia*), the primordial unity of *yin* and *yang* (*taiyi*), and the six heavenly stems (*liuren*). Furthermore, applicants were tested on divination methods and permutation studies. In the exam six options were suggested, which had to be finished within two exam days.

Yuan dynasty

The famous Temujin Chinghiz Khan (1162?–1227) claimed his origins came from a divine force of *tengri*, which gave him the right to rule everything below the sky. He appeased the heavens with rituals, such as sprinkling mare's milk to honour his ancestors or sacrificing sheep to alleviate illness in his family. Practices and theories of divination were also central to Mongolian beliefs, and it was from this perspective that Chinghiz Khan's successor, the Great Khan Möngke (1251–1259), promoted the study of the heavens and invited the Persian astronomer Jamal al Din Muhammad ibn Tahir ibn Muhammad al Zaydi (n.d.) to his court. A Buddhist monk Liu Bingzhong (1216–1247), who excelled in mathematics and astronomy, is believed to have spurred Khubilai's initial interest in calendar practices and theories, which Chinese advisors evidently encouraged.

Khubilai Khan's multi-ethnic and multi-cultural court connected Chinese astronomy to the larger context of the Pax Mongolica, and this was a two-way exchange. Chinese astronomers travelled west and became active, for instance, at the newly built observatory of Persian Maragha (located in modern Iran) of the Ilkhanat founded in 1258 by Khan Hülägü. Conversely, Persian, Muslim, and Indian experts surged into the Yuan capital Beijing, bringing astronomical instruments, astrolabes, diagrams of armillary spheres, sundials, and terrestrial and celestial globes to the court.

Khubilai established the Institute of Muslim Astronomy (*huihui sitian jian*) in 1271—before the final defeat of the Southern Song in 1279—and the Institute operated parallel to the Chinese Astronomical Bureau until the early Qing dynasty in the seventeenth century. It had about forty staff, including administrative personnel, teachers, and pupils and mainly concentrated on the Islamic astronomical system (*huihuili*), creating Persian *zij*-calendars in Beijing. In 1279, court astronomers began to construct a new armillary sphere and a wooden gnomon to establish a new 'season granting system' (*shoushili*), and only a decade later these instruments were cast in bronze. Guo Shoujing (1231–1316), the Khan's Chinese astronomer, made use of Islamic concepts such as adjusting times of phenomena to their longitude. He also used trigonomic-geometrical methods to converse between ecliptic and equatorial coordinates, but he did not adopt the Islamic *zijes*.

Guo also planned prototypes of a tall gnomon and a simplified instrument for daily use but as Sivin has shown: 'old equipment and prototypes yielded most of the data for the season-granting system that the Yuan History reports'. Such was the Mongols' appreciation for gold that they even gilded the bronze armillary sphere of the *huangyou* period, keeping the instrument as a requisite and emblem. This was despite the fact that Guo Shoujing's amended armillary sphere's mechanism had proven superior, solving many of the earlier problems.

Ming dynasty

Guo Shoujing's calendar remained the most accurate in the history of Chinese mathematical astronomy for the following four hundred years and, thus, far into Ming rule. Mongolian and Persian astronomical knowledge also stayed central to the Ming dynastic approach to heaven. The first Ming emperor Zhu Yuanzhang, however, changed the political agenda of heavenly observation by assigning the Astronomical Bureau the task of looking out for strange events and anomalies (*bianyi*) with respect to the stars, planets, conjunctions, wind, clouds, fog, or dew. This meant that every astronomical bureau had to be equipped with an observation platform.

The first director of the Astronomical Bureau, Liu Ji (Bowen, 1311–1375), soon distilled some general bureaucratic advice on this task in his 'Secret Policies of Astronomy' (*Tianwen milüe*). Liu's literary accounts provided working tools alongside the bureaucratic records themselves, and were constantly cited in later throne memorials to demonstrate authority or persuasion. Liu required the Astronomical Bureau's astrological notes to chronologically list records of any auspicious symbols (*xiangyi*) in the 'Veritable Records of the Ming Dynasty' (*Ming shilu*), as did the later senior grand secretary (*shoufu*) of the young Wanli emperor (1563–1620) Zhang Juzheng (1525–1582) while ruling the Ming state.

The first imperial observatory of the Ming had been built south of Nanjing for the Chinese bureau, but was soon handed over to the Islamic branch, when the Chinese bureau moved to Jiming mountain, northwest of the city. Even though this was a new location, Zhu Yuanzhang requested the old instruments from the Yuan Astronomical Bureau in 1368. And, when Beijing became the Ming capital again in 1421, under the third Ming emperor Zhu Di (reign name Yongle, 1402–1424), he used the Yuan

Shadow definer.

observatory first, although probably without instruments. In 1437 the vice director of the Astronomical Bureau requested a wooden replica of the Yuan instruments in Nanjing and had them cast in bronze for a new star-observing platform (*guanxing tai*) that had been built on top of the Beijing city wall near modern Jianguo Gate, where they still can be found today.

The shift of latitudinal position should have created issues: Beijing and Nanjing vary by 7°, and at least at one location instruments had to be adjusted. Instead, memos since the 1480s note damages, blaming the artisans for having messed up. Calibrations, however, are not discussed. One reason for this might be that the mandated responsibility of observing anomalies did not necessarily require quantitative methods.

Zhang Shen (n.d.) was the first observer to raise the issue of precision with regard to the armillary sphere in 1502. He went so far as to oversee the construction of new wooden prototypes, but any proposal for new casts was deferred. Thirteen years later, the wooden samples were presented to the court, only to be confiscated directly by the Ministry of Rites (*Li bu*) for day-to-day use. Meanwhile, the bronze instruments at the

Shadow definer.

city gate had become imperial artefacts and, as such, they could not be modified without imperial authorization, so they became inaccurate, unused, and decrepit.

The Jesuit, Matteo Ricci (1552–1610) saw the instruments at the Nanjing Observatory in 1600 and noted in a report that they were set at 36°, which was inaccurate for both Beijing and Nanjing. He used this as an opportunity to illustrate the backwardness of Chinese astronomy. Ricci and his successors attempted to gain access to the court, producing and displaying astronomical instruments, celestial globes, sundials, and optical glasses to provoke the elite's curiosity and interest in their religious agenda. As Xu Guangqi advanced to the position of Minister of Rites (*libu shangshu*), he recommended Niccolo Longobardo (1559–1654) and Johann Terrenz (1576–1630) for appointment to the Calendrical Bureau. Giacomo Rho (1593–1638) and Johann Adam Schall von Bell (1591–1666) followed, guaranteeing European Jesuit knowledge a place in Chinese studies of heaven far into the Qing dynastic period.

Earth—Agriculture and Hydraulics

Song dynasty

Armillary spheres or globes were both instruments of investigation and, simultaneously, visual aids, allowing scholars to comprehend the configurations and workings of heavens on a material level. Because of this, the Song state employed models alongside written and visual accounts, to disseminate knowledge across its manifold regions in fields such as agriculture, while scholar-officials increasingly edited accounting information and guidelines for their fellow colleagues on paper. One early reference to the use of models in the Song state for such purposes mentions a cattle epidemic that hit Songzhou and Haozhou in the autumn of 997. In an attempt to alleviate the region's shortage of cattle during the harvest and to prepare the fields for the next growing period, official Wu Yuncheng (918–?), advisor to the prince (*taizi zhongyun*), had provided funds to buy new cattle from the Jiangnan and Huainan regions, but their delivery was delayed. Realizing that there would be no immediate relief, Wu recommended the introduction of foot ploughs (*tali*). Such ploughs had originally been promoted by the first emperor of the Song, Taizong, for land clearance in regions such as Huaichu which not only lacked cattle but were also scarcely populated. The vice director of the palace library (*mishu cheng*) and assistant of the history institute (*shiguan*), Chen Yaosou (961–1017), immediately responded to Wu's request, dispatching messengers with wooden models of the foot plough to Huainan.

Illiterate farmers could grasp the gist of this modelled form and had not much use for written compendia. But scholar-officials also took notice of mundane practical tools in their writing. Zhou Qufei (*jinshi* 1163), for instance, describes the foot plough in his twelfth-century travelogue for his fellow scholars: 'It has a form like a spoon and must be about six *chi* long. At one end it has a horizontally placed log of approximately one *chi* length, enough to be grasped by two hands. From the middle of the plough stick on the left side a short handle is attached, so that the farmer can step on it with his left foot.'

Zhou recognized the value of this foot plough, because the cultivation of new lands to increase agricultural productivity continued to be a priority for Song rulers while their people were being forced southwards. Zhou, however, was clearly targeting his fellow scholars, not farmers. Books were a tool to raise the official-scholar's awareness of the mundane issues and commoner's knowledge that he required in his daily work.

Many such initiatives to spread knowledge about agricultural tools were assisted by imperial attempts to promote new crops and agricultural methods. Early in the eleventh century (1011), emperor Zhenzong (968–1022) had introduced rice varieties such as yellow fast-ripening rice (*huanglu*) for paddy farming and the drought-resistant champa (*zhancheng dao*) for dry farming regions to feed the Song's growing population. Local officials such as Chen Fu (1076–1154) continued these advances, publishing a comprehensive 'Book of Agriculture' (*Nongshu*) in 1149. Chen, who claimed to draw on personal experience, was evidently aiming his message at his fellow literati colleagues or, more specifically, local literati-officials and rural literate elites, reminding them insistently of the importance of agriculture for society and state. Chen may have hoped that his compilation would gain recognition from higher echelons and help him to gain promotion, although he wisely refrained from explicitly expressing such aims in his book.

In his writing, Chen explained his expertise by noting his colleagues' ignorance about some issues he considered vital for agricultural tasks: seri- and riziculture, and water buffalo husbandry. In fact court institutions were still built in accordance with northern agricultural systems, in units which produced cattle, horses, sheep, and poultry, as well as bees for honey and silkworms for textile manufacture, while court food culture began to embrace exotic elements of the south. Scholars at the court would have consulted Jia Sixie (n.d.) who wrote a book on 'Essential Techniques for the Welfare of the People' (*Qimin yaoshu*) in the sixth century after living in China's north, where crop farming and husbandry were dominant. Because of this experience, Jia Sixie included information about horses, *niu* (in this case referring to northern cattle and oxen), sheep, pigs, chickens, swans, geese, and fish.

Chen Fu's focus reflects the changing circumstances of Song life, whereby the Song population began to toil in the southern lands because the Jurchen Jin in the north forced them away from their previous resources. Chen Fu and his colleagues discovered the benefits of a southern Chinese climate, and of some regions that the earlier generations had considered untamed and wild. The literati thus compiled guides on rearing silkworms, on growing and using mulberry trees, silk spinning, and tea production. Plants such as the tong-oil tree and the peony received dedicated scholarly attention, as did local food cultures, revealing how diverse China had become under imperial rule.

Bureaucratic measures were also taken to prevent or react quickly to veterinary epidemics that might threaten a whole region's agricultural base. Laws on the sale, trade, and transportation of oxen and horses were reformed to respond to the spread of diseases. Such legal restrictions about cattle reveal that the Song lacked both locomotive and nutritional energy, which might explain this era's interest in water-driven devices, as well as water and windmills. Contemporary calculations on human and animal force stress how one ox could replace four to five men working with foot ploughs, and that foot ploughs were therefore less efficient. However, in cases of epidemics when yields had to be brought in immediately, foot ploughs provided a viable (and often the only) alternative. In 1139 the state again resorted to using models to alleviate the devastation wrought by a vast veterinary plague that had hit southern Jiangnan.

By that time the Song state's policy for controlling animal epidemics was already well advanced. It had put a forewarning system in place that obliged all officials to regularly report on their region's epidemics, which is one reason why the historical records give the impression of an era struck by disasters. While cattle and horses could roam freely in the northern plains, herding had become the major form of rearing in the southern regions as local offices began to rent cattle and water buffalos to farmers who could not afford to keep a dray animal of their own.

Equine husbandry modelled the structure of a veterinary bureaucracy that aimed to facilitate agricultural production as much as military campaigning. Stables had to be maintained across the empire to quarantine diseased equine livestock. The Song state also appointed veterinary wardens and doctors across central and local institutions, and its imperial agency prescribed medication from the court.

The importance that Song statecraft and intellectual thought assigned to large livestock is omnipresent also in this era's sculpture and illustrative arts. Artists produced nostalgic representations of horses as a political critique of politics and in reminiscence of the northern lifestyle. Painters elaborated native, exotic, or mythical birds, beasts, fish, or plants, or depicted the novel images of water buffalos grazing between mulberry trees and paddy fields. Art was one form through which the people mediated coming to terms with their changing environment. In poetry, scholars reflected on their material world taking up forms of oral communication and memorization such as the chants used by carpenters, weavers, or silkworm breeders in their daily work.

Lou Shu (1090–1162), a county magistrate near the imperial capital Hangzhou, for instance, elaborated in a set of forty-five verses the gendered ideal of 'Tilling and Weaving' (*Gengzhi shi*) (twenty-one verses on rice cultivation and twenty-four on silkworm rearing) which Wang Gang, a military officer from Jiangxi, published a century later in 1237 with accompanying woodblock illustrations. The illustrated version of Lou Shu's poems became a cultural icon and its representation of an ideal agricultural landscape was endlessly reproduced by Ming and Qing emperors.

Water was a central resource for the expansion of riziculture during the Song era, but also a problem. Artistic images reflect the presence of water control in the minds of

Opposite: Large livestock.

the Song elites. One reason for this was that the capital, Bianjing, was situated in a basin at the crossing of two rivers and was thus constantly under threat of being flooded. In the same year that Su Song proposed his clock, Shan E handed the court an elaborate report on the expansion of waterways around the Changjiang and Wujiang rivers, the Wuyan dam, and the Jiazhu river basin. Court scroll paintings of the period promote the deeds of the mythical emperor and icon of hydraulic engineering the Great Yu and celebrate hydraulic projects, irrigated landscapes, canal and bridge building, as well as water mills and lifting devices.

Many Song state projects rationalized large-scale water conservation and irrigation projects that had started in the Tang period. Water flows were visible in landscape paintings, while those literati who were interested in a realistic depiction took a particular interest in motifs such as water mills and ships, which became the major source of transportation for people and goods in the south. Canals were built to provide an inland connection between north and south, thus avoiding piracy along the coastal routes.

In both shipbuilding and hydraulic engineering, the state included modelling in its bureaucratic processes. Scholars appreciated models for instructional purposes and as a way to negotiate design with the artisans. The prefect of Chuzhou, Zhang Xue (?–1138): 'wished to construct a large ship (*da zhou*), but his advisers were not able to estimate the cost. (Zhang) Xue therefore showed them how to make a small model vessel (*jiao yi zao yi xiao zhou*), and then when its dimensions were multiplied by ten (the cost of the full-size ship) it was successfully estimated'. By this period, the Song state had standardized methods of scaling up in fields such as architecture, and had even compiled a handbook of 'Building Standards' (*Yingzao fashi*) for palace construction.

In 1158 the Jurchen Jin state, which had forced the Song people south in 1127, designed a state bureaucracy similar to that of the Song. Zhang Zhongyan (active around the 1130s–1140s), working for the Jurchen Jin, produced a model to communicate his design for a ship to the illiterate artisans.

Zhongyan made a small boat (*xiao zhou*) with his own hands, which was several (tens of) *chi* long. It all fitted perfectly together, from bow to stern, without any need for glue or lacquer. He called this his 'demonstration model' (*gu zi mao*), which could be used for teaching purposes, and the astonished artisans showed him the greatest respect for this achievement. At this point, it seems the bureaucrat for once showed artisans how to build ships.

As models and miniatures began to be put to new use in Song burial rituals, scholars also created models to describe issues of larger landscape modelling. Around the same time as the shipbuilders Zhang Zhongyan and Zhang Xue, Tang Zhongyou (1138–1188) constructed a model to detail the construction of a bridge on one of the most important routes crossing the river in three places close to the cities of Tiantai.

———————————

Opposite: Drawing of a ship.

預備大黃船

船面自頭至稍七
丈玖尺叁寸

船底自頭至無板
處伍丈貳尺肆寸

無板處稍貳丈貳
寸

頭闊玖尺深伍尺
肆寸

中闊壹丈伍尺深
陸尺貳寸

梢闊壹丈肆尺深
肆尺貳寸

伏獅左右各長一丈五寸

十四

寶
珠
亭
伐

水
殿

盧康

Later sources would elaborate Tang Zhongyou's engagement in bridge construction in an anecdote about a responsible official's natural reaction to immoral behaviour caused by inappropriate circumstances. Tang arrived in the morning but had to wait because the ferry system was busily transporting travellers, traders, animals, and goods to the other side. Tang had to endure seeing 'immoral customers who had lost all decent self-restraint' and thus decided to ask the local dignitaries why there was no bridge at such an important junction.

Tang Zhongyou: 'created a wooden model which he placed in a wooden pond. Then he added water to imitate the effects of a flood and carefully observed the water's advance and retreat as he opened [the basin]'. Subsequently, so the source continues, Tang arranged for corvée labourers to begin the construction work in spring, continuing until the ninth month. Two embankments were constructed east of a local landmark, the Huanghua pavilion. Tang stipulated the wall was to be strongly built with huge rocks, unlike the usual brick construction. The bridge itself consisted of three levels: 'to break the flow of the water. If there was overflow on the southern embankment the water would end up in a wooden shore which could hold 115 *xiao* [of water]'.

Tang Zhongyou was, like his peers Su Song and Zhu Xi, a multi-talented official who used the full repertoire of opportunities in his era to serve his people and state. As well as bridge building, his historical reputation was based on an elaborate historiographic discussion of administrative and governance issues which included the entire gamut of publication methods: his 'Illustrations and Treatises of Emperors and Kings throughout the Generations' (*Diwang jingshi tupu*) draws together numerous exquisite texts including the correspondent contemporary charts, illustrations, and maps that his eras had produced on astronomy and astrology, legal, fiscal, and military practices and theories.

Yuan dynasty

An interplay between Mongolian nomadic and Chinese sedentary agricultural practices spread throughout Asia, affecting and modifying economic and social life during the Yuan era. In conquering the Song population, the Mongolian Yuan kept Song state bureaucratic structures in the southern region largely intact, although taxes increased. Tenancy relations were established on private and governmental land, with farmers often paying 50 per cent of their crop yield as rent. In the north, the Yuan extensively confiscated agricultural lands for grazing purposes, meaning that grain then needed to be transported from the south to the north.

While water was abundant in the south, the north of the empire often suffered from drought. When Khubilai Khan made Khanbaliq (Cambaluc; known in Chinese as Dadu or Tatu, or 'Great capital') his capital, he asked his official and astronomer Guo Shuojing to harness the rivers and canals in the region to provide the city with sufficient water. In the Yuan period officials thus acquired and displayed multiple talents, as had their Song predecessors.

When Wang Zhen completed his 'Book on Agriculture' (*Nongshu*) in 1313 he was able to draw on a growing textual archive of specialized studies and essays on sericulture and farming, breeding methods for ducks, goldfish, and horses, and literary repositories on diverse plants. Wang also quoted profusely from both Jia Sixie and Chen Fu. Wang explained that his book on agriculture was needed to alleviate the effects of disasters and of the long war between the Song and Yuan, both of which, he believed, were still inhibiting agricultural productivity. However, Wang's book did not just evoke an ideal past. It was innovative in many regards, as it combined both northern and southern methods. Furthermore, Wang interspersed his own comments describing his own observations and personal experience amongst Jia's and Chen's texts.

The most important novelty of Wang Zhen's books was his combined use of texts and illustrations. In his 'Illustrated Register of Farm Implements' (*Nongqi tupu*), he paired technical drawings of machinery and implements with explanatory remarks on their construction and use, delivering 'templates of action' for state officials who were obliged to travel across the empire to take up their diverse appointments. Wang Zhen's approach reflects the increasing technical possibilities of the Yuan era, and a maturing of scholarly explorations of visual and textual means to augment their strategies and ideas.

Like Jia Sixie, Wang included a chapter on animal husbandry in his agricultural study. The sections on sheep and pigs mostly consist of quotes from the *Qimin yaoshu*, and the section on cattle and water buffalo was mostly reproduced from Chen Fu. Wang also copied Chen Fu's remarks on negligent lack of hygiene in cattle farming, paying respect to the changes necessitated by the move south, even though the Yuan state now allowed large livestock to roam freely in the north. Wang Zhen wished to preserve the multiplicity of agricultural practices and theory of his generation whilst simultaneously offering advice on new or alternative methods which could be used in times of crisis or disaster. Wang paid close attention to didactics, and consulted with carpenters to understand the mechanism and design of water lifting devices as well as simple tools such as harrows or seed-drills, even probing into soil preparation himself in order to be able to explain agricultural techniques to his peers.

Wang was genuinely interested in spreading the knowledge of agricultural methods to his colleagues who often had to transfer from post to post throughout the empire. In due course, Wang also supported Yuan state initiatives to promote the farming of new crops and plants. For instance, he explained in detail a two-roller ungeared cotton gin which processed fibre that the Mongolian state had introduced to China. This was part of the Yuan state's active encouragement to farmers to begin cotton production, which it had been promoting since 1289. By 1296 the take-up of cotton farming was such that the state could include cotton in its tax system, although it continued to promote cotton farming by keeping the tax rate lower than that for silk or hemp.

Ming dynasty

From the end of the Ming dynasty to the beginning of the Qing, scholar-officials increasingly developed the literary library of local agricultural practice. 'Shen's Book of Agriculture' (*Shen shi nongshu*) and Zhang Luxiang's (1611–1674) 'Supplement to the Book of Agriculture' (*Bu Nongshu*), for instance, described agricultural achievements in north Zhejiang province. Both books describe agricultural tools and their use in routine activities like soil preparation, irrigation, manure transportation and use as fertilizer, as well as crop protection. Like their predecessors, they also covered the growing of crops and plants, forestry, animal husbandry, and the household textile industry. A growing number of leaflets 'supporting agriculture' (*quannong*) and household encyclopaedia were published to spread the knowledge of new crops such as sweet potatoes, maize, and tobacco.

Agriculture diversified constantly, particularly in Jiangnan which had rich soil. Farmers and officials experimented with different varieties of rice and farm tools, fertilizer use, and cropping systems. Merchants and officials improved their awareness of the main varieties and sub-varieties of rice, especially intermediate and late-yielding rice, selecting and propagating new varieties to suit an urbanizing society. Officials and farmers promoted the use of tools including the rake (*tieda*) and a leveller (*tiandang*) used to level the field, as both facilitated rice paddy production. The three- or six-tooth rake was key to wheat planting in ridges, which was crucial to double crops of rice and wheat in the paddy fields.

Farmers also experimented with diverse fertilizers, from human manure to oil cakes, and mixtures of each, to produce two or three crops per year, such as rice in summer and dry land crops such as wheat, beans, and rapeseed in winter. At the end of the Ming period, Xu Guangqi included a detailed list of the different kinds of manure then in use in his 'Complete Books of Agricultural Politics' (*Nongzheng quanshu*, 1625–1628), explaining the way to use them, popularizing as well as innovating a basic though substantial technology of Chinese agricultural practice which would enable the state to feed its growing population.

As in the earlier Song and Yuan periods, Ming era agricultural practice was closely intertwined with the management of waterways. Early Ming approaches to hydraulic engineering built on Guo Shoujing's and Yu Ji's (1272–1348) ideas of water construction from the northwestern region.

The effects of hydraulic reorganization often became evident in changing crop patterns that spurred another reformation of taxes and the corvée system that the Ming had inherited from the Yuan. Zhu Yuanzhang, the first Ming emperor, claimed that a continuation of the Yuan practice of inherited professions in the fields of crafts and agriculture, coupled with a quota system for corvée labour for the state-owned production of silk, salt, and porcelain, as well as mining, minting, and large-scale construction work was beneficial for both the state and the common people, since it prevented corruption and made the state's demands more predictable.

While the quota system was implemented, Zhu Yuanzhang's son and officials during the reign of the Yongle emperor soon began to unbalance the system in the area of luxury goods such as silk by increasing court demands. In fact, the system could only work as long as desires and demands remained static and the duties of maintenance and repair were continuously fulfilled. Despite the aim for constancy, new tributary schemes, disasters such as floods or droughts, or even fluctuations in local crop patterns or harvest yields presented constant challenges.

In the fifteenth century, officials in the Lower Yangtze delta attempted to replace a crisis-response approach in the field of hydraulic management with ongoing programmes of continuous repair and maintenance to spread financial investment over generations. This new policy enabled local areas to raise funds and mobilize labour in compliance with other local needs, thereby reducing the burden of such works for the local community. However, this system also meant that, over time, the scope of projects shrank and disputes about water resources or disagreements about maintenance duties emerged. Localities increasingly turned away from any financial or organizational engagement beyond their regional remit. This meant that large-scale schemes were endangered, leading to frequent incidents of Yellow River disasters occurring during the Ming dynasty which required the attention of central government. A number of famous ministers and other members of the Late Ming court, including Xu Guangqi and the Jesuits, thus became involved in dyke construction, river regulations, and canal building.

The importance given to water regulation in the Ming era literary understanding of agriculture demonstrates how the emphasis had shifted towards sericulture and rice farming. Visual and material culture reflects this change in the absence of large livestock such as cattle, horses, donkeys, and camels. Similarly, state bureaucratic structures changed, turning their attention to silk, food crops, and crafts such as porcelain production.

Man: medicine and health care

Medical education and institutions existed as early as the Han dynasty, yet much of what is reported catered exclusively to the court and elite. With the Song institutionalization of medicine, the concept of 'Confucian doctors' (*ruyi*) emerged—people who combined scholarly erudition and moral values with medical expertise. Medical learning was quickly integrated into the Hanlin academy (*Hanlin yuan*), the highest school of learning in the state.

Song medical approaches were inclusive and contextualized, rather than isolating an illness or the human body alone, as the bureaucratic landscape of this era vividly reflects. Institutional structures in equine care were similar to those of human care, especially regarding pharmaceutical treatment and directives intended to control epidemics. The state-produced literature mainly aimed at enhancing self-treatment and prescriptions and thus was often pharmaceutical in nature. Bodies were believed to function in similar ways so much of what was written about humans was meant to function equally for animal care. Song scholars perceived veterinary and human diseases as cases of disorder with the natural harmony of nature or the world, evidenced by natural disasters and ultimately caused by misguided human behaviour. Certainly humans had to actively participate in the treatment and act morally to facilitate healing. Doctors, hence, were less interested in the ingredients of medication as they were in human ecology, asking whether people were morally weak, or lacked sufficient understanding of natural principles, and so acted carelessly. Individual illness, epidemics, and catastrophes were considered to be a matter of knowledge, morality, and action.

Institutional settings of hygiene and medical care from the tenth to seventeenth century reflected the belief in man's ultimate accountability, with their emphasis on surveillance, prevention strategies, knowledge, and education in both human and veterinary care. It is important to note that any parallels between state structures for veterinary and human medicine were not purely a result of, for instance, a seeming Song fondness of elaborate bureaucracy—the institutions were similar because the medicinal concepts and approaches corresponded.

In both the human and veterinary medical sector, Song dynastic institutions emerged piecemeal, resulting from lively debates on public hygiene and medicinal care. This emergent character of Song health care is evident in the historical development of the court's knackery which began as an institution that de-boned, skinned, and cut the meat, muscles, and intestines of dead horses, oxen, camels, and donkeys and became a local court office which selected officials as veterinary doctors and taught commoners who were deployed in military and civil veterinary care across the country.

Craftsmen (*gongjiang*) of the directorate, auxiliary officials of the palace (*qin congguan*), officials of the imperial coachman (*qimazhi junshi*), and guardsmen (*xiangbu*) were thus asked to prepare the intestines and meat, and to 'repair' (*xiang bu suojiao*, literally 'mend') the bones so that the new metropolitan and provincial graduates and officials (*suo jiao di guan xuexi*) could study them as teaching aids. Recording of these actions was mandatory. The investigators had to meticulously report all these procedures, including providing illustrations (*tu*) 'for later inspection' (literally: to be kept in reserve *yigong biyong*). Institutional approaches from the Northern and Southern Song era indicate a genuine desire for standardization and synergy, which scholars appear to have believed was the only way to implement specific measures across the large territory. An impressive material manifestation of the normalization efforts of Song medical scholars was a bronze cast of the human body, which was used as a didactic tool to teach *qi*–body flows and train army doctors who accompanied military troops or were installed in offices across the country.

Textual debates of the time reflect an equally deep concern about standards, which many officials considered the only way to manage the manifold needs and demands of the empire. An imperial pharmacy was established by Wang Anshi in an attempt to spread the use of prescribed formulas for those who could not afford a doctor or individual medical care. A medical branch was installed in the Hanlin academy, producing regular updates of the imperially commissioned *materia medica*, which was disseminated across the imperial state system.

Texts were central to Song imperial attempts to standardize medical treatment and the state launched several campaigns to publish formulas and concoctions. In 974,

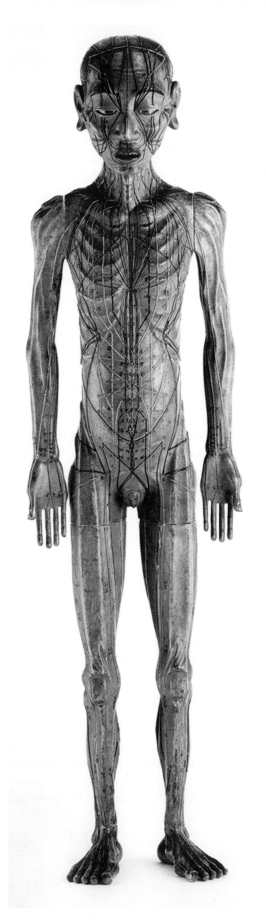

Acupuncture model.

Lu Duoxun (925–996) was commissioned by the imperial court to publish his 'Newly Determined *Materia Medica* of the Kaibao Era' (*Kaibao xin xiangding bencao*), proposing, for instance, the Vietnamese Sophora root (*shandou gen*): 'as an efficient treatment against horse jaundice, fever, and coughing. It also kills worms'. Worms, that is insects or their larvae, were considered to be key culprits of livestock illnesses and equally dangerous to humans. Once a medical method of prevention or treatment proved successful, the information was reproduced and disseminated across the empire.

These and later works elaborate the ingredients, tastes, smells, colours, and classificatory characteristics of medicines. The 'First Part on Supreme Herbs' (*Caobu shangpin zhi shang*) in the classification of medicinal herbs for emergencies of the Daguan era (*Daguan jingshi denglei beiji bencao*) describes the *fuji zigen* in clear detail as 'bitter in taste, it is cold and not poisonous', providing a guide for its use by non-skilled patients as well as for doctors who did not know the plant.

Grounded in the early days of dynastic rule and conceptualized during the Qingli reforms of the 1040s many ideas and efforts of the first century came to fruition mainly towards the end of emperor Shenzong's (1048–1085) reign. The Song state's approach to medicine after 1076 must be viewed in the context of the economic reform and sociopolitical upheaval initiated by the grand councillor Wang Anshi. The Northern Song state continuously pondered reform and refinement of institutional structures, which the Southern Song state and the Jurchen Jin era then implemented. During the Jin era, for instance, the 'imperial medical section' (*taiyishu*) was re-named the imperial medical bureau (*taiyi ju*), managing and configuring medical education autonomously. The Jin court divided its three hundred students into nine categories (*ke*). In 1140, medical specialists also formed part of the ministerial parliament (*shangshu sheng*) and the ministry of personnel (*libu*), with their main responsibility being to care for the emperor's health.

The Jin also installed a training and examination system, testing candidates for medical court service regularly every third year. Those who did not pass the examination were allowed to continue their studies and work as doctors for the commoners, but not for the emperor (or court?). Medical theories followed the Northern Song traditions, particularly the works published during the Huizong era (1111–1118), and classic texts from previous periods were also integrated.

While Southern Song doctors found the idea of an imperial pharmacy still difficult to comprehend, prescriptive medicine received great prominence during the Jin era. Apprentices trained by working alongside practising doctors. The appointment of doctors was institutionalized at the level of the capital, the district, and the market place. Just like the court system, the local medical system had distinct branches (*ke*) and relied on reference works such as the 'Discussion of Recipes from Xuanming' (*Xuanming lunfang*) or the 'Patterns Preventing Cold-illnesses' (*Fanghai zhige*). Practitioners such as Liu Zongsuo (n.d.) scrutinized special illnesses such as those called 'hot-nature illnesses' (*rexing bing*) in his tract 'Discussion of the Heat of Fire [phase]' (*Huore lun*).

Yuan dynasty

By the end of the Song era, medical schools had been established throughout the empire, as the Yuan gazetteer compiler Wang Yuan (n.d.) reveals in his record of the local medical system that existed in his lifetime. Wang emphasized the contrast between a Song literary approach to health care and the Mongolian veneration of *yin-yang* learning through language and medical schools, whilst noting that the Song tradition of medicine had also valued acts such as incantations, chants, and ritual healing.

Khubilai Khan's successors recruited diviners or physicians for their armies and government while conquering Eurasia. Not all of these conscriptions were enslavements; in fact, experts were granted privileges such as tax exemptions and advantageous posts in the powerful imperial academy of medicine. In contrast to the Song system, in which medical officers were ranked as low as 6, the Mongols ranked medical officers as high as 2, akin to a ministerial position. In medicine, as in astronomy, the Mongolian Yuan doubled appointments in all institutions, partnering a Han Chinese medical office (*taiyi yuan*) with a Muslim 'office of broad grace' (*guanhui si*) to promote Muslim medical treatment.

When it came to professional expertise, Khubilai Khan ignored ethnicity, employing Chinese doctors such as Xu Guozhen (n.d.), who originated from a generations of physicians, as well as Muslim practitioners such as Ngai-Sie Isu'a (1227–1308). Both men were consulted on general political matters as well as on medicine. The Mongols were probably influenced by the highly developed medical institutions in the Islamic world of that time and had learned to respect medical experts. Enormous hospitals had been established throughout Islamic countries since the ninth century.

Central state records reveal the workings of court institutions that mainly serviced the court and capital region, showing how the Mongol rulers left the details of medical care to the locals, who consulted temple schools. Xu Heng (b. 1209), son of Xu Guozhen and the most influential proponent of neo-Confucian schools at Khubilai's court, advised young men to enter medical service in temple schools rather than by civil service examination, as had been done during the Confucian era. In these temple schools, students learned from role models rather than from books.

As the Yuan promoted medical learning in order to eradicate fake physicians and illiterate practitioners, more credence was given to texts than to individual skills. In 1269, ministers suggested that: 'from now on people of the various medical departments, shall each work in their speciality, when they see a patient they shall rely on the medical canons to determine the symptoms, prescribe medicine, apply acupuncture, or burn moxa'. Reiko Shinno suggests that such medical schools probably instructed first-generation practitioners, in contrast to medical education within kinship structures, handed down from father to son.

Ming dynasty

Public health during the Ming era became a local issue again as doctors increasingly reverted to thinking in cases, understanding illnesses as singular contextualized events. In such instances, the huge effect of bureaucratic ordering on scholarly understandings of nature, heaven, Earth, and men were particularly evident, as the idea of exemplars in this era included equal weighting to political, philosophical, and social thought.

The Ming state became increasingly indifferent towards medical education and treatment, leaving such tasks to the locally or privately sponsored efforts of philanthropists, instead of officials. Whereas Han officials during the Yuan period installed medical care in the temple structures of the three progenitors' cult, allowing each to learn from the other, the Ming neglected such temples' functions as healing centres, promoting instead orthodox Confucian doctrines of medical learning at the court. Charity pharmacies and medical bureaus were maintained until the reign of Xuande (1399–1435). By the 1560s local structures had become dilapidated and competent doctors tried to avoid working in any of the ill-paid and run-down local medical bureaus.

Knowledge, Learning, and Bureaucracy

From the Song to the Ming era bureaucratic structures considerably influenced approaches to knowledge codification and dissemination, and paper bureaucracy grew.

Song scholarship shaped the realms and formats of knowledge, as the literati defined their own social and political identities as the ruling elite, embracing literature and the arts even more than their predecessors, the aristocratic Tang rulers. However, with the invention of print and paper, scholars also became increasingly anxious about the origins of knowledge and lineages of thought. Antiquarianism, that is, the study of the 'good issues of ancient times' (*haogu*) flourished, focusing on the three dynasties preceding the Han, 'script on bronzes and stones', i.e. epigraphy (*jinshi xue*), the 'origin of characters' (*ziyuan xue*), logography (*wenzi xue*), and 'stele inscriptions' (*beiming*). In these fields, scholars established the basis for the field of evidential study (*kaozheng*) that would reach maturity during the Qing dynasty. In unearthing and tracing old knowledge, scholars attempted to understand rituals and morals through the preservation or recovery of objects. With the advance and commercialization of printed matter, scholars therefore became increasingly critical about the truths of the past as they were represented in contemporary print. Raising doubts about the reproduction of ancient knowledge acted as a catalyst for studies of the nature of things and affairs, heaven, men, and Earth.

Technical inventions and materials became increasingly visible in print, facilitated by daily bureaucracy. The literati pondered the relevance and potential of things and 'established methods' (*lifa*), policies and operation plans (*jinglue*), or guidelines (*gui*) to manage society, state, and knowledge flows. As officials and moral leaders of society, men such as Zeng Gongliang (998–1078) and Ding Du (990–1053) hence recorded the recipe for gunpowder in the year 1044. Their tract on the 'Essentials of Military Affairs' (*Wujing zongyao*) introduced weapons such as the fire-carpet (*huotan*) to a civil elite, for use in wars against the Liao and, later, the Jurchen Jin, and Mongolian Yuan. Many of these weapons had already been in use for some time before they appeared in literary accounts. This is a general pattern of the records of innovation of the Song era.

When they seized the capital Bianjing, the Jurchen acquired a number of Chinese technologies, like Su Song's astronomical clock tower. Although the Jin failed to reconstruct Su Song's clock, they skilfully used gunpowder weapons in attacks against the Mongolians who made inroads into Jin territory in the thirteenth century. Fire guns (*qiang*) helped to repel a Mongolian attack on Bianjing in 1232, but it did not save the Jin from defeat, as the Mongolians quickly adapted, employing the new weaponry with superior military skills. In 1236, for instance, cannons were widely in use.

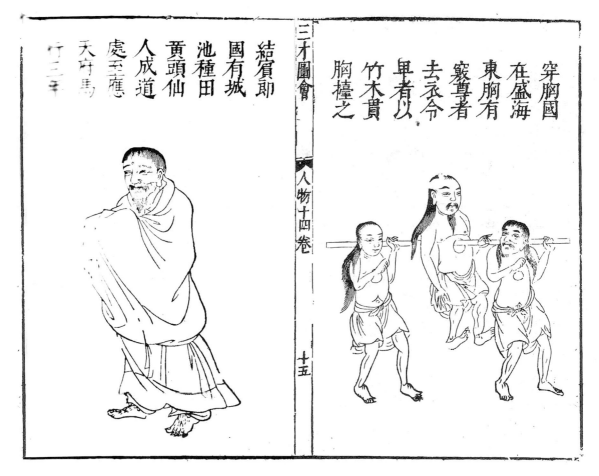

A south-facing chariot.

Knowledge in books survived even when the practical ability to put this knowledge into reality had vanished because scholars in China were interested in tracing the origin of things and preserving ancient ideas. For instance, the 'Official History of the Song' records that in 1027 the painter Yan Su (960?–1040?) (who also served in several local and central positions during his lifetime) attempted to build and improved a south-facing chariot (*zhinan che*). A second attempt was made by the artist Wu Deren in 1107, who attempted to build and improved a south-facing chariot (*zhinan che*). Even though the texts described the vehicle's structure and form, the knowledge of how to create it was lost.

Historians and scholars from the Tang period, and increasingly from the Song onwards, traced the origins of things and affairs in literary accounts as a way to remind the emperor (and themselves) of the complexity of knowledge, as a way to elucidate critical issues in order to comprehend, and therefore to rule, the world properly. These studies, building lineages of dependent inventions for the production of things and affairs (including institutions, habits, ritual, and moral concerns), also established trust in novel ideas and innovations, connecting everything new to the past. By the Ming period, the catalogue of studies on the origins of things had grown considerably.

Learning from the ancients or from the original writing remained a persistent aesthetic debate throughout the tenth to seventeenth centuries, crossing dynastic boundaries of the Song, Liao, Jin, Yuan, and Ming states. Ancient histories provided evidence of things that worked, and were reliable and effective. Accordingly, Zhao Bingwen (1159–1232), who carried out the position of high minister while the Jin dynasty rose and fell, proposed the Tang dynasty as the model for honesty and kindness. Conversely, the poet Yuan Haowen (1190–1257), who also lived during the Jin rule, advocated Song Confucian aesthetic categories and classical poetry, while Li Chunfu (1177–1223) advocated 'original writing', disregarding the ancient words completely. Scholars from the tenth to the seventeenth centuries wrote and published texts, but they also applied models, sketches, artefacts, and other means to preserve and innovate diverse 'cultures of knowledge' in the many regions of Song, Liao, Jin, Yuan, and Ming dynastic rulership.

5 The Scientific Revolution

JOHN HENRY

ALTHOUGH still contested by historians who prefer to emphasize the continuities underlying all historical change, the 'Scientific Revolution' has become the accepted designation for the period during which something recognizably like modern science emerged. While this term is arguably a misnomer—since the 'revolution' took about two centuries to accomplish—there can be no denying that the human endeavour to understand the natural world underwent such radical changes beginning in the sixteenth century that by the end of the seventeenth century there had been a complete sea-change. It was the fully comprehensive nature of these developments, and the truly remarkable achievements arising from them, which have led to the period being seen as one of revolution.

Between 1500 and 1700 the world picture shifted from a geocentric finite cosmos of nested heavenly spheres which allowed no empty space, to a heliocentric solar system in an infinite universe that was void except where it was dotted with stars. A prevailing belief in a qualitative dichotomy between the heavens and the Earth (the heavens being conceived as completely different from the Earth) gave way to the belief that planets were like the Earth, and stars like the Sun, and the acceptance of the universal applicability of the laws of nature. Moreover, the laws of nature were no longer simply conceived as mere regularities (bees make honey, the Sun rises in the East), but became codified for the first time as precise statements about how specific aspects of the world work, and were seen as capturing causal relationships between phenomena and as having predictive power. There arose new theories of motion, of the generation of life and its organization, a revised human anatomy, and a new physiology. This period also saw the introduction of the experimental method into what had previously been an essentially contemplative 'natural philosophy', and a new belief that mathematical analysis, in spite of its clearly abstract nature, could be used to help in understanding experimental results, and in understanding the physical world more generally. The previously contemplative natural philosophy also underwent dramatic change when it embraced the idea, previously confined to practitioners of mathematical and occult arts, that knowledge of the natural world should be put to use for the benefit of mankind. Going hand in hand with these changes was the emergence of new forms of organization and institutionalization among those with an interest in studying the natural world; in particular it was a period which saw the formation of societies devoted to the understanding of nature. The Scientific Revolution took place in Western Europe and, although in large part its starting point was the knowledge of the natural world first developed in ancient Greece and subsequently transformed by Islamic scholars and then by medieval Christian scholars, it went far beyond what these earlier civilizations and others such as the Chinese had achieved. Although science has now become an international enterprise, it is still fair to say that, for example, a Japanese Nobel Prize winner in physics is a practitioner of Western science (in the same way that a Japanese concert pianist, or violinist, plays Western music). The story of the Scientific Revolution, therefore, is also, in part, the story of the rise of the West. Certainly, it is a story of the origins of how knowledge of the natural world, what we now call scientific knowledge, became so prominent in Western, and subsequently world, culture.

The literature on the Scientific Revolution is vast and wonderfully rich and complex. What's more, it is still growing. What follows is a necessarily selective account of those aspects of the revolution which have acquired the highest profile in the historiography of science.

How It All Began: The Renaissance and the Scientific Revolution

It is easy to see that the Scientific Revolution, like the Protestant Reformation, constitutes an important part of the wider changes in intellectual authority that were characteristic of the period known to historians as the Renaissance, and so it can be said to share the same general causes as this major change in European history. A full account of its causes would, therefore, have to encompass the decline of the old feudal system and the rise of commerce, together with the concomitant rise of strong city-states and national monarchies during a period of increasing decline of the Roman Catholic Church and the Holy Roman Empire. In a brief account like this one, however, we must confine ourselves to those initiating causes which can be seen to have had the most direct effects. But before turning to these, it is also worth noting that the Renaissance itself should not be seen as having its own prior identity which enables us to see the Scientific Revolution as a mere spin-off of this larger movement— the changes that constituted the Scientific Revolution, no less than the changes that constituted the Protestant Reformation, were part and parcel of the changes which have led historians to see the Renaissance as a major turning point in history, and helped to make the Renaissance what it was.

Many of the changes in the Renaissance stemmed from the invention of printing, and the new exploitation of the magnetic compass and gunpowder (which were evidently known long before in the East), all of which had major cultural and economic repercussions. These can also be seen to have had a direct bearing upon developments in, and attitudes toward, natural knowledge. Printing, and the manufacture of paper (also adopted from the East), enabled the dissemination of knowledge as never before; and the compass and gunpowder demonstrated just how useful knowledge of the natural powers of things could be.

The magnetic compass made it possible for mariners to strike out across the open sea, instead of hugging coastlines, and played a significant role in the Renaissance voyages of discovery. But these voyages had another highly significant effect on subsequent thinking. It is not true that educated thinkers before the discovery of the New World by Christopher Columbus (1451–1506), assumed that the Earth was flat. The ancient Greeks knew that it was spherical, and the influential early Church Father, St Augustine (354–430), cautioned Christians against taking the Bible literally when it implied the world was flat, because they would only embarrass themselves in the eyes of pagans. Even so, throughout the Middle Ages the idea developed that the sphere of the Earth was floating in a larger sphere of water, with only one hemisphere of the Earth above the water. This idea developed out of starting assumptions taken from Aristotle (384–322 BC), the ancient Greek thinker whose authority came to dominate medieval natural philosophy. There was, however, an alternative view, promoted by Claudius Ptolemy (*c.*90–*c.*168), a later ancient Greek writer, who had become the major authority in the technical subjects of astronomy and astrology. Ptolemy's *Geography* (or *Cosmography*, as it was translated) became newly available to the Latin West in 1406 and it was clear from this that he simply believed in a single 'terraqueous globe'—that is to say, the Earth was simply a single sphere whose surface was made up of land masses and oceans or seas; there was no extra sphere of water, in which the Earth was floating. Ptolemy's *Geography* on the one hand encouraged the great navigators of the Renaissance period to believe it might be possible to circumnavigate the globe (without becoming lost in the supposed larger sphere of water, in which the globe was held to be floating), while on the other hand the Renaissance circumnavigations of the globe showed that Ptolemy was correct, and the supposedly Aristotelian theories developed by medieval philosophers were wrong.

But there were a couple of other major effects of these voyages. Exploration of the New World and other parts beyond Europe gave rise to an increasing awareness of cultural relativism. This was especially remarkable in the case of China, where a highly advanced civilization had no connection whatsoever to the Judaeo-Christian religious tradition which had previously been regarded as synonymous with civilization. Furthermore, it showed that the traditional wisdom (again deriving from Aristotle's authority, but in this case also supported by Ptolemy), that life could not survive in the antipodes, was misconceived.

This brings us to another major aspect of the Renaissance which was to have far-reaching repercussions on the understanding of the natural world. Some of the political and economic changes in the Renaissance manifested themselves in the emergence of powerful and wealthy individuals or families (exemplified most clearly, for example, by the ruling families in the different city states of Renaissance Italy—the Medici in Florence, the Sforza in Milan, and so on) who could demonstrate and enhance their standing by acting as secular patrons to artists and scholars. It was these secular patrons who enabled the visual arts to flourish in this period—which is, of course, one of the most prominent features of the Renaissance. Equally important for our purposes, however, was the patronage of scholars, who were encouraged to

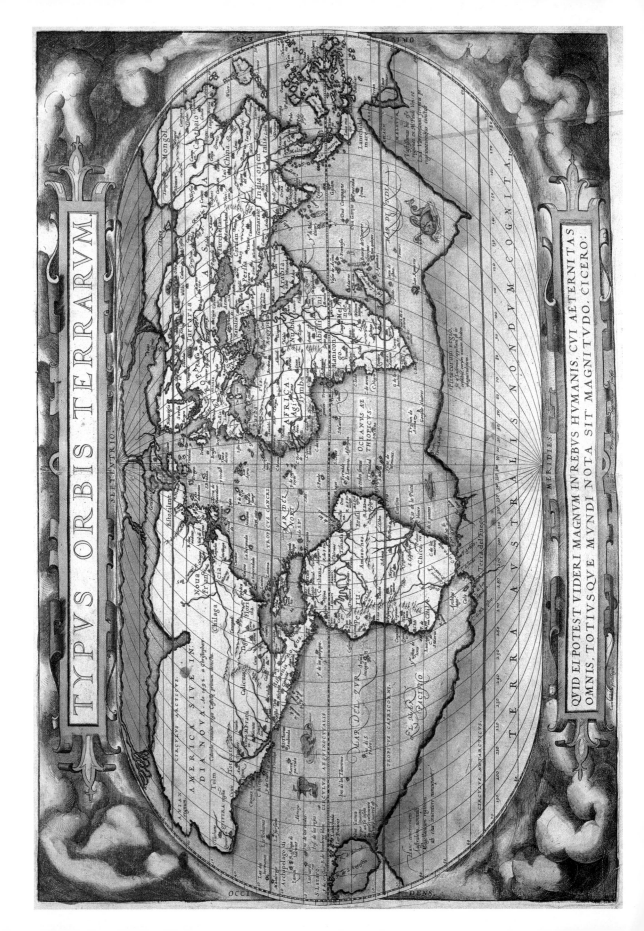

gather libraries of great books for their patrons. The result was a re-discovery of the writings of many ancient philosophers, and the concomitant realization that Aristotle was not *the* philosopher, as medieval philosophers and theologians referred to him, but was merely *a* philosopher. Furthermore, the discovery of books *about* ancient philosophy, most notably *The Lives of the Philosophers*, by Diogenes Laertius (*fl.* second century AD), made it clear that, for the ancients themselves, Aristotle was by no means the most admired philosopher.

It is important to note here that the curriculum in the Arts Faculties of all the universities throughout Europe was largely based on the natural philosophy of Aristotle. Since all students had to become masters of arts before they could proceed to study in the so-called higher faculties (of theology, law, and medicine), this meant that the most highly educated people throughout Europe were thoroughly steeped in Aristotelian natural philosophy. Their investment was now looking increasingly ill-founded. The discovery of writings by other philosophers, including Plato (*c.*427–347 BC), the Neoplatonists, Stoics, and Epicureans, provided a rich fund of alternatives. Eclectic attempts to combine the best features of the newly available ancient philosophies met with some success in moral and political philosophy, but were less successful in natural philosophy. One alternative, therefore, was to switch allegiance from Aristotle to Plato, or some other ancient thinker. Other Renaissance philosophers, however, perhaps more disorientated or more dismayed by the overthrow of traditional intellectual authority, tended increasingly to reject recourse to any authority and turned to personal experience as the best means of acquiring knowledge of nature. An important influence here was the fact that one of the revived ancient philosophies, which was seen to be popular among the ancients themselves, was scepticism—with its built-in rejection of authority.

The voyages of discovery, and the re-discovery by scholars of alternative philosophies to that of Aristotle, increasingly led to a rejection of human authority as a valid source of knowledge. This in turn had repercussions in religion, and these fed back into negative attitudes to authority. When Martin Luther (1483–1546) rejected the authority of the Pope and the priest in religion, urging instead that every man should be his own priest in the priesthood of all believers, he encouraged the faithful to read the Bible for themselves. In our secular world, we might regard this merely as an affirmation of the authority of the Bible. But for Luther, the human authority of the Pope, or the local priest, was coming between the believer and the source of truth.

Opposite: The *Theatrum Orbis Terrarum* (*Theatre of the Globe of the Earth*) is usually credited as the first printed Atlas, compiled by Abraham Ortelius (1527–1598), using maps which had been produced earlier by other cartographers. This map of the world is based on the map of Gerard Mercator (1512–1594), first produced in 1569, using Mercator's newly developed system of projecting the three-dimensional globe onto a two-dimensional surface. Mercator's Projection was especially useful for navigation because courses on a constant compass bearing appeared as straight lines. The development of cartography was a major aspect of the Scientific Revolution, showing the impact of the voyages of discovery on the development of mathematics, and consequently the beginnings of increased social standing of mathematicians. Abraham Ortelius, *Theatrum Orbis Terrarum* (1570).

So, to read the Bible for oneself (something which was forbidden to the ordinary believer at the time) was to reject authority and to go to the source. The natural world was often regarded as God's other book, and just as the faithful were now expected to read the Book of Scripture for themselves, so it seemed to devout natural philosophers that God could be served by reading the Book of Nature. Natural philosophy had come to be seen, in the Middle Ages, as the hand-maiden to the so-called 'Queen of the Sciences', theology. Before the Renaissance there was only one theology, represented by Roman Catholic orthodoxy, and only one natural philosophy, represented by the scholastic Aristotelianism developed by the medieval schoolmen. Apart from all the other effects of the fragmentation of Western Christianity after the Reformation, there was its effect on reinforcing the rejection of human authority and emphasizing the need to go to the source of truth. In the case of natural philosophy this simply meant the natural world itself. Where natural philosophers had once sought the answer to any question about the natural world in the books of Aristotle, this was now seen to be indefensible. The alternative was to study the natural world itself. Simple and obvious though this might seem to us, it represents a major and crucially important shift in the history of human endeavour.

The time was ripe for the development of a new experiential or empiricist approach to the understanding of the physical world. This new attitude was clearly exemplified by the radical Swiss religious, philosophical, and medical reformer known as Paracelsus (1493–1541). He not only wrote reformist works, developing a uniquely original system of medicine, but he also explicitly defended his new approach on empiricist grounds. In an announcement of the course he intended to teach at the University of Basle in 1527, for example, he rejected 'that which those of old taught' in favour of 'our own observation of nature, confirmed by extensive practice and long experience'. It is easy to see, from the vigorous reaction against Paracelsus, that his rejection of the authority of Galen (c.130–201) and Avicenna (980–1037), who played the same authoritative roles in medicine as Aristotle and St Thomas Aquinas (1225–1274) did in natural philosophy, was seen by many not just as an attack on the traditional medicine sanctioned by the colleges of physicians throughout Europe, but as an attack on the colleges themselves, and even as an attack on the Church and State, of which the colleges were seen as representatives. Just two years after Charles I (1600–1649) was beheaded outside the Palace of Whitehall, the Royal College of Physicians in London was denounced as a 'Palace Royal of Galenical Physick'. When the sectarian herbalist Nicholas Culpeper (1616–1654) translated the College-sponsored *Pharmacopoeia* into English for the first time (1649) he made an even more damaging comparison: 'Papists and the College of Physicians will not suffer Divinity & Physick to be printed in our mother tongue.'

In spite of the radical and subversive nature of much of Paracelsus' teaching, many aspects of his new system of medicine were embraced by and absorbed into revised versions of more traditional medicine. One of the reasons for this was undoubtedly the perceived empirical success of Paracelsianism, so lending support to the efficacy of this new methodology. Doctors, after all, were professionals seeking to make a living from fee-paying patients; if new therapies seemed to lead to new cures, it was inevitable that these would be sought out by patients, and provided by practitioners. But increased empiricism took hold even in areas of medicine which had no immediate bearing on therapeutic success. The reputation of Andreas Vesalius (1514–1564) was not just based on his superbly illustrated anatomical textbook, *De humani corporis fabrica* (*On the Structure of the Human Body*, 1543), a book which surely shows the crucial importance of printing, but also on his new method of teaching. Where previously anatomy lecturers read from one of Galen's anatomical works, while a surgeon performed the relevant dissections, Vesalius dispensed with the readings and performed his own dissections, talking the students through the procedure and what it revealed. Shortly afterwards (1594), his university, at Padua, constructed the first purpose-built anatomical lecture theatre, with steeply raked tiers of seats, enabling all students a clear and not-too-distant view of the cadaver.

Paracelsus, and no doubt others even outside the ranks of his followers, failed to see any value to the physician in the 'anatomy of corpses', but eventually the heightened level of anatomical study was to have practical benefits. A number of new discoveries by Vesalius and his successors at Padua, as well as their emphasis upon the importance of comparative anatomy for the understanding of the human body, was to lead to William Harvey's discovery of the circulation of the blood.

Harvey (1578–1657) was a student at Padua between 1597 and 1602, and continued with the kind of anatomical study he learned at Padua upon his return to England. Where other medical schools focused only on the human body, the Paduan emphasis upon the value of comparative anatomy allowed Harvey to perform vivisections on dogs and other animals. This enabled Harvey to see beating hearts in action, and to go far beyond what was possible by simply inspecting dead human hearts. Harvey was able to show that the Galenic assumption of two separate systems in the body—the veinous system originating from the liver, and the arterial system originating from the heart—was completely wrong, and that there was only one system taking blood out from the heart to all parts of the body via the arteries, and bringing it back by the veins, to be re-circulated endlessly. Although resisted at first, Harvey's experimental demonstrations of his discovery (published in 1628) were so elegant, and his audience so used by now to the relevance of experiment in revealing truths about nature, that his theory soon became accepted. This immediately meant that the whole system of Galenic physiology had to be recast. The result was a marshalling of effort by anatomists and physiologists throughout Europe, leading successively to numerous new discoveries.

Vesalius claimed that he had discovered over two hundred errors in Galen's anatomical works; he blamed these on the fact that Galen was forbidden by his Roman overlords to dissect human bodies. This is understandable, but it is much less understandable to us that these errors were not noticed before the sixteenth century. Vesalius made it clear that the reason for this was precisely the slavish adherence to ancient authority that was now rejected by Vesalius and other Renaissance thinkers:

Contemporary anatomists are so firmly dependent upon I-know-not-what quality in the writing of their leader that, coupled with the failure of others to dissect, they have shamefully reduced Galen's writings into brief compendia and never depart from him—if ever they understood his meaning—by the breadth of a nail. Indeed, in the prefaces of their books they announce that their writings are wholly pieced together from Galen's conclusions and that all that is theirs is his. ... So completely have all yielded to him that there is no physician who would declare that even the slightest error had ever been found, much less can now be found, in Galen's anatomical books, although it is now clear to me from the reborn art of dissection ...

A Revolution within a Revolution: The Copernican Revolution

There can be no denying that a major aspect of the Scientific Revolution was the reform of astronomy introduced by Nicolaus Copernicus (1473–1543) in his *De revolutionibus orbium coelestium* (*On the Revolutions of the Heavenly Spheres*, 1543). Although its impact was at first limited to the technical community of astronomers, it eventually led not only to a revised cosmology, but also to a revised physics. It is easy to see that this too followed upon the Renaissance voyages of discovery, and the re-discovery of ancient philosophy. The very first chapter of Copernicus' book affirms that the Earth is a single terraqueous globe, as described in Ptolemy's *Geography*; problematic as Copernicus' theory was for his contemporaries, he must have recognized that it was easier to claim Ptolemy's terraqueous globe was in continual motion than to claim that a sphere of water, in which the sphere of the Earth was floating with one hemisphere above the surface of the water, was in continual motion.

But if Ptolemy's *Geography* made it easier for Copernicus to defend his position, it was the re-discovered ancient writings that inspired him to forge that position. It was well known that the technical astronomical account provided in its most complete and useful form by Ptolemy in the *Almagest* (as it was known based on the title given by Islamic scholars from whom it came to the Latin West) was not really compatible with the nested homocentric spheres which constituted the Aristotelian cosmos. Ancient cosmology took it for granted that heavenly motions were perfectly circular, perfectly uniform (that is to say the motions maintain a constant speed), and centred on the Earth. In fact, as we now know (thanks to discoveries made during the Scientific Revolution), the planets speed up and slow down as they, along with the Earth, orbit the Sun, and their orbits are elliptical not circular. Ptolemy's mathematical ingenuity enabled him to describe a set of values defining the motions for each of the planets

Opposite: The University of Padua had the first purpose-built anatomy lecture theatre, opened in 1594. The steeply raked galleries enabled a reasonably close view of the proceedings even to a large audience. Andreas Vesalius had begun the practice (around 1540) of using dissection to demonstrate to students what could actually be seen in the human body (as opposed to using dissection merely as an opportunity to point to anatomical features described in the authoritative anatomy texts written by the ancient Greek writer Galen). Subsequent professors followed his lead, and that led to the building of this specialized lecture theatre. It can be seen as the embodiment of the shift in the Scientific Revolution from reliance on ancient authority to a new emphasis upon discovering knowledge through one's own observations and experiences.

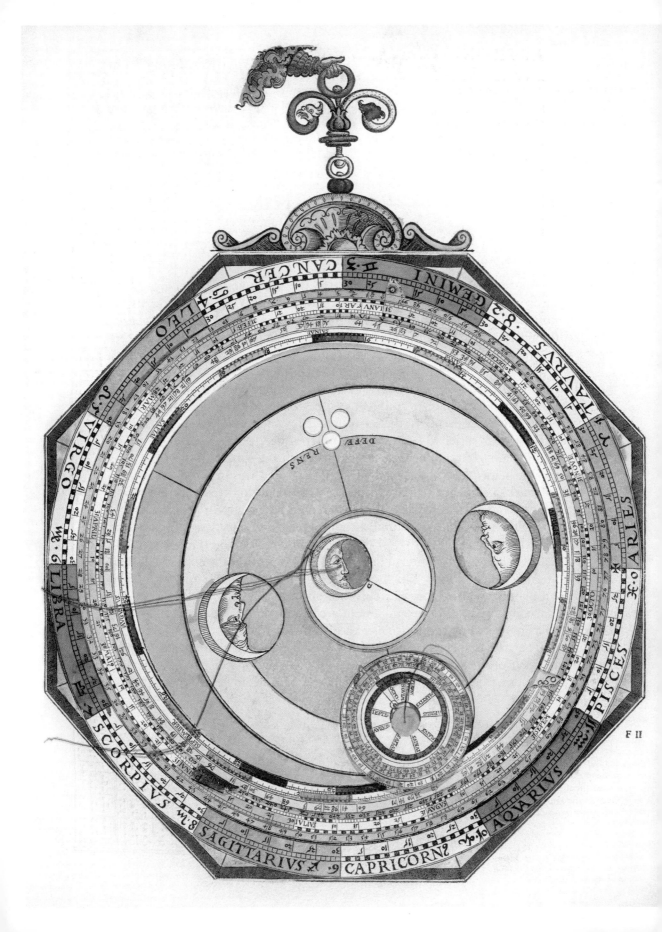

F II

which, for the most part, conformed to uniform circular motions. But the planetary motions had to be envisaged not as centred on the Earth, but as moving on imaginary circles, called epicycles, which in turn moved in circles around a larger circle, the deferent, which encircled the Earth, although centred on a point in space some distance away from the Earth.

It was by no means clear to astronomers, much less to philosophers, whether Ptolemy's eccentric wheels within wheels could be considered compatible with the simple homocentric cosmology. Furthermore, Ptolemy's account did not offer a coherent system, but merely a set of values for dealing with each planet in turn. For Copernicus, the result was what we might recognize as a Frankenstein's monster. It is, he wrote,

> just like someone taking from various places, hands, feet, a head, and other pieces, very well depicted, it may be, but not for the representation of a single person; since these fragments would not belong to one another at all, a monster rather than a man would be put together from them.

Copernicus took it upon himself to provide a new astronomy which was compatible with cosmology, but in so doing he had to change not only astronomy but cosmology as well. It seems clear that he took it upon himself to do this, at least in part, because he wished to restore what he saw in ancient writings as a close alliance between astronomy and cosmology, before Ptolemy forced them apart. We need not go into the complexities, and anyway Copernicus was not entirely successful because he too clung to the premise that heavenly motions must be uniform and perfectly circular, but by putting the Earth among the planets he arrived at a system in which the order of the planets from the Sun could be established by geometry, and that order coincided with the order suggested by the lengths of the planetary orbits (Mercury's being the shortest, Saturn's the longest). The order was not new (except for the fact that the Earth was now in between Venus and Mars, the position previously occupied by the Sun)— the length of orbit had been used by Ptolemy to provide the order of the planets, but this was merely conventional, and Ptolemy had no other means of confirming the order. The conformity of Copernican geometry with the conventional order was sufficient to convince Copernicus that his theory was correct. Few of his immediate contemporaries saw it the same way, however. After all, a geometrical nicety, essentially an aesthetic point about the conformity of geometry with tradition, is a small thing to weigh against the extraordinarily unlikely claim that the Earth is in motion.

Opposite: Peter Apian's folio volume, with hand coloured plates, was a collector's item from its first appearance. More than just a book, the plates included attached paper volvelles, which could move to demonstrate the motions of each of the planets. This page shows how the epicycle of the Moon moves on its deferent around the Earth. The 'heavenly spheres' in Ptolemaic (and Copernican) astronomy did not refer to the actual planets (the spots of light visible in the sky), but to the supposed spheres completely surrounding, and rotating around, the Earth. Each sphere had to be thick enough to accommodate the epicycle Ptolemy attributed to its particular planet. The epicycles of Venus and Mars were calculated to be very large, and so the spheres of these two planets had to be especially thick. It was only after the appearance of Johannes Kepler's *Astronomia nova* (1609) that planets came to be regarded as moving independently on orbits that were merely geometrically defined. Petrus Apianus (1495–1552), *Astronomicum Caesareum* (1540).

STELLÆBURGUM five OBSERVATORIUM SUBTERRANEVM, A TYCHONE BRAHE Nobili Dano: IN INSULA HVÆNA, EXTRA ARCEM URANIAM, EXTRVCTVM CIRCA ANNVM M D LXXXIIII.

Amstelædami, Joanne Blaeu excudebat

Tycho Brahe's Stellaeburgus or Stjerneborg (Castle of the Stars) as depicted in the *Atlas major* (1663) of Joan Blaeu (1596–1673). Between 1576 and 1580 the Danish astronomer Tycho Brahe built a major astronomical observatory on the island of Hven in the straits separating Denmark from Sweden. Although equipped with the best possible sighting instruments, their accuracy was discovered to be compromised by movements caused by wind and weather. Accordingly, Tycho immediately built a new observatory, on an adjacent site, which he called his 'Castle of the Stars'. Described as 'subterranean', the instruments were evidently mounted in pits and covered by shutters or rotating domes when not in use, and were able to provide the accuracy Tycho desired. Tycho could not accept the motion of the Earth, and so did not accept Copernican theory, but he knew that Ptolemaic astronomy was seriously flawed. He believed that the correct way to reform astronomy was to gather accurate observations of the movements of the heavenly bodies, and Stjerneborg was equipped with instruments specially designed to allow unprecedented accuracy. Johannes Kepler was later able to establish that Mars (and the other planets) did not move in perfect circles, but on elliptical orbits, by using Tycho's observations of Mars.

Ptolemaic astronomy was in such disarray by the sixteenth century (its inaccuracies having accumulated over the centuries) that astronomers eagerly embraced Copernicus' new system. For the most part, however, they continued to believe it was incompatible with Earth-centred cosmology, and saw it just as an improved mathematical model with no basis in truth. Even so, the new attitude toward observation, and testing age-old claims against physical reality, made other inroads into astronomy and cosmology. The leading astronomer, Tycho Brahe (1546–1601), believed that accurate observations would enable the reform of astronomy without having to assume the motion of the Earth. A wealthy member of the Danish nobility, he equipped his palace, Uraniborg, with impressive sighting devices and other astronomical instruments, and established himself as an observer of unprecedented accuracy. But he was also fortunate enough to witness the rare event of what is now called a supernova (an exploding star) in 1572. Visible even in daylight, the phenomenon had to be regarded by Aristotelian thinkers as atmospheric. But Tycho took it upon himself to measure the parallax of this new light in the sky—such measurements enable an estimate of the distance from Earth, and so could establish whether this was a sublunar, or a heavenly, phenomenon. Tycho was able to establish, beyond doubt, that this was indeed a new star in the heavens. He subsequently used the same techniques to establish that comets, which again he was lucky enough to be able to observe with the naked eye (telescopes were not yet invented), were also located above the Moon. These results were highly significant because they disproved the Aristotelian claim that the heavens are perfect and unchanging, and that, consequently, new stars are impossible, and comets, like meteors, must be atmospheric (or meteorological) phenomena, taking place below the Moon.

The culmination of this kind of astronomical observation was reported in Galileo's *Siderius nuncius* (*Message from the Stars*, 1610), following his use of the telescope, an instrument which had recently been invented in the Netherlands for commercial purposes. Galileo (1564–1642) reports seeing effects which suggested to him that there are seas, mountains, and valleys on the Moon, and that it is, therefore, not merely an aetherial light in the sky, but is a massive body exactly like the Earth. Galileo left unsaid the implication that if this massive body can be acknowledged by everyone to be moving through the heavens, then the same could be true of the Earth. Galileo also showed that Jupiter has its own satellites, just as the Earth has its Moon. Again, this was important because the Earth seemed anomalous in the Copernican system (in which everything went around the Sun, except for the Moon). Finally, Galileo showed that there are innumerable stars which are invisible to the naked eye and only become visible when viewed through a telescope. This addressed another objection to the Copernican theory. According to the new theory the stars must be inconceivably farther away from the Earth than in the old geostationary world system because they show no parallax—that is to say, they show no movement when observed six months apart, or when observed from opposite sides of the Earth's vast orbit. In response to this, supporters of Copernicus had already introduced the suggestion that perhaps the universe is infinite in extent, with the stars scattered through it; Galileo's telescopic view of many more stars than were visible to the naked eye gave more credence to this view.

Galileo took the Copernican theory out of the hands of a few specialist astronomers and a few maverick natural philosophers, and brought it to the attention of all

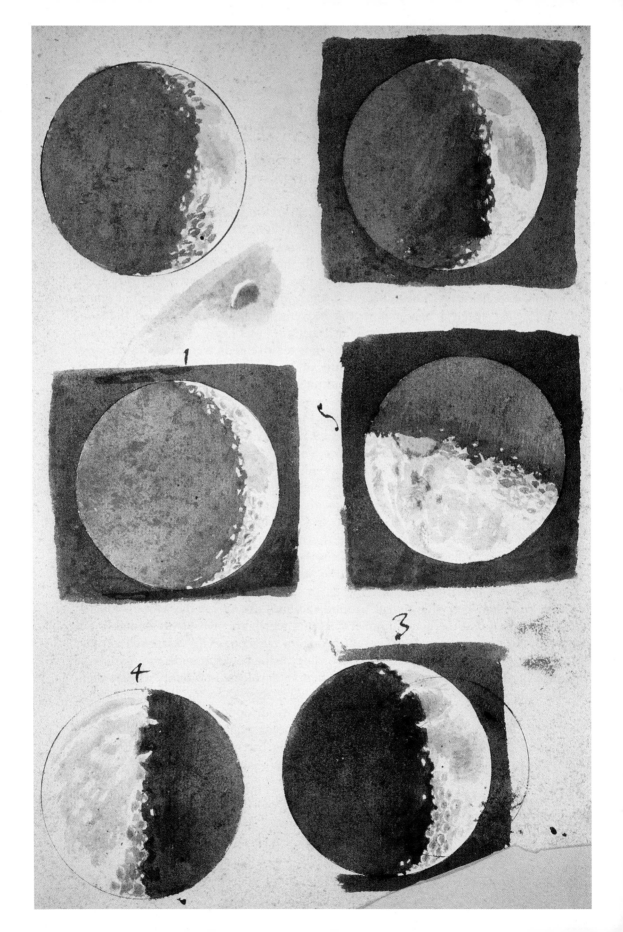

educated readers. Furthermore, his determination to show the truth of Copernican theory led him into conflict with his church, a *cause célèbre* which made the Copernican theory impossible to ignore. But Galileo's arguments in support of Copernicanism were complemented by the careful and ground-breaking mathematical astronomy of Johannes Kepler (1571–1630). Using Tycho's extremely accurate observations of Mars, and believing he should not be satisfied with approximate agreement (because he believed Tycho had been sent by God to help him reform astronomy), Kepler eventually broke with the tradition of celestial circularity and established that the planets move in ellipses, and speed up and slow down in a predictable way as they perform their orbits. The seal was finally set upon Kepler's work when Isaac Newton (1642–1727) showed, in his *Principia mathematica* (1687), that these planetary movements followed from the universal principle of gravitation.

Copernican theory had implications far beyond astronomy. We have already seen that it led to theories of infinite space, but this in turn led to new speculations about the possibility of void space, or vacuum. Aristotle rejected the existence of vacuum as a contradiction in terms, but the vastness of space between the orbit of Saturn, the outermost planet, and the nearest stars, which Copernican theory demanded, suggested that the space must simply be empty. Accordingly, the possibility, or not, of vacuum became a growth area in natural philosophy after Copernicus. The first air-pump, for creating artificial vacua, was invented in 1650 by the Burgomaster of Magdeburg in Germany, Otto von Guericke (1602–1686). Experiments with air-pumps soon took on a life of their own and led to major reforms of scientific knowledge in a number of areas, but the fact remains that this experimental research grew out of the Copernican revolution. Guericke makes it perfectly clear in his book, *Experimenta nova . . . de vacuo spatio* (*New Experiments on Empty Space*, 1672) that his main concern in beginning his experiments was to provide support for the Copernican theory. Remarkably, Guericke also invented the first generator of static electricity (a ball of fused sulphur which developed a static charge when rubbed) to provide further evidence in favour of Copernicus. Guericke showed how his sulphur globe attracted various small objects which then remained on its surface as the globe was spun round. 'Now we can see', Guericke wrote, 'how the sphere of our Earth holds and maintains all animals and other bodies on its surface and carries them about with it in its daily twenty-four hour motion.' The electrical phenomena Guericke described attracted immediate attention and, again, led to major new understandings of the natural world, but these new developments also arose initially from Copernicanism.

Opposite: Galileo improved upon the newly invented telescope, used to indentify ships coming into port, and turned it to the heavens. Here is a page from his notebooks where he first made ink sketches of his observations of the Moon (1609). Galileo interpreted the light spots just inside the dark part of the Moon as mountain tops, just catching the Sun's light. Similarly, he interpreted the dark marks just inside the illuminated part of the Moon as valleys still lying in shadow. The clear implication was that the Moon was not a perfectly spherical orb of a notional heavenly 'fifth element', but was just like the Earth. This and other telescopic discoveries that he made (many stars invisible to the naked eye, moons of Jupiter, phases of Venus, spots on the Sun) provided circumstantial evidence in favour of the Copernican theory, though they were insufficient to prove its truth.

The importance of the Copernican stimulus to a new understanding of space—infinite and empty—can be seen in Isaac Newton's concept of absolute space. Newton's laws of motion presented in his *Principia mathematica* are only valid on the assumption that there is a real, unchanging, non-interacting, infinite empty space which provides the arena in which all bodies move and act upon one another. Unlike Aristotelian space, in which there are so-called natural places for each of the five elements, and in which natural movements are completely different in different regions (straight line, up and down, natural motions below the Moon; circular natural motions above it), Newton's space has to be everywhere the same throughout its infinite expanse, undifferentiated and unaffected by the bodies in it, in short an absolute space. When Newton wrote his *Mathematical Principles of Natural Philosophy* he began by defining the technical terms and concepts he was about to use. In spite of the crucial importance of the notion of absolute space to his physics, however, he did not have to define it. 'I do not define absolute space', he simply wrote, 'as being well known to all.' By 1687 it was indeed well known (although not unanimously accepted—some natural philosophers still held out for a notion of space as only relative to the bodies occupying it), but only as a result of the up-take of the Copernican theory, and the demands of this system for a reform of previous views of space.

Moreover, the Copernican theory led to new work on the theory of motion. According to Aristotle, nothing can move unless it is moved by something, and if the mover ceases to operate, the motion will also cease. But if the Earth is in motion, as Copernicus says, what keeps it in motion? This became a major problem for supporters of Copernicus. William Gilbert (1544–1603), a London-based physician and would-be reformer of natural philosophy, hit on the idea of using magnetism to explain the motion of the Earth. Realizing from the work of earlier students of magnetism that the Earth itself was a giant magnet, Gilbert argued that since magnets can spontaneously move themselves, they must have a soul—Aristotle himself had conceded that animate things could move themselves—and if the Earth is a magnet it must have a soul, and so can move itself. Gilbert's animistic ideas were embraced by later thinkers, though usually only after turning his notion of the magnetic soul into some sort of magnetic force or principle (no longer seen as animate, merely occult). Kepler adopted this idea to explain how planets can move in ellipses, and his theory of a magnetic force operating between the Sun and the planets was easily replaced in Newton's *Principia mathematica* by the attractive force of gravity. Galileo, meanwhile, was developing his own account of how the Earth kept in motion, and his theory can be seen to lead, culminating again in the work of Newton, to the principle of inertia. This principle rejects the Aristotelian necessity for a continual mover, arguing that once movement is initiated it will continue until something intervenes to stop it. Galileo believed that such perpetual motion needed to be circular—once something is set in motion in a circle it will continue to move that way indefinitely. But Descartes and others pointed to motion in a straight line as the natural form of motion, and Newton set the seal upon it, and simultaneously showed that planetary motions could only be deviated from motion in a straight line and bent around into an ellipse as the result of a continuously acting force—gravity. Although Galileo was wrong about inertial motion, he had much better luck with his associated experimental investigations of free fall, and was able to replace the Aristotelian assumption that bodies fall faster the heavier they are with his law of free fall which states that all bodies fall (disregarding any interference from the medium through which they fall) at the same rate.

Replacing Aristotle: From Occult Philosophies to the Mechanical Philosophy

Copernicanism was not the only innovation to have such wide-ranging consequences. Dissatisfaction with the Aristotelian system, accumulating as a result of scholarship based on ancient alternative philosophies and increasing discoveries that revealed the fallibility of his system (from confirmation of the terraqueous globe to the appearance of the new star in 1572, and so forth), led to increasingly cynical dismissals of his natural philosophy.

This included a rejection of his hylomorphic theory of matter and the concomitant attempt to explain all material phenomena in terms of the four (so-called) *manifest* qualities of hot, cold, dry, and wet. Hylomorphism refers to the Aristotelian assumption that bodies are made of matter (*hyle*) and form (*morphe*)—matter cannot exist without form (try imagining a lump of matter which has no shape whatsoever); and it makes no sense to talk of the form of nothing, so form has to be imposed on some matter. To explain the countless different varieties of body (from metal, to wood, to fluff, or vapour and beyond), Aristotle took a reductionist line and assumed that all could be explained in terms of different combinations of four elements (earth, water, air, and fire). These four elements embodied four fundamental, irreducible, qualities (fire was hot and dry; earth, cold and dry; water, cold and wet; air, hot and wet). Aristotle wished to explain all the changes of bodies in terms of these four manifest qualities—that is to say, four qualities which were obvious and easily detected by the senses (especially by touch). Unfortunately, Aristotle was forced to recognize that not all properties of bodies could be reduced to these four qualities. Although something smooth, for example, could be assumed to have water in its composition—since wetness seemed to the touch to correlate with smoothness; and a drink that made one hot must have air or even fire in its composition—what could we say about a magnet's ability to attract a piece of iron? This ability did not seem to be explicable in terms of any of the four qualities. Wetness might account for stickiness, and so the adherence of the iron to the magnet might be seen as the result of a binding wetness, but what made the iron move toward the magnet? In some cases fire seemed to draw things toward it—air and other light things—but no fire could pull in a piece of iron. There was nothing for it but to accept the evidence provided by experience that magnets had an *occult* quality, and this quality could not be reduced to the manifest qualities.

Aristotle, and his medieval scholastic followers, avoided recourse to occult qualities as much as possible—seeking to explain everything in terms of the manifest qualities. But Renaissance philosophers found themselves increasingly resorting to occult

qualities. Medicinal plants were said to operate by their manifest qualities. Some plants were cooling (either when eaten, or when applied as a poultice) and could be used to combat fever; others were sudorific or diuretic and could be used to dry out someone who was judged to be suffering from an excess of moisture in their system. But plants that were known in the local lore of a particular region to have healing effects, but which were not mentioned by Aristotle or his ancient follower and specialist in plants, Theophrastus (*c.*371–*c.*287 BC), or the later Dioscorides (*c.*AD 40–90), could not always be seen to work by one of the manifest qualities. This was exacerbated by the increase of medicinal plants brought into Europe from the New World. Where there was no ancient authority to indicate how a plant exerted its healing effect, natural philosophers increasingly resorted to occult qualities.

Subsequently, the most ambitious Renaissance philosophers tried to develop their own systems of natural philosophy, which they hoped would replace the increasingly untenable Aristotelian system. At first, these would-be replacement systems all relied heavily upon occult principles. Presumably, these innovators had all noticed the increasing usefulness of occult qualities, and saw them as an under-exploited aspect of Aristotelianism, which potentially could be used as the basis for a new philosophy. This new attitude to the occult was undoubtedly encouraged by the re-discovery of ancient writings attributed to the Greek god, Hermes Trismegistus (who was assumed to be a real sage, whose wisdom had been acknowledged to be so great that he had been deified by the ancients). Although essentially Neoplatonic, and therefore rather religious in tone, these re-discovered writings were linked to various alchemical and other magical texts which were also attributed to Hermes Trismegistus. The result was that secular Renaissance thinkers were able to overcome the strictures of the Church against magic (which for the Church was always associated with the activities of demons), and claim that *natural* magic (a magic based not on demonic activity but on the natural occult powers of things—occult powers bestowed upon these things by God at the Creation) was part of the oldest wisdom known to man. Underlying these claims was a belief that Adam had known all things, and that after the Fall, his wisdom was successively forgotten from generation to generation. But the further back in time the scholar could go, the closer he came to recovering Adamic wisdom. Hermes was widely seen, in the Renaissance, as a contemporary of Moses, and a pagan sage who had not yet forgotten everything Adam had known.

Opposite: Hermes Trismegistus depicted on the pavement in Siena Cathedral (1488). Various unknown Neoplatonic writers of late antiquity tried to give authority to their writings by claiming they were written by 'Thrice-great Hermes', the Greek messenger god. A number of alchemical works were attributed to him, and then, in the Renaissance, a number of spiritual or religious works, also supposedly by Hermes, were discovered. The latter were post-Christian and alluded to some Christian beliefs (most notably the Holy Trinity), but because Renaissance scholars assumed Hermes must have been a contemporary of Moses, these Hermetic writings seemed to show the religious prescience of this pagan 'god' (regarded by the Christians as a deified sage). The result was a belief in an ancient uncorrupted wisdom, shared by pagans as well as those in the Judaeo-Christian tradition, and ultimately deriving from the wisdom of Adam, the first man. The Hermetic writings came to be seen as the key to recovering this ancient wisdom, and led to a belief that magical traditions were part of this ancient wisdom.

The first translator of the Neoplatonic Hermetic writings from Greek into Latin was Marsilio Ficino (1433–1499), who developed his own occult system of philosophy, *De vita* (*On Life*, 1489). This proved to be extremely influential and inaugurated a vogue among the most ambitious Renaissance thinkers for developing alternative systems of philosophy. At first all those who attempted to develop alternative new philosophies relied to a greater or lesser extent on occult or magical approaches. The leading figures here include Giovanni Pico della Mirandola (1463–1494), Pietro Pomponazzi (1462–1525), Cornelius Agrippa (1486–1535), Paracelsus, Girolamo Fracastoro (c.1477–1553), Jean Fernel (c.1497–1558), Girolamo Cardano (1501–1576), Bernardino Telesio (1509–1588), Francesco Patrizi (1529–1597), Giordano Bruno (1548–1600), William Gilbert (1544–1603), Francis Bacon (1561–1626), and Tommaso Campanella (1568–1639). Jean Fernel took up a position that was close to Aristotelianism but with a much greater role allowed for occult qualities. Francesco Patrizi, by contrast, rejected Aristotelianism almost entirely and developed a philosophy which was much closer to Neoplatonism. Others took up positions somewhere along the spectrum between occult Aristotelianism and more Ficinian Neoplatonism. Although some, such as Fernel and Paracelsus, proved influential and gathered followers, none of them were able to persuade the majority of their educated contemporaries that they had arrived at the true philosophy, worthy of replacing Aristotelianism.

Then, at the beginning of the seventeenth century, a new and powerful alternative to occult philosophies emerged. It is possible this radically different approach arose from modes of thought that were comparatively common among mathematical practitioners, but the state of historical research at present only allows us to say that two leading mathematical thinkers independently developed an entirely kinematic approach—that is to say, they tried to explain all physical phenomena exclusively in terms of bodies in motion. These two radical innovators were Galileo Galilei and René Descartes (1497–1650). There was a third innovator who developed a kinematic physics, Isaac Beeckman (1588–1637), and he actually introduced Descartes to this new way of doing physics, but unlike Descartes, he never fully developed his natural philosophy and never published his ideas, so he had no subsequent effect on developments (except indirectly, through Descartes). A fourth thinker who tried to develop an entirely kinematic physics was Thomas Hobbes (1588–1679), but his system of physics was a derivative combination of ideas from Galileo and Descartes.

Galileo was as robustly opposed to magical ways of thinking as he was to Aristotelianism, and it looks as though he conceived the possibility of developing a kinematic physics when he realized that Copernicus' moving Earth might provide an explanation of the tides—an explanation which did not involve the occult influence of the Moon. The association between the Moon and the tides had been known since ancient times and was assumed to be due to occult influence. A moving Earth, however, suggested to Galileo that the oceans might be slewing about as the Earth rotates (the details are complex but Galileo drew an analogy with the fresh water brought across the lagoon to Venice in barges—if a barge's motion changed, the water would shift and rise up either at the front or the back of the barge). Galileo's investment in this idea was so great that he even saw the tides as proof that the Earth must be in motion, as Copernicus said. Accordingly, he made his theory of the tides the mainstay of his attempt to prove the truth of Copernican theory in his *Dialogue on the Two Chief World Systems* (1632)—this was the same work in which he introduced the idea, mentioned earlier, that once a body is set moving in a circle it will continue to do so indefinitely, unless something stops it (again, this was crucial for his kinematic physics). In his final work, *Discourses on Two New Sciences* (1637), he developed his kinematics further, and even contrived to explain acceleration in free fall without recourse to the occult idea of gravity (by restricting discussion of acceleration due to gravity to the fact that 'equal increments of speed' are given to the body—there is no discussion as to why the speed increases).

Meanwhile, the French mathematician, René Descartes, had developed a much more comprehensive and systematic kinematic physics. Descartes was about to publish this in 1633 when he heard of the condemnation of Galileo by the Inquisition for his promotion of Copernicanism. Since Descartes's system also depended on the assumption that Copernicus was correct, he suppressed his work, and his fully developed system was not published until 1644, as *Principia philosophiae* (*Principles of Philosophy*). Descartes's system is rich and complex, but in its essentials it combined an atomistic philosophy with a physics based on the centrifugal forces generated when a body is made to rotate (think of whirling a stone around in a sling—the stone tends to move away from the centre of rotation). Atomism was one of the ancient philosophies that had been rediscovered by Renaissance scholars, and sought to explain everything in terms of atoms moving, colliding, combining, and dispersing. This may seem close to kinematics, but the most prominent of the revivers of atomism, Pierre Gassendi (1592–1655), for example, assumed that atoms had their own internal energies and principles of movement. Accordingly, Gassendi has been seen as closer to the tradition stemming from Ficino, and his atomism has distinct occult elements built into it. This was not the case with Descartes; if he included unexplained principles of activity in his earliest thinking, he was soon able to excise them and develop a completely kinematic system in which everything was explained in terms of the motions of invisibly small particles, their collisions, and the transfer of motion from one particle to another in those collisions. It was with regard to the transfer of motion that Descartes developed his three laws of nature (the third of which was supplemented with seven rules of collision)—the first explicit statement of precise laws which could then be used to analyse and predict the behaviour of inanimate bodies.

Galileo's and Descartes's attempts to develop systems of philosophy which did not rely on occult principles were immediately recognized as potentially useful, and proved highly influential. Cartesianism came to be known as the mechanical philosophy because all physical change was explained in terms of the intermeshings, friction, and collisions of material parts. Descartes even extended this to living things, so that plants and animals (and even human bodies—though he did acknowledge the existence of an immortal soul in humans) were seen to be like automata. This represented a

major change in theories of life. For Aristotle only living things were capable of self-movement (including growth in the case of plants), and this ability denoted the presence of a soul in the self-moving, living thing (albeit merely a vegetative soul, or an animal soul). But Descartes pointed out that clocks and other automata (he lived in an age when ingenious craftsmen had been able to make remarkable automata) could move themselves, and yet nobody inferred from their movements that they had souls. Descartes dismissed all but the rational, immortal soul of human beings (which he retained on religious grounds), and living things were as much machines as anything else in his mechanical philosophy.

Descartes's system was so fully worked-out, so comprehensive in its coverage (in principle at least), that it began to supplant Aristotelianism in universities—especially in the Netherlands. In his native France, Cartesianism was perceived as a threat to sound religion and for a while it was proscribed by the Crown, but it seems fair to say that the succeeding generation of French natural philosophers, and in particular those who constituted the newly founded Académie des Sciences in 1666, were predominantly Cartesian. This is not to say that the flaws in Descartes's system were not noticed (the system was more ingenious than it was workable), but it was undeniably the only system of philosophy which was capable of replacing the fully comprehensive Aristotelian system, lock, stock, and barrel—no other thinker presented a system that even came close. At a time when Aristotelianism was regarded as no longer tenable, Cartesianism was widely regarded as the only philosophical system capable of taking its place.

In Britain, the Cartesian system did not fare so well. Natural philosophy in England was strongly affected by the philosophy of Francis Bacon, the first major philosopher to appear in England since the Middle Ages. Bacon tried to develop his own occultist (it has been described as semi-Paracelsian) natural philosophy but his main impact derived from a programme he developed for reforming natural philosophy. It was always recognized that occult qualities could only be discovered, or understood, as a result of experience: nothing about the appearance of a magnet enables us to predict its effect on iron, but we learn it easily when we see the magnet interacting with a piece of iron. Some of the earlier thinkers in the occult tradition, such as Paracelsus, Fernel, and Cardano, had been explicit about the need to base natural philosophy on experience (and of course this went hand in hand, as we have seen, with the anti-authoritarian aspects of Renaissance thinking), but Bacon was the first philosopher

Opposite: Astronomical Clock Prague Old Town Hall (1410). As well as showing the position of the Sun in the zodiac, the phases of the Moon, and various other astronomical and calendrical details, this clock provides an hourly spectacle of the twelve apostles, death, and other figures moving by clockwork. Another astronomical clock, embellished with various automata, was completed in 1574 at Strasbourg Cathedral. When the so-called mechanical philosophy, developed by René Descartes and his followers, began to replace Aristotelian natural philosophy, these clocks and other automata came to be seen as crude imitations of the 'clockwork universe' made by God. For Robert Boyle, writing in 1685, the world was 'like a rare Clock, such as may be that at *Strasbourg*, where all things are so skilfully contriv'd, that the Engine being once set a Moving, all things proceed according to the Artificers first design' (*Works of Robert Boyle*, vol. 10, p. 448).

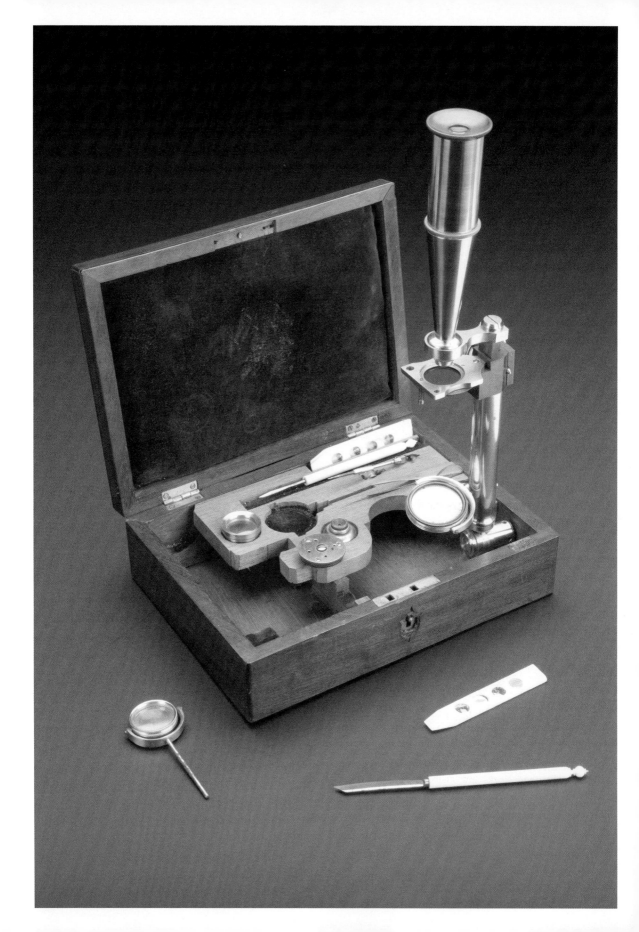

to provide a philosophical defence of experience and a reasoned account of the importance of what became known as the experimental method. This method, and the Baconian approach more widely, was taken up by the succeeding generation of English thinkers, and the new philosophy as it developed in England came to be called the experimental philosophy.

Descartes's system was for the most part presented as a rationalist philosophy, advancing logically from initial premises. When Descartes did present seemingly experimental evidence, he interpreted it in a way that suited his preconceived doctrines. He appropriated Harvey's experimental demonstration of the circulation of the blood, for example, but made it serve his own mechanistic conclusion, rather than the more occultist conclusion of Harvey. Harvey's experiments had clearly shown that the contraction of the heart was its active stroke, and concluded the heart had an unexplained ability to repeatedly contract. This was too occult for Descartes, who insisted the heart was so hot it could explosively vaporize incoming blood from the lungs, thereby inflating the heart and sending the blood out into the great artery. It followed from this that the *expansion* of the heart must be its active stroke, and Descartes implied that is what Harvey's experiments showed. Accordingly, English readers of Descartes tended to be suspicious of his work and continued to offer their own experimentally based version of natural philosophy. It was important to English thinkers that their experiments could be presented as being performed in a theory-free way (without theoretical preconception), just as the experimental method had been developed by Bacon in his *Novum organum* (*New Organon*, 1620, echoing the traditional title given to Aristotle's collected works on logic—*Organon*).

The culmination of the English approach can be seen in Isaac Newton's *Philosophiae naturalis principia mathematica* (*Mathematical Principles of Natural Philosophy*, 1687). Newton's incomparable abilities as a mathematician enabled him to explain the motions of the planets, as determined by Kepler, on the assumptions that planets seek to move indefinitely in straight lines (the principle of inertia), but are continually attracted to the Sun with a force that varies inversely as the square of the distance between the Sun and the planet (the universal principle of gravitation). Accepting the Baconian tradition, Newton felt no compunction to provide the kind of mechanistic account of how gravity operated on the planets that Cartesians demanded. Cartesians explained gravity in terms of the downward push of continually descending invisibly small particles, but for Newton these supposed streams of particles were hypothetical entities with no evidence that independently verified their

---

Opposite: While experimenting with lenses to create his telescope, Galileo also developed an instrument for magnifying the very small (1609). His *occhiolino* was renamed 'microscope' by later members of the Accademia dei Lincei ('Academy of the Lynx-eyed'), a scientific society of which Galileo was a member. The microscope came to prominence later in the century when a number of discoveries were made, and various works of microscopy were published (for example Robert Hooke's *Micrographia* of 1665). The revived atomist philosophy, and the mechanical philosophy, promoted its use. Henry Power, in his *Experimental Philosophy* (1664), even hoped that microscopes might be improved so that we could see 'the constant and tumultuary motion of the Atoms'.

existence. Newton's gravity, like the attractive power of the magnet, was simply confirmed on the one hand by everyday experience, and on the other by the fact that the motions of bodies affected by gravity could be analysed and predicted in terms of Newton's mathematical laws of nature (which replaced Descartes's three laws, and proved much more useful). 'And it is enough', Newton wrote, 'that gravity really exists and acts according to the laws that we have set forth and is sufficient to explain all the motions of the heavenly bodies and of our sea.'

It is important to note, however, that English natural philosophers did not have a monopoly on experimentalism. Galileo, whose studies of motion were conducted with carefully designed experiments rolling balls down inclined planes, or with experiments with pendulums, for example, helped to establish a strong experimental tradition in Italy. This is most easily seen in the work of the Accademia del Cimento, founded in 1657 by a number of Galileo's followers, as an institute for pursuing experimental philosophy. Even in the Netherlands and France, where Descartes's rationalist philosophy had most success, experimental approaches were used to test his claims. One of the most famous and far-reaching examples of this was the set of experiments performed by Blaise Pascal (1623–1662) to establish, contrary to Descartes, the possibility of void space, and the role of atmospheric pressure in various phenomena. Following on from the claims of Evangelista Torricelli (1603–1647), a student of Galileo's, that a column of mercury in a glass tube whose lower end is immersed in a bath of mercury is held up by the pressure of the atmosphere on the surface of mercury in the bath, Pascal reasoned that the lesser atmospheric pressure at the top of a mountain would only be capable of supporting a shorter column. Pascal's hypothesis was confirmed by conducting experiments during an ascent of the Puy de Dôme in central France, and he went on to assert that the space above the mercury in the tube was a vacuum, publishing the results of this and other experiments in *Experiences nouvelles touchant le vide* (*New Experiments Concerning the Void*, 1647). Experiments on void space were subsequently devised and conducted at the Royal Society by Robert Boyle (1627–1691) and Robert Hooke (1635–1703), using an air-pump that was an improvement on Otto von Guericke's, so it is clear that, from Torricelli to the Royal Society via Pascal and Guericke, experimentalism was a European movement.

Attempts to replace the Aristotelian system, then, began with recourse to occult notions and philosophies of nature that were based to differing extents on assumptions about the occult qualities and powers of things. From Fracastoro's *De sympathia et antipathia rerum* (*On the Sympathy and Antipathy of Things*, 1546) to Newton's declared aim in the Preface of the *Principia* to explain all phenomena in terms of 'certain forces by which the particles of bodies, by some causes hitherto unknown, are either mutually impelled toward each other . . . or are repelled and recede from each other', natural philosophers tended to assume that bodies had unexplained (though God-given) principles of activity which accounted for many of the phenomena of nature. Within this larger picture we can see the efforts of Galileo, Beeckman, Descartes, and Hobbes to rescind all use of unexplained powers and principles of activity and to invoke only passive bodies in motion as their explanatory principles. This exclusively kinematic movement ultimately failed (although Descartes had many followers into the eighteenth century), but it was powerful enough to change the character of the more occult and dynamic natural philosophies. Where newly proposed natural philosophies before Galileo and Descartes were overtly occult, afterwards the new philosophies tended to be dynamic versions of mechanistic philosophy, in which assumptions about the unexplained powers and active principles of things were justified in terms of experience. Although writing in the Preface to the *Principia* about gravitational attraction, an action at a distance that neither Galileo nor Descartes could have accepted for a moment, Newton still felt able to present it as part of mechanical philosophy: 'I wish we could derive the rest of the phaenomena of Nature', he wrote, 'by the same kind of reasoning from mechanical principles.'

The New Philosophies and Society: Patronage and Pragmatism

Intellectual developments, no matter how recondite and abstruse, do not take place in a vacuum, with no reference to anything but other intellectual developments. In the foregoing account, even though we were focusing primarily upon intellectual developments, to understand those developments we had to allude to the wider context, starting with the broad context of the set of historical changes known as the Renaissance. But there were a number of more specific social and political changes which made the Scientific Revolution what it was.

If the rise of magic was made possible by its newly acquired respectability after the recovery of the Hermetic corpus, its adoption in practice owed more to its promise of pragmatic usefulness than to any Hermetic doctrines. The same concern for the pragmatic uses of knowledge can be seen in the increasing attention paid by scholars and other elite groups to the techniques and the craft knowledge of artisans. The increasing importance of mining and metallurgy in the economy of Europe, for example, drew the attention of practically minded intellectuals. The first printed account of Renaissance mining techniques, including instructions on the extraction of metals from their ores, how to make cannons, and even how to make gunpowder, was the *De la pirotechnia* (1540) of Vanuccio Biringuccio (1480–1539). Written in Italian by a mining engineer who rose to the rank of director of the papal arsenal in Rome, it was evidently intended as an instruction manual for others working in similar circumstances to Biringuccio himself. This can be compared with the *De re metallica* (1555) of Georgius Agricola (1494–1555). Agricola was a humanist scholar who taught Greek at Leipzig University before turning to medicine. Practising in a mining area, and initially interested in the medicinal uses of minerals and metals, he soon developed a compendious knowledge of mining and metallurgy. The fact that the *De re metallica* was published in Latin shows that it was aimed at an audience of university-trained scholars, not at miners or foundry-workers. Furthermore, the book's numerous editions and wide dissemination throughout Europe show that Agricola did not misjudge the audience.

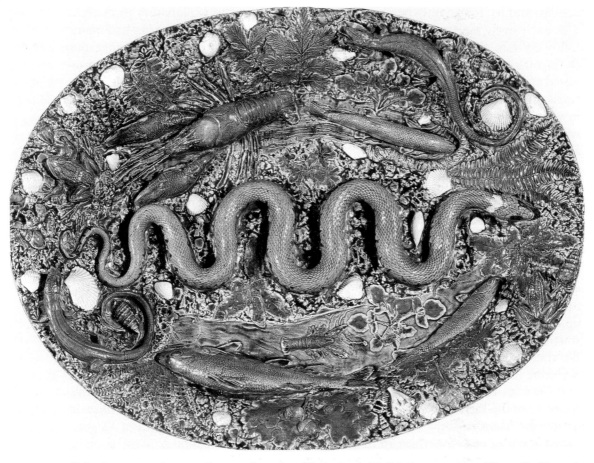

Palissy's success and renown (he was a favourite of Catherine de Medici, Queen consort of Henry II of France), can be seen as indicative of the move towards practical and pragmatic knowledge in the Scientific Revolution, and in the increased acknowledgement that artisans and craftsmen could contribute to learning. His rusticware plates show his own interest in natural history, and the corresponding interest of his wealthy customers. He was also able to give public lectures to wealthy patrons in Paris, drawing upon the knowledge of minerals, rocks, soils, underground water sources, and other subterranean phenomena which he had acquired in pursuit of improving his craft as a potter. He insisted that theory must be based on proven practice, and was able to expose a number of established theories as false.

Similar stories could be told in other areas. Although only a potter, Bernard Palissy's (1509–1590) efforts to reproduce Chinese porcelain made him famous. He was able to give public lectures in Paris on topics in what we would think of as mineralogy, geology, hydrology, and agriculture. In 1580 he published his fund of knowledge in *Discours admirables*, never missing an opportunity to extol the virtues of practice over theory. The introduction of the magnetic compass, one of the major practical innovations of the Renaissance to rank alongside printing and gunpowder, effectively led to a new science of magnetism. It was noted that compass needles did not simply point north, and various attempts were made to put their other movements, notably variation and declination, to use for determining longitude, or latitude when cloud or fog obscured the heavens. Here, for example, the discovery of declination by a retired mariner and compass-maker, Robert Norman (*fl.* 1590), was taken up and expounded by William Gilbert in the first thorough study of electrics and magnetism, *De magnete* (*On the Magnet*) of 1600.

Other scholars remained content with talking in general terms of the importance of craft knowledge. The Spanish humanist and pedagogue, Juan Luis Vives (1492–1540), acknowledged the importance of trade secrets in his encyclopaedia, *De disciplinis* (*On the Disciplines*, 1531). Francis Bacon, similarly, wanted to include the knowledge and techniques of artisans in a projected compendium of knowledge which was to form part of his *Instauratio magna* (*Great Restoration*), a major reform of learning. Bacon's influence in this regard can be seen not only in various groups of social reformers in England during the Civil War years and the Interregnum, but also in the Royal Society of London for the Promotion of Useful Knowledge, one of the earliest societies devoted to acquiring and exploiting knowledge of nature. The Society made a number of repeated attempts, using specially produced questionnaires, to ask its members to return information about local craft techniques and artisans' specialist knowledge in and around their places of residence. The idea was to produce a 'history of trades' to supplement the usual natural histories.

We have already mentioned the role of secular patrons in stimulating the arts, and in effectively promoting the re-discovery of ancient philosophies other than Aristotle's, by establishing personal libraries and employing scholars to build them up. During the Middle Ages the only patron for major paintings was, effectively, the Church, and this inevitably showed in the subject-matter and style of the resulting paintings. Secular patrons, however, wanted (sometimes at least) secular subject-matter and more realistic representations—and the resulting effect on the art of the Renaissance is wonderfully obvious. But secular patrons also affected the way the natural world was studied, and how that knowledge was put to use. Essentially, the concern of the secular patron was with the pragmatic usefulness of knowledge, although in some cases that usefulness might amount to nothing more than the aggrandizement of the patron—a confirmation of his wealth and power.

The earliest groupings of investigators of nature all seem to have been brought together by wealthy patrons, particularly by sovereigns and princes. Indeed the royal courts must have been one of the major sites for bringing together scholars and craftsmen, which we have already seen was one of the characteristic features of the Scientific Revolution. The amazingly elaborate court masques and festivals conceived in order to publicly display the magnificence and glory of the ruler required a huge team of facilitators. Learned scholars would devise appropriate themes, combining traditional notions of chivalry and honour with more fashionable lessons taken from newly re-discovered classical stories, while architects and engineers would design the elaborate settings intended to illustrate the moral themes and a vast array of other artisans and craftsmen would be brought together to make it all a breath-taking physical reality. It is hard to imagine a comparable site during the period for the

The Ambassadors (1533) by Hans Holbein (*c.*1497–1543). This portrait of two ambassadors to the court of Henry VIII includes highly accurate and detailed depictions of various scientific and mathematical instruments. The symbolism of these items is undoubtedly complex and multi-layered, but it is clear that contemporary viewers of the painting would be expected to recognize these instruments as indicative of new ways of thinking about the world and the heavens, based on observation and measurement rather than ancient authority. The books depicted on the lower shelf, near the lute, are a work by the Protestant reformer, Martin Luther, and an instruction manual on commercial arithmetic by Peter Apian (1527)—modern works, indicative of recent changes. Even the anamorphic skull prominent at the front of the painting, and which can be seen correctly when viewed from a particular position, shows the artist's command of mathematics. The perspective in the painting is perfectly rendered (consider, for example, the tiled floor, and the lute), but the artist can go even farther and construct an image which can only be properly viewed from a specific viewing point.

creative collaboration of scholars and craftsmen. Unless, of course, it was one of the many sites where the arts of war demanded such collaborations.

If festivals and wars were only occasional affairs, the offer of more long-term patronage to alchemists and other natural magicians, engineers, mathematicians, natural historians, and natural philosophers was obviously done with the aim of increasing the wealth, power, and prestige of the patron. Usually this meant that the patron was most concerned with some practical outcome from the work of these servants of his court. Even in the case of seemingly more remote and abstract physical discoveries, it is possible to see such practical concerns in the background. When Galileo, professor of mathematics at the University of Padua, discovered the moons of Jupiter with his telescope and named them the Medicean Stars, after the ruling Medici family of Florence, he was bargaining for patronage by offering celestial and quasi-divine significance to Duke Cosimo, as well as putting him onto the star maps. But he did not stop there. By trying to produce tables of the motions of the moons of Jupiter, which he hoped would provide a means of determining longitude at sea, Galileo was potentially turning his discovery into one of the utmost practical benefit, from which the Medici could hardly fail to gain.

Mathematical practitioners in general tended to benefit from the pragmatic interests of secular patrons. The result was a rise in social and intellectual status which enabled mathematicians to make bolder claims for their subject than was previously possible. Mathematics was held in low regard by Aristotle, who insisted that natural knowledge must be grounded on causal explanations—if the causes of a phenomenon were known then that phenomenon was properly understood. Mathematical analysis of a particular phenomenon or set of phenomena could not offer the kind of causal explanations demanded by Aristotle and his later scholastic followers and so was regarded as of little use to contemplative natural philosophy. The more pragmatic concerns of secular patrons, however, could often benefit from mathematical analysis. Furthermore, mathematicians could also legitimately claim that their results were usually certain and therefore reliable—which could seldom be said for results acquired through Aristotelian speculation.

A new confidence in mathematics can be seen, for example, in Copernicus' presentation of his new astronomical and cosmological theory. Although Copernicus could offer no explanation at all as to how the Earth could move, he was convinced nevertheless of the physical truth of his theory. It was as if Copernicus was willing to accept the truth of his theory not, as an Aristotelian would have demanded, on physical and causal grounds, but simply because the mathematics pointed to its truth (by providing a geometrical demonstration of the order of the planets which fitted the order suggested by planetary periods, and by explaining how all the planets seemed to incorporate in their own motions the annual motion of the Sun—since this was actually the annual motion of the Earth, projected, as it were, onto each planet). When Copernicus published his *De revolutionibus* in 1543, not all mathematicians had such confidence in the power of their own discipline, but by the time Newton published his demonstration of the *Mathematical Principle of Natural Philosophy* in

1687, the battle had been won, and mathematical approaches to an understanding of the physical world have been regarded as essential ever since. Although Copernicus was not himself in the employ of a wealthy patron with pragmatic concerns, it seems hard to deny that his confidence in the power of mathematics was affected by the general increase in status of mathematics and mathematicians that was a major outcome of secular patronage in the Renaissance.

The political potential of natural knowledge was a major reason for Francis Bacon's concern to reform the means of acquiring knowledge and of putting it to use, as described in his various programmatic statements and illustrated in his influential utopian fantasy, *New Atlantis* (1627). The most prominent feature of Bacon's utopia is a detailed account of a research institute, called Salomon's House, devoted to acquiring natural and technological knowledge for the benefit of the citizens. Bacon repeatedly sought patronage for his so-called 'Great Instauration', first from Elizabeth, then from James I, although neither were to agree to it (even though Bacon did become, briefly, James's Lord Chancellor). But Bacon's posthumous fame was such that Charles II of England and Louis XIV of France did recognize the political potential of enhanced knowledge of the natural world and offered their patronage to what were to become the leading scientific societies in Europe, both of which were explicitly modelled on Salomon's House. In the French case at least, thanks to Jean-Baptiste Colbert (1619–1683), the controller general of finance, the Académie Royale des Sciences (1666), with its salaried fellows, can be seen effectively as an arm of the State. The Royal Society, founded in the year of the Restoration of the English monarchy (1660), never gained more than nominal support from an administration that was preoccupied with more pressing matters. It had to be much more apologetic, therefore, in its attempts to demonstrate its usefulness to the State. Even so, it can be seen from the propagand-izing *History of the Royal Society of London* (1667), by Thomas Sprat, and other pronouncements of the leading Fellows that the most committed members of the Society, at least, saw their experimental method as a means of establishing truth and certainty and so ending dispute in philosophy. This, in turn, was presented as a model which could be used to bring an end to the religious disputes which had divided England since before the Civil Wars, and to establish order and harmony in the State. The existence, to say nothing of the success, of the Académie and the Royal Society shows that the new natural philosophy was far more directly concerned with political matters than the natural philosophy of the medieval period.

But these were by no means the only new institutions devoted to the study of the natural world. Indeed, Bernard de Fontenelle (1657–1757), secretary of the Académie Royale des Sciences from 1697, referred to a 'new Age of Academies'. In some cases the group was called together by a wealthy patron with an interest in natural knowledge and its exploitation. One of the earliest of these was the group of alchem-ists, astrologers, and other occult scientists brought together at the court of Rudolf II (1552–1612) in Prague; another was the Accademia dei Lincei (Academy of the Lynxes), founded by the marchese di Monticello, Federico Cesi (1585–1630). The evident attractiveness of such collaborative enterprises can also be seen in the astonishing interest shown by scholars all over Europe in the Brotherhood of the Rosy Cross, whose intended reforms of learning, based on alchemy, Paracelsianism, and other occult ideas were announced in two manifestos which appeared in 1614 and 1615. In fact, to the disappointment of those like René Descartes (1596–1650) who tried to make contact with them, the Brotherhood seems to have been as fictitious as Bacon's Salomon's House. The manifestos were written by a theologian and alchemist, Johannes Valentinus Andreae (1586–1654), seemingly as a clarion call to like-minded reformers; but he seems to have rapidly abandoned the idea of making the Brotherhood into a reality when he saw the nature of the interest it generated. If Rosicrucianism came to nothing, however, Bacon's vision, as we have already seen, was to have a profound effect.

The self-consciously reformist attitudes of the early scientific societies, and their public pronouncements of their methods and intentions in journals and other publications, mark them out as completely different from the universities. It used to be said that the universities during this period were moribund institutions, completely enthralled by traditional Aristotelianism, and blind to all innovation. This has now been shown to be completely unjustified, and the important contributions of some members of university Arts and Medical Faculties to innovation in the natural sciences has been reasserted. Nevertheless, it seems fair to say that it was usually individual professors who seemed innovatory, not the institutions to which they belonged. If there were exceptions to this it was in the smaller German universities, where the local prince might hold greater control over the university by his patronage. A number of such universities introduced significant changes in their curricula. In particular, the introduction of what was known as *chymiatria* or chemical medicine (embracing Paracelsianism and rival alchemically inspired forms of medicine) as a new academic discipline radically transformed a number of German universities. Similarly, interest in the potential practical benefits of natural magic and other forms of occult philosophy at the Kassel court under Moritz of Hessen-Kassel (1572–1632) led to the appointment of occultist professors at the University of Marburg, not just in the Medical Faculty, where *chymiatria* became prominent, but in all the other faculties as well. Even so, for the most part it remains true to say that the European universities in general seemed slow to change and institutionally committed to traditional curricula, even if individual professors might seem innovatory. In the case of the new academies or societies, however, the institutions themselves were innovatory, and they had a much greater effect on changing attitudes to natural knowledge.

Another important feature of the interest of wealthy patrons in natural marvels was the development of what were called cabinets of curiosities, collections of rarities and oddities from the three kingdoms of nature: mineral, vegetable, and animal. Originally envisaged, perhaps, as nothing more than spectacles symbolizing the power and wealth of the collector, the larger collections soon came to be seen as contributing to natural knowledge, providing illustrations of the variety and wonder of God's Creation. Cabinets of curiosities became, in the phrase of Samuel Quiccheberg (1529–1567), overseer of the *Wunderkammer* of Albrecht V of Bavaria (1550–1579), theatres of

Domenico Remps, *A Cabinet of Curiosities* (*c.*1689). This painting nicely illustrates a small cabinet of curiosities, of the kind that might have been acquired by a collector of limited means. It includes natural objects, and unusual artefacts, and even merely pictures of interesting natural phenomena. The collector would try to acquire background knowledge of each of the items, which could then be passed on to interested visitors. Cabinets were not just decorative, therefore, but were intended to contribute to knowledge and understanding of the world. Interest in forming such collections led to networks of correspondence between collectors, and even to the emergence of local brokers or suppliers in remote and (therefore) exotic parts of the world. The largest collections soon outgrew any cabinet, and ultimately became the basis for the earliest museums.

wisdom. Quiccheberg's plan for organizing such collections was published in 1565 as the *Inscriptiones…theatri amplissimi,* and proved influential among curators of cabinets for over a century. The curator of Archduke Ferdinand of Tyrol's (1529–1595) collection, Pierandrea Mattioli (1500–1577), became one of the leading naturalists of the age. Focusing particularly on the botanical specimens in the collection, Mattioli greatly superseded the work of the ancient authority on botany, Dioscorides, in his influential *Commentaries on Dioscorides* (1558). Part of the success of this work derived from the accurate illustrations, supplied by craftsmen also under Ferdinand's patronage.

The larger and more successful collections soon became early tourist attractions, drawing gentlemanly visitors on their 'Grand Tours'. Perhaps more significant for the spread of natural knowledge was the fact that acquisition of new specimens for the collections demanded extensive networks of interested parties communicating with one another about the latest discoveries and where to acquire them. Eventually, of course, these collections and their obvious pedagogical uses were to inspire the formation of the more publicly available botanical gardens, menageries, and museums. Indeed in some cases, the larger collections formed the nucleus of the first public museums. The collection of the Tradescant family, acquired by Elias Ashmole (1617–1692), formed the nucleus of the Ashmolean Museum in Oxford, while Sir Hans Sloane's (1660–1753) collection provided an impressive beginning for the British Museum in London.

The New Authority of Natural Philosophy

The historical record strongly suggests that modern atheism appeared in the late sixteenth century—certainly the word was coined at this time—and it is usually attributed to the revival of ancient scepticism (and other effectively atheistic ancient philosophies, such as Epicureanism), to the fragmentation of the Christian Churches (and the concomitant weakening of their authority) after the Reformation, and to the rise of the new philosophy. However, the role of natural philosophy in the decline of religious belief is by no means clear and simple. Certainly, there was no intrinsic reason why any of the new natural philosophies should have turned contemporaries away from religious belief. There can be no denying, however, that those who were disposed toward atheism were easily able to, and did, appropriate the latest thinking about the natural world, and used it to serve atheistic purposes.

As a result of the 'Galileo affair', as it was known to contemporaries, the Copernican theory is often seen as emblematic of the supposedly inimical relations between science and religion. But, the Roman Catholic Church took no measures against the Copernican theory, which had been published in 1543 and dedicated to the Pope, until 1616, when it made a hasty ruling against it. The perceived need to make this ruling arose because of what the Church authorities in Rome saw as embarrassing public controversy between Galileo and his enemies. The subsequent development of the affair can also be seen to owe more to Galileo's tactless unconcern for scoring points off his rivals and turning them into implacable enemies than it did to any supposed enmity between science and religion. Certainly, the circumstances leading to Galileo's condemnation were so unique and specific to him and his actions that no general conclusion should be drawn about any supposed incompatibility between the scientific approach and religion. Galileo was not condemned for being a Copernican, he was condemned for writing a book promoting it when he had (allegedly) promised the Congregation of the Index (the Inquisition) that he would never discuss it.

Witches at their Incantations (*c.*1646) by Salvator Rosa (1615–1673). Accounts of witchcraft, ghosts, lycan-thropy, and other occult phenomena played a significant role in the Scientific Revolution. The period coincided with the emergence of atheism and irreligion, and atheists often appropriated the latest ideas in natural philosophy to support their views. This was easy to do, especially with the advent of the mechanical philosophy, because natural philosophers were accustomed to trying to explain natural phenomena in terms of what were called secondary causes. God was the *primary* cause, but it was agreed that he rarely (if ever) directly intervened in the natural world, choosing to act instead by means of physical *secondary* causes. Atheists found it easy, therefore, to exclude God and explain phenomena solely in terms of the secondary causes. A number of devout natural philosophers tried to counter this trend by asserting the truth of ghost stories, tales of witchcraft, and so forth. These seemed to confirm the existence of a spiritual world, and therefore of God—witches were reputed to get their powers from Satan, but there could be no Satan without God. These attempts proved short-lived (and may in fact have contributed to the decline in belief in witchcraft), and were replaced by natural theology—the attempt to prove the existence of God by showing that the intricacies of nature could not have happened without an intelligent designer.

Indeed, if we consider all the major contributors to the Scientific Revolution, it is easy to see that they were usually devout believers, and in some cases, such as Robert Boyle and Isaac Newton, they even made important contributions to contemporary theology. But no matter how devout the originator of naturalistic innovations might be, contemporaries with their own atheistic agenda could always seize those innovations for their own purposes. Descartes's conception of specific laws of nature, for example, led him into profound theological speculation. He was aware that inanimate bodies cannot 'obey' laws of nature and assumed that the laws must be somehow enacted and maintained by God. It was easy, however, for subsequent generations of thinkers to dismiss what Descartes saw as irremovable theological implications of the laws of nature, and to simply regard the laws as built into a godless nature. But the atheists were too hasty, and even today there are vigorous debates among philosophers of science about the precise status of laws of nature and how they can have meaning. Similarly, although Descartes himself insisted on the reality of the immortal soul, it was easy for his less devout followers to dismiss it, just as he had dismissed the vegetative and animal souls, and to insist that men, no less than the other beasts, were merely machines.

Devout natural philosophers were appalled to see their ideas put to irreligious uses and there was a strong movement toward what is called natural theology—a theology based not on Scripture but on showing what seems to be the intricate design of the natural world, and its dependence, therefore, on a supreme designer and creator. But the momentum of secularization could not be stopped. Natural theology gave rise to what was called deism (accepting, on naturalistic grounds, the existence of a divine Creator but rejecting the validity of Scripture), and weakened the authority of the Churches. Churchmen saw deism as not far removed from atheism, and the history of increasing secularization proved them to be right.

The many successes of the natural philosophers of the Scientific Revolution raised the authority of natural knowledge, and meant that it could be used to rival the long-held authority of the Church. But, scientific knowledge does not speak for itself, any more than Scripture does, and the authority of scientific knowledge was wielded in different ways by different thinkers. Even so, it is a major and final legacy of the Scientific Revolution that, in the succeeding age, the Age of the Enlightenment, it was scientific doctrine, especially Newtonian doctrine, that shaped theory and practice, not just in the natural sciences themselves, but in the new 'Sciences of Man', embracing moral, psychological, political, economic, and sociological thought.

6 Enlightenment Science

JAN GOLINSKI

THE eighteenth century has not always been recognized as a significant era in the history of science. A few decades ago, historians tended to see the period as over-shadowed by the towering genius of Sir Isaac Newton. 'Newtonianism' was regarded as the key to an age that remained in thrall to Newton's accomplishments in astronomy and natural philosophy. The whole century was viewed as a prolonged calm in the wake of the previous century's Scientific Revolution and before the breaking storm of the Industrial Revolution to follow. Those who saw the history of science in terms of such 'revolutions' found no serious candidates until the end of the period, and then maybe only in chemistry. What seemed like a whole century of steady diffusion of already-established ideas—'normal science' without any significant interruption—held little interest from this point of view.

This attitude has changed, partly because of new approaches to the history of science and partly because of the emergence of the Enlightenment as a central theme in the cultural history of the eighteenth century. Less emphasis is now given to revolutionary intellectual transformations, which are generally held to have been less instantaneous and monolithic than they previously seemed. Science is now more frequently viewed as a set of practices than as a structure of ideas. Hence, more attention is given to its connections with technology, communications, and trade. What now emerges when we look at the eighteenth century is a busy scientific culture, populated by many individuals who used to be overlooked in the search for the great minds of the age. Scholars now attend not just to the writers of significant scientific texts, but to compilers and plagiarists, printers and booksellers, and to the readers of these works, which included women and children as well as men. Alongside the circulation of texts, we follow the maps and drawings, instruments and artefacts, specimens and commodities—all of the media by which knowledge is distributed. We recognize not just natural philosophers (precursors of the later 'scientists'), but artisans, engineers, craftspeople, navigators, architects, naturalists, entrepreneurs, and many others. Swept up in this ferment of knowledge were not just the leading thinkers of the age but all the people who read books and periodicals, attended schools and public lectures, collected plants and minerals, designed and built machines, and discussed scientific ideas.

The new view of the history of science coincides happily with a new emphasis on the Enlightenment, the term being used not simply as the label of a period but to name a process of cultural change ('enlightenment' with a small *e*). The process was one of accelerated circulation and exchange, in which communications media proliferated and a larger portion of the population was brought within range of the traffic in information and opinion. It was experienced in different forms in different places, stimulating various degrees of opposition against political and religious authorities, and generating countervailing forces of reaction. It was centred in Western Europe, but touched also on Asia and the Americas, coinciding as it did with a particular phase of European peoples' expanding influence over the rest of the world. The focal themes of study and debate included the process of cultural development itself, which was the subject of historical inquiries and comparative studies of contemporary societies elsewhere. These discussions were often premised on a notion of 'human nature', the supposedly universal foundation of human character, which reflected the prevailing assumption that the natural world as a whole was governed by uniform laws. Notions of human nature were built upon studies in the physical and natural sciences, which thus informed debates about human capabilities, cultural progress, and the proper ordering of society. The French mathematician Jean-Baptiste d'Alembert captured the atmosphere of pervasive and boundless intellectual inquiry in 1759, when he wrote:

from the principles of the secular sciences to the foundations of religious revelation, from metaphysics to matters of taste, from music to morals, from the scholastic disputes of theologians to matters of trade, from the laws of princes to those of peoples, from natural law to the arbitrary laws of nations, . . . everything has been discussed and analyzed, or at least mentioned.

Nations and Communications

The Enlightenment was identified as a process of cultural change in the most famous attempt to define it, published toward the end of the period it conventionally names. Immanuel Kant's essay 'What Is Enlightenment?' was contributed to the *Berlinische Monatsscrift* in 1784 in response to a question raised in the journal the year before. The brief text has been argued over repeatedly ever since, and subjected to fiercely opposed interpretations by twentieth-century philosophers. Kant identified enlightenment with 'mankind's exit from its self-incurred immaturity'. He pointed to its open-ended character with his declaration that his contemporaries lived not in an enlightened age but rather one of enlightenment—one in which cultural development was continuing rather than having been completed. Kant's essay also drew attention to the centrality of free discussion, in what he called the 'public' domain, as a condition of enlightenment; though he also acknowledged that individuals' freedom of action could legitimately be restrained in respect of their official duties. Those who held office as administrators, clergy, soldiers, and so on, could be required to obey their orders without question, though they should also have the freedom to discuss whatever they wished in their public capacity. This definition of the realm of free debate informed the later analysis of the 'public sphere' by the twentieth-century German social theorist Jürgen Habermas, who used the term to cover a range of institutions and media arising during the period. Habermas's public sphere has come to be seen as a central feature of the Enlightenment, perhaps most perfectly embodied in the flourishing newspapers, periodicals, clubs, and coffeehouses of eighteenth-century England. The public sphere comprised mechanisms for evaluating and criticizing the political structures of civil society, as well as for circulating information. It made people self-conscious about the process of cultural development in which they were participating.

The public sphere also provided the distinctive context for Enlightenment science. One of the defining features of scientific discourse in the period was its openness to the public. Printed materials were addressed to a readership that was (at least in principle) international and unbounded. Experiments were performed before witnesses and (again, in principle) were replicable anywhere. Even the closely guarded secrets of artisans and craftspeople were exposed to unprecedented public view, in such publications as the multi-volume *Encyclopédie* (1751–1772). The sociability that marked the public sphere underpinned communications in at least seventy formally constituted learned societies during the era. These ranged from the national scientific academies sponsored by monarchs in London, Paris, Berlin, and St Petersburg, to provincial institutions in such towns as Dijon, Mannheim, Bologna, and Spalding, or in colonial outposts such as Boston, Philadelphia, Saint Domingue (Haiti), and Batavia (Jakarta). Below this level were thousands of informal clubs, self-organizing societies, circulating libraries, and gatherings in inns and coffeehouses. Lectures and discussions were held in these places, improvement schemes launched, newspapers read and their contents debated. The activity was fed by the production of printed materials in large quantities. The main centres of book publication included London, Edinburgh, Paris, Amsterdam, and Leipzig. Censorship and copyright protections were patchy and often unenforced. The centres of 'piracy' or unauthorized reprinting included Dublin and Philadelphia. And the book trade successfully evaded the policing of international borders. Books that were supposed to be banned in France, for example, flooded into the country from the Netherlands, Switzerland, and England. Periodicals also circulated internationally. The leading scientific academies published their proceedings, such as the *Philosophical Transactions* of the Royal Society of London, and the *Mémoires* of the Académie Royale des Sciences in Paris. And scientific material also turned up in many other publications, including almanacs and general-interest periodicals such as the *Ladies' Diary* (from 1704).

A new consciousness of cultural geography was commonly a part of the experience of enlightenment. Nationhood was an important dimension of political identity in some places, but not in all. France was politically unified by absolutist rule, though its provinces preserved a degree of independent cultural life, as the numerous academies of learning in provincial towns testified. Great Britain was also politically unified after the union of English and Scottish parliaments in 1707, but Scotland retained its cultural identity and still possessed independent religious and legal institutions. Economic development and the growth of towns in the Scottish Lowlands and the

English regions supported energetic cultural and scientific life in those areas. The Lunar Society of Birmingham drew together an influential network of industrialists and men of science in the English Midlands. Germany and Italy, though linguistically unified, remained politically fragmented until the nineteenth century. Particular cities emerged as foci of enlightenment: Berlin, Göttingen, Naples, and Milan, for example. The British colonies in North America were initially drawn into closer cultural relations with the homeland through the flourishing North Atlantic trade. Then, from 1776, the American Revolution broke those ties, launching the growth of an independent cultural life in the former colonies that joined to form the United States.

Aside from the national level, the Enlightenment unfolded also as an international and a regional phenomenon. European intellectuals sometimes thought of themselves as united in a 'republic of letters', since they were linked by networks of correspondence and the trade in publications. Latin and French were the universal languages of learning and diplomacy, and some continental intellectuals began to learn English. On the other hand, the experience of particular regions—even those often thought of as peripheral rather than central—could be a spur to cultural change. Examples of such peripheral centres include Edinburgh and Naples. In both places, urban elites confronted what they regarded as more backward neighbours in close proximity. The contrast was even starker in North America, where colonial Creoles who thought of themselves as polite and refined rubbed shoulders with Native peoples they denigrated as 'savages'. Striving to emulate what they perceived as more enlightened centres of culture, members of these local elites differentiated themselves from their more primitive neighbours. At the same time, intensified trade within Europe and across the Atlantic created another stimulus to enlightenment. Viewing the situation from Edinburgh in 1742, the Scottish philosopher and historian David Hume likened the condition of Europe in his time to the golden age of classical Greece, with its numerous rival city-states. He concluded that: 'nothing is more favourable to the rise of politeness and learning, than a number of neighbouring and independent states, connected together by commerce and policy'.

Cosmic Machinery

It is a mistake to subsume the whole of eighteenth-century science under the category of Newtonianism, but there is no denying the celebrity attached to Newton's name, especially in the first few decades. His great work on mechanics and celestial dynamics, *Principia* (*Mathematical Principles of Natural Philosophy*, 1687), was the foundation of the fame he began to accrue in the second half of his long life. Publication of his second important book, *Opticks* (1704, with significant revisions in subsequent editions), followed shortly after his assumption of the presidency of the Royal Society of London. That position, and the knighthood he received in 1705, brought with it significant powers of patronage in the world of learning, which Newton was not shy of using in his later years. His death in 1727 led to burial among the English kings and queens in Westminster Abbey, to memorials, and indeed to little short of apotheosis in paintings, engravings, and poetry. François-Marie Arouet (Voltaire) witnessed Newton's funeral while in exile in London, and later celebrated his scientific accomplishments in *Lettres philosophiques* (1734), and *Éléments de la philosophie de Newton* (1738). While Newton was hailed as a hero of the Whig party and the Latitudinarian faction in the Church of England, Voltaire made him a symbol of religious toleration and anticlericalism in France. Technical work in Newtonian mechanics, by d'Alembert and by Pierre-Louis Moreau de Maupertuis, along with Voltaire's campaign, eventually led to the recognition of Newton's natural philosophy in the ranks of the Académie. The marquise Gabrielle Émilie du Châtelet, whose command of mathematics was more profound than that of her lover Voltaire, also contributed by translating Newton's *Principia* into French.

Notwithstanding the magnitude of his accomplishment, however, Newton's work set many problems for his successors. His writings offered various, sometimes conflicting, opinions on philosophical issues, and the arguments over their interpretation lingered for decades. In his model of the cosmos, Newton built upon the work of René Descartes, who had made the Copernican arrangement of the Solar System the basis of a comprehensive cosmology in the mid-seventeenth century. Newton incorporated Johannes Kepler's laws of planetary motion and showed how to derive them mathematically from fundamental laws of mechanics and gravitation. This opened the way for further work on the orbits of the planets, the Moon, and comets. But Newton left unresolved the issue of whether this system could sustain itself. He suggested that the planets' orbits would tend over time to decay, requiring restoration by divine action through the agency of comets. In the second decade of the century, a public exchange of letters between the German philosopher Gottfried Wilhelm von Leibniz and Newton's surrogate Samuel Clarke thrashed out the question of what this implied for philosophical theology. For Clarke, as for Newton, active principles of divine agency must be recognized at work in the natural world. The alternative was an entirely self-sustaining universe, the vision proposed by the philosophers Thomas Hobbes and Benedict Spinoza, which—according to Clarke—was a short step from atheism. To Leibniz, on the other hand, any such divine intervention implied a flaw in the original design of the cosmos. The most perfect creation, like the best-made clock, would run by itself, obeying without deviation the uniform and regular laws established by the creator. In the course of subsequent decades, opinion swung toward Leibniz's side of the debate. By the end of the century, the French mathematician Pierre-Simon de Laplace was reputedly declaring that he had no need of the God-hypothesis in his own work on celestial dynamics. Thanks to his work and that of others, it was even possible to envision how the Solar System could have been formed through the action of gravity on a huge cloud of matter. Newton had denied that this could have occurred, insisting that the motions of the planets clearly displayed the vestiges of the divine act of creation. But, by the end of the century, the discovery of

Opposite: *The Apotheosis of Isaac Newton* (George Bickham, 1732). Museum of the History of Science, Oxford.

PRÆCLARISSIMUS·ISAACUS·NEWTON·EQUES.

ECCE PHILOSOPHORUM PRINCEPS

See the great *Newton*, He who first Survey'd   *O wondrous Man!* in whom the heav'nly Mind,
The Plan, by which the Universe was made;   Shines forth distinguish'd, and above Mankind;
Saw Natures Simple, yet Stupendous Laws,   Whilst here on Earth, how humble, wise and good!
And prov'd the Effects tho' not explain'd the Cause.   In Heav'n a Star of the first Magnitude.

London Published May 8th 1787 by Robert Sayer 53 Fleet Street.

such heavenly clouds in the form of nebulae suggested that the cosmos could have developed entirely through the action of uniform natural laws. These laws could be interpreted as a mode of providential activity, but they did not have to be.

As we shall see later in the chapter, a similar tendency to extend the reach of uniform natural laws occurred in other areas of the sciences. Newton had proposed the existence of various kinds of active principles, which supplemented the inertial tendencies of matter itself and were identified with God's activity in the material world. In the enigmatic 'Queries' he added to his book *Opticks*, the principles were applied to explain phenomena in the domains of chemistry, electricity, and physiology. But later investigators often detached these powers and forces from the theological meaning Newton had given them. Rather than agents of divine action, they came to be seen as inherent properties of matter itself. Newton had insisted that matter had no inherent powers of activity; but, from the 1740s on, such powers were increasingly imputed to it. If this was a kind of Newtonianism, then, it was one of which the founder would vehemently have disapproved.

A Philosopher giving that Lecture on the Orrery, in which a Lamp is put in Place of the Sun (Joseph Wright, 1766). Derby Museums and Art Galleries.

One setting in which Newtonian science was presented to a public audience was depicted by Joseph Wright of Derby in his 1766 painting, *A Philosopher giving that Lecture on the Orrery, in which a Lamp is put in Place of the Sun*. The image has been chosen many times as an emblem of Enlightenment science, since it shows spectators literally bathed in the light of knowledge. They are seen receiving knowledge through their senses—through vision above all—by collectively witnessing a demonstration-experiment. Newtonian natural philosophy appears as consistent with the empiricist epistemology of Francis Bacon and John Locke, and with the practices of collective witnessing that had been adopted by such groups as the Royal Society in the seventeenth century. The object of the spectators' scrutiny is an orrery, a mechanical model of the Solar System. Such models were widespread in the era, made by clockmakers and other craftsmen, and ranging in size from the largest versions (mounted vertically on the stage of a theatre), to the tabletop device shown by Wright, to mechanisms the size of clocks or pocket watches. Because of their intricate craftsmanship, orreries displayed the wisdom of the divine design in the cosmos at large. The foresight of God's creation was mirrored by that of the orrery's maker, and the regularity of His providential laws was demonstrated by the constancy of the clockwork. Phenomena such as comets and eclipses, which had mystified or terrified people in earlier ages, could be shown to arise from the normal motions of the heavenly bodies. Thus was the message of enlightenment following from scientific discovery conveyed. And yet, it may well be that the spectre of doubt raised by the Leibniz-Clarke debate hovered over such displays. Like the mechanical replicas of humans and animals that fascinated people at the time, these artefacts suggested that the natural world was itself a kind of machinery. That being so, spectators might be tempted to consider whether the cosmic machine really *needed* a God. Once the clockwork had been made and wound up, was there any evidence that the maker stayed around? Or could the universe be thought of as running entirely by itself?

Wright's painting also provides some clues about the settings in which these questions were pondered. Wright portrayed not the members of a learned academy but an assembly of mixed sexes and ages, probably a family or group of friends in a domestic interior. Evidently, by the 1760s, the practices of collective witnessing had diffused well beyond the limited circles in which they had begun. In England, they had in fact been associated with popular expositions of Newton's work since the first two decades of the century. William Whiston began teaching in London coffeehouses soon after being sacked from his position in Cambridge in 1711; he had been Newton's successor in the Lucasian chair until his ousting for religious heterodoxy forced him to seek a new career. Within a few years he was joined by the French Huguenot refugee John Theophilus Desaguliers, who combined work as Newton's experimental assistant at the Royal Society with public lecturing in London. In the 1740s, Benjamin Martin pioneered a new lifestyle for scientific lecturers by taking to the road as an itinerant. In his first few years of travels, Martin visited Bath, Birmingham, Chester, and Shrewsbury, announcing his arrival in advance in newspaper advertisements, and collecting subscriptions from those who wished to hear him. He supplemented his

GABRIELLE·EMILIE·DE·BRETEVIL·MARQUISE·DV·CHASTELET·LORRAINE·

income from lecture fees by offering his writings and instruments for sale. Although large orreries might be beyond the financial means of many middle-class consumers, other devices such as telescopes, microscopes, and barometers were more affordable; they brought at least a basic level of scientific experimentation into many bourgeois households. Martin's career-path was followed by hundreds of others, in Britain, France, Ireland, Germany, and North America, in the remaining years of the century and into the following one. Adam Walker, who trod this path in the north of England, Scotland, and Ireland in the 1760s and 1770s, exhibited a large vertical orrery, which he called the 'eidouranion', suitable for theatrical display. James Ferguson, a Scottish lecturer, author, and instrument-maker, who designed several orreries himself, has been tentatively identified as the original of the philosopher in Wright's painting.

Women and children participated alongside men in these scenes of Enlightenment public science. The year before Newton published his *Principia*, Bernard de Fontenelle had invited women into the audience for natural philosophy in his *Entretiens sur la pluralité des mondes* (1686). The work treated the cosmology of Descartes in a series of conversations between a male philosopher and an aristocratic lady. Discussing what it would be like to view the Earth from elsewhere in the Solar System, the characters explore the novel perspective subsequently experienced by viewers of the orrery. Fontenelle's elegant text translated issues of the Earth's place in the Copernican cosmos into the idiom of polite discourse; it initiated a genre of mixed-sex dialogues on scientific topics. John Harris's *Astronomical Dialogues between a Gentleman and a Lady* (1719) was an early example in English, later joined by Benjamin Martin's *Young Gentleman and Lady's Philosophy* (1744). The most celebrated text of this kind after Fontenelle's was Francesco Algarotti's *Il Newtonianismo per le dame* (*Newtonianism for the Ladies*, 1737), soon translated into English and other languages. Algarotti mostly covered Newton's theory of light and colours, spicing up the exposition with snatches of poetry, extended metaphors, and flirtatious banter between the protagonists. In 1761, the English publisher John Newbery put out an exposition of Newton's science for children, supposedly written by the young boy 'Tom Telescope', who entertained and instructed his peers with lectures on matter and motion, optics and astronomy.

Thus, women and even children had a recognized place in certain kinds of public science. Books were specifically addressed to them; it was taken for granted that they would be involved when science entered the home. Women also heard public lectures on the sciences, even though they were almost entirely barred from universities. In Paris, they attended the courses of Jean Antoine Nollet from the mid-1730s, learning about mechanics, optics, and electricity. Nor was the marquise du Châtelet the only woman with a high level of technical expertise. In Bologna, Laura Bassi held a university chair and lectured on Newtonian physics for nearly thirty years. The mathematician Maria Gaetana Agnesi followed Bassi onto the faculty of

---

Opposite: Portrait of Émilie du Châtelet (unknown artist, undated). Private collection, Château de Breteuil.

Portrait of Margaret Bryan and her daughters (William Nutter, 1797). National Portrait Gallery, London (NPG D8491).

the University of Bologna. Maria Winckelmann Kirch worked at the Berlin Observatory in the first decade of the century. And later Caroline Herschel discovered comets and nebulae while working as an astronomer near London with her brother William. There is no denying, however, that women faced significant obstacles in gaining recognition among the scientific elite. Du Châtelet was obliged to publish some of her works under the name of a male colleague. Bassi had to teach mostly from her home, or occasionally at the university from behind a curtain. Kirch and Herschel were treated as assistants to the men they worked with. The institutions of learning remained barred to women long after the end of the age of Enlightenment; in fact, they were not admitted to the ranks of the leading scientific academies until the mid-twentieth century. Because of their exclusion from formal scientific institutions, it has often been forgotten that women participated significantly in Enlightenment public science.

Forces, Fluids, and Feelings

Static electricity—produced by rubbing glass or amber, which then attracts or repels bits of thread or paper—was a phenomenon known since ancient times. In the early decades of the eighteenth century, it began to be possible to charge objects reliably by machines. Electricity became an object of systematic investigation and assumed a place in Enlightenment public science. Francis Hauksbee the elder, a London instrument-maker and experimenter to the Royal Society, designed a generator with a spinning glass globe rubbing against wire brushes. Newton showed an interest in Hauksbee's experiments and claimed they demonstrated the existence of the subtle expansive fluid or 'ether', which he believed surrounded all bodies. Other investigators referred to the 'effluvia' thrown off, and perhaps also drawn in, by objects when they became charged. In the early 1730s, the experimenter Stephen Gray demonstrated the phenomenon of conduction, showing that some materials could transmit an electric charge over a distance even if they did not themselves appear to be charged.

Nollet was one of the most entrepreneurial of the experimental lecturers who brought these phenomena into the public sphere. In his courses, held in Paris from the mid-1730s, he showed a variety of electrical effects to genteel audiences of men and women. Individuals who stood on insulating plates could be charged by a conductor attached to an electrical machine, or a boy could be charged while suspended on ropes from the ceiling. The boy could then have sparks drawn from him, or could attract and repel pieces of paper. In such an experiment, electricity offered an alternative to light as a metaphor for the communication of knowledge from person to person. As in the scene of the orrery depicted by Wright, scientific knowledge was transmitted in sociable gatherings of mixed sexes and ages. There might even be a suggestion of eroticism in the physical contact between men and women as the sparks fly. In some experiments of the time the suggestion was made quite explicit, with a young woman who had been charged up being discharged by an 'electrical kiss' from a male spectator.

N. le Sueur Invenit.   R. Brunet fecit.

In 1746, electrical experimentation was enhanced by a (literally) shocking new device: the Leiden jar, discovered by Pieter van Musschenbroek, a professor at the University of Leiden. The jar was capable of holding a very substantial charge, provided it was charged up while the outside of the glass was connected to the Earth. It could then be discharged by touching the metal pin inserted through the cap. The shock was so strong it could be passed along a lengthy chain of people holding hands, and felt by each one. The problem was to explain how the jar worked, and the most plausible solution was offered by the Philadelphia printer and journalist Benjamin Franklin in his *Experiments and Observations on Electricity* (1751). Franklin proposed that a subtle and expansive electrical fluid pervaded normal matter. If the fluid was present in a body in more than the usual amount, the body was positively charged; if there was a deficiency of fluid, the body was negatively charged. The theory offered explanations for the phenomena of charging, conduction, and neutralization. It suggested that the Leiden jar should be understood as a container for the electric fluid, which accumulated on the inside and drove away fluid from the outside of the glass. When the negatively charged exterior was connected with the positively charged interior, a strong discharge resulted. Franklin gave his theory a direct practical application by showing that bolts of lightning were nothing other than massive electrical discharges between storm clouds and the ground. He campaigned for the erection of conducting poles on buildings, to lead the electrical fluid to ground during thunderstorms and protect against damage.

The Leiden jar and Franklin's lightning conductors entered the repertoire of commercial lecturers. The Baptist minister Ebenezer Kinnersley, a friend of Franklin's, toured the eastern seaboard of North America and the West Indies lecturing on electricity in the late 1740s and 1750s. Kinnersley presented electrical powers as part of the providential order of nature and emphasized the utility of lightning conductors. Like Nollet's demonstrations, however, his lectures bore multiple meanings. Ostensibly celebrating the benefits of rational knowledge, such shows also stimulated the imagination of viewers by displaying the wonders of nature. They elicited feelings from those who witnessed them, and even more directly from those who allowed electricity to pass through their bodies or were treated with it for various ailments. In these ways, electrical demonstrations found their place within the culture of sensibility, which took hold after the middle of the century in Western Europe and colonial North America. Electricity was experienced sociably; it directly elicited emotions and bodily passions. It thus provided a medium and a metaphor for the sympathetic feelings that were supposed to bind people together in civilized society. In some performances it stood in for the natural powers of attraction between men and women. An image published by Benjamin Martin in 1755 captures some of these emotional aspects of electrical demonstrations. It shows a tabletop electrical machine

Opposite: Jean-Antoine Nollet's electrical experiment with a boy suspended from the ceiling (1746). Frontispiece to Nollet, Essais sur l'électricité des corps (Paris, 1746).

being contemplated by a man, woman, and child, indicating how electricity complemented a heightened awareness of affective relationships within the family and in society at large.

This emotional and aesthetic dimension was an important aspect of Enlightenment public science. In his *History and Present State of Electricity* (1767), the English natural philosopher and preacher Joseph Priestley wrote that the pleasure of studying nature resembles that of 'the sublime, which is one of the most exquisite of all those that affect the human imagination'. The Leiden jar presented an obvious example of the combination of pleasure and fear that writers on aesthetics identified with the experience of the sublime. Priestley argued that similar emotions could be elicited in other fields of science too. During his work to produce and characterize different types of gases (or 'airs'), in the late 1760s and early 1770s, he frequently referred to the surprise and awe he experienced. He derived emotional satisfaction from revealing the providential order of nature, as when he discovered that air which had been degraded by humans or animals breathing it could be restored to goodness by the action of plants. (In modern terms, Priestley had uncovered the process of photosynthesis, by which plants convert carbon dioxide into oxygen, reversing the process of animal respiration.) To him, the discovery revealed a previously unknown aspect of God's benevolence in the creation, an occasion for legitimate wonder and astonishment. Priestley presented his findings to his readers, and encouraged itinerant lecturers to display them in public, in order to share the experience as widely as possible. The fact that such things were being disclosed at this time was taken as a further indication of the workings of divine providence. It showed the advance of general enlightenment, with the consequent dawning of true religion and political liberation that Priestley expected to follow. Study of the natural sciences thus both revealed God's plan in nature and served the providential cause of human enlightenment, according to Priestley. Not surprisingly, he regarded it as the noblest and most sublime activity.

Collecting and Exploring

Both the metaphor of the machine and the preoccupation with the powers of matter shaped the ways in which living things were understood during the Enlightenment. In the seventeenth century, some physicians and natural philosophers began to think of the human body in mechanical terms. The operations of muscles and bones were analysed on the analogy with pulleys and levers, and the flow of blood and other bodily liquids was conceived in terms of fluid dynamics. The so-called 'iatromechanical' approach was particularly associated with the Italian mathematician Giovanni Borelli and the anatomist Marcello Malpighi. The tradition survived into the eighteenth century, bolstered by the success of the mechanical philosophy and encouraged by the fascination with automata. In the 1730s, the French inventor Jacques de Vaucanson

---

Opposite: Benjamin Franklin Drawing Electricity from the Sky (Benjamin West, *c.*1816). Philadelphia Museum of Arts.

Plate XXV.

A *New* ELECTRICAL MACHINE *for the* TABLE.

made several celebrated moving figures, including a flute-player and a duck that seemed to digest its own food. A decade later, the French physician Julien Offroy de La Mettrie published a notorious book suggesting that human beings were nothing more than such mechanical devices, albeit much more elaborate ones than human craftsmen could make. La Mettrie's *L'homme machine* (1748) was published anonymously in Leiden, and it purported to explain not only human physiology but also the operations of the passions and the intellect in terms of the motions of solids and fluids in the body. The book was condemned by ecclesiastical authorities as an attack on the doctrine of the soul and on conventional morality. A copy was burned in Paris by the public executioner, and the author was fortunate to be given refuge in Berlin by king Frederick II of Prussia.

Notwithstanding La Mettrie's scandalous work, prevailing opinion around the middle of the century was moving away from the view that living things could be explained in mechanical terms. An influential voice on this question was Georges-Louis Leclerc, comte de Buffon, director of the royal botanical gardens in Paris, whose huge multi-volume work on natural history began to appear in 1749. Buffon scornfully dismissed attempts by Cartesians and Newtonians to analyse animals and plants as mechanical systems. Instead, he insisted that living beings manifested powers of matter that were specifically organic. Although Buffon was coy about taking the next step, the supposition opened the door to a materialist philosophy, in which vitality itself was reduced to the inherent properties of matter. Denis Diderot, editor of the *Encyclopédie*, was willing to embrace this conclusion in such works as *Pensées sur l'interprétation de la nature* (1754) and the unpublished *Le rêve de d'Alembert* (written in 1769, but not printed until the nineteenth century). Diderot pointed to experimental research on muscular irritability, the spontaneous generation of life from non-living matter, and the regeneration of detached body parts in simple organisms like the freshwater polyp or hydra. To him, the implication was clear: matter had an inherent capacity to organize itself into the form of living things. There was thus no need to suppose a special act of creation to produce each species. Simple forms of life would be constantly arising from non-living matter, and evolving over the generations into more complex or advanced forms, perhaps even including human beings. Diderot did not spell out the process of evolution in detail; his aims were primarily moral and political rather than scientific. He saw materialism as a weapon to be used against the Catholic Church in France, and he shamelessly promoted a naturalistic system of morality to replace that based on religious doctrine.

Materialist speculation eventually fed into the scientific theories of evolution that emerged in the nineteenth century. The connecting links were turn-of-the-century evolutionists such as Jean-Baptiste Lamarck and Erasmus Darwin. But, in the

---

Opposite: A New Electrical Machine for the Table (Benjamin Martin, 1755). From Martin, *The General Magazine of Arts and Sciences* (vol. 1, London, 1755).

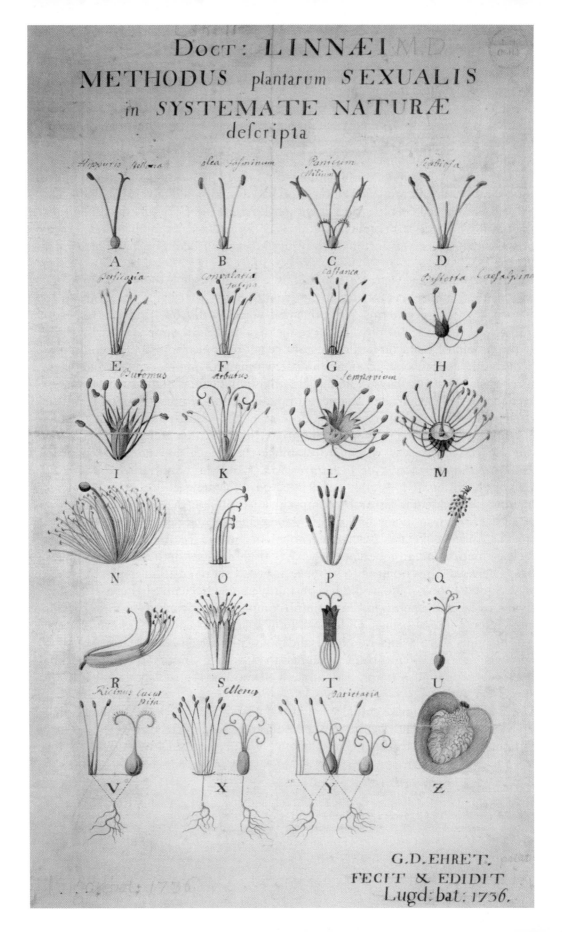

eighteenth century, materialism remained confined to small circles of free-thinking intellectuals, mainly in France. The dominant mode of study of living things was quite a different enterprise, namely natural history. This was an activity of describing, naming, classifying, and collecting the products of all three kingdoms of nature: animals, plants, and minerals. It viewed the natural world as manifesting a systematic order, which was assumed to be the divine plan of creation and unchanging over time. Natural history was a participative process that enrolled people in all parts of the world. Its most prominent proponent was the great Swedish naturalist Carolus Linnaeus (Carl von Linné). Linnaeus devised a hierarchical classification scheme, first set out in his *Systema naturae* (1735, with eleven further editions during his lifetime), in which all living things were categorized and assigned an appropriate name. The system gave rise to the Latin names of plants and animals still used in biology today. Linnaeus himself classified and named nearly 8,000 species of plants and nearly 4,500 species of animals. For plants, he created a system of classification based on their reproductive organs, specifically the number of stamens (male organs) in their flowers. Among animals, he originally designated six categories: quadrupeds, birds, amphibians, fish, insects, and worms. In the tenth edition of *Systema Naturae* (1758–1759), he replaced quadrupeds with the new category of mammals, choosing as the defining feature of this class the mammary glands that the mothers use to nourish their young.

Linnaeus sought to make natural history a global enterprise, recognizing its potential economic benefits for the nations that pursued it. In his botanical garden at Uppsala, he tried to naturalize exotic species that he believed could contribute to Sweden's prosperity, including tea bushes, mulberry trees, and rhubarb. He exchanged letters and specimens with a wide circle of correspondents, and he dispatched a series of disciples on voyages of exploration throughout the world. Pehr Kalm travelled to Russia and North America in the 1740s; Pehr Löfling explored South America in the 1750s; Daniel Solander sailed on Captain James Cook's first Pacific voyage in 1768–1771; and Anders Sparrman joined Cook's second voyage in 1772–1775. Linnaeus's worldwide supply network and intensive experiments on transplanting species were imitated by the major European powers as they expanded their global influence. The Jardin du Roi in Paris was supplied by networks established by French explorers. On his return from Cook's first voyage, Joseph Banks took over the management of the Royal Botanic Gardens at Kew, and secured specimens from subsidiary institutions at Botany Bay, the Cape of Good Hope, and elsewhere. A similar arrangement linked the royal gardens in Madrid with facilities in Mexico City, Bogotá, and Lima. The networks were founded to serve imperial interests, but they allowed individuals in colonial outposts to become partners in the creation of scientific knowledge. Men, women, and even slaves, associated with the European settler communities in many parts of the world, seized the opportunity. John Bartram,

Opposite: Linnaei Methodus Plantarum Sexualis (Georg Ehret, 1736). Natural History Museum, London.

a naturalist in Philadelphia, exchanged plant specimens with Peter Collinson, a London merchant. Jane Colden, the daughter of a physician living on Long Island, New York, catalogued more than three hundred species of American plants. Gramman Kwasi, a freed slave in Dutch Surinam, discovered a valuable medicinal root that was reported to Linnaeus, who gave Kwasi's name to the tree from which it came.

Natural history was one of the main scientific purposes of the exploratory voyages of the period. By the 1760s, European exploration of the Pacific was well under way, with British, French, and Spanish ships venturing into the ocean, and Russian travellers approaching it overland. Banks and Solander sailed on Cook's *Endeavour* in 1768, equipped at Banks's personal expense with supplies of bottles, cases, and preserving spirits to collect insects, birds, and plants. Their collecting activities at each of the ship's destinations—Tahiti, New Zealand, Australia, and the East Indies—yielded what was judged an unprecedented harvest for natural history by no less an authority than Linnaeus himself. The British enterprise was spurred on by competition with French mariners, especially Louis Antoine de Bougainville, whose ship *Boudeuse* circumnavigated the world in 1766–1769. Bougainville's naturalist Philibert Commerçon, and his female assistant Jeanne Baret (who shipped out disguised as a man), described and collected specimens throughout the voyage, including the South American plant *Bougainvillea*, which later became familiar in many other parts of the world. Bougainville also collected a human being from Tahiti, Ahutoru, who became a celebrity in France before dying on the voyage that was to return him to his homeland. Cook followed the same practice, picking up on his first voyage a Tahitian sage and medicine man called Tupaia, who was able to translate the languages of other Pacific Islanders and even New Zealand Maoris, until he died in Batavia. On Cook's second voyage, another Tahitian, Omai, was brought back to London on the *Adventure*. He became famous in London society, having his portrait painted, visiting parliament and the opera, and being introduced to king George III. A well-known painting by William Parry shows him in the company of Banks and Solander. He is shown as a handsome and dignified individual, garbed in a costume that evokes classical antiquity, and treated by his European friends with respect, though he is also apparently an object of curiosity. The picture indicates the Enlightenment fascination with the lives of people who were labeled as primitives or 'savages'. Omai was returned to Tahiti on Cook's third voyage in 1777.

Natural history—including its anthropological dimension—was viewed by Europeans as a means to promote territorial expansion, colonization, and trade. In Tahiti, Banks observed the nutritious qualities of the breadfruit plant, and conceived the scheme of transplanting it to the West Indies to feed the slaves working on sugar plantations. This was the purpose of William Bligh's ill-fated voyage on the *Bounty* in 1787–1789. After that voyage succumbed to the infamous mutiny, Bligh completed the task successfully on the *Providence* in 1792–1793. And natural history was not the only science practised on such voyages. In fact, European ships of the period have been called 'floating laboratories', since they served as mobile sites for various kinds of scientific experimentation. Franklin made use of his floating laboratory on each of his

Omai, Sir Joseph Banks, and Daniel Solander (William Parry, *c.*1775–1776). National Portrait Gallery, London (NPG 6652).

eight crossings of the Atlantic; he measured atmospheric conditions and ocean currents, observed birds and marine life, and practised navigational techniques. In 1768, he published the first chart of the Gulf Stream, converting mariners' lore into public knowledge and making it available for all sailors on the North Atlantic routes. Oceanic navigation also demanded expertise in astronomy. Determination of a ship's longitude at sea required mastery of sophisticated astronomical methods, such as measuring the position of the Moon or the motions of the satellites of Jupiter. Although John Harrison's improvements in the design of chronometers in the 1760s and 1770s promised to give mariners accurate means of timekeeping at sea, by which longitude could be found, astronomical methods remained an indispensible complement to even the best available timepieces.

Some of the voyages of the time, including Cook's first venture to the Pacific, were motivated by the need to improve astronomical measurements. The two transits of Venus during the eighteenth century, in 1761 and 1769, spurred the imperial powers, especially Britain and France, to mount expeditions to observe them. By timing exactly the movement of the planet across the face of the Sun from several locations, it would be possible to determine the Earth's distance from the centre of the Solar System, which was the unit by which all other distances and motions of the planets were calculated. In 1761, the French sent observers to the Indian Ocean and to Siberia. The British Astronomer Royal, Nevil Maskelyne, assumed responsibility for observations on the island of St. Helena in the South Atlantic. Less blessed by fortune, Charles Mason and Jeremiah Dixon, who had been dispatched by the Royal Society to Sumatra, were obliged to observe the transit from the Cape of Good Hope, since their destination had been occupied by French forces. By 1769, on the other hand, the Seven Years' War was over, and a degree of scientific cooperation between Britain and France resumed. Cook was able to view the event in Tahiti, and the French astronomer le Gentil was allowed by the British to observe from Pondicherry in southern India. Other observers viewed the transit from locations worldwide, including Yakutsk, Saint Domingue, and Baja California.

At the same time as knowledge was accumulating about the natural world through the massive collecting enterprise of metropolitan institutions, geographical knowledge of the globe was also being concentrated in the same locations. Latitudes and longitudes of thousands of places worldwide were measured with greater accuracy. Coasts and harbours on distant continents were charted by European navigators. Ocean currents and trade winds were recorded on printed charts. European countries were mapped with unprecedented accuracy, and the grid of latitude and longitude was extended across the North American continent as the basis for territorial divisions and urban planning. An early project was that of Gian-Domenico Cassini, astronomer at the court of Louis XIV, who launched a triangulation survey of France, resulting in a new map of the nation in 1693. Now the king could contemplate his realm and run his finger over its outline, though the map conveyed the disappointing message that the country was actually smaller in area than had previously been thought. Cassini also compiled the observations of travellers throughout the world to produce a map of the whole globe, centred on the North Pole, which he had carved into the floor of the Paris Observatory. His project inspired many other attempts at mapping in the following century, as the triangulation method was improved with more precise surveying instruments. The English surveyor and military leader William Roy led a triangulation survey of the Scottish Highlands after suppression of the Jacobite rebellion in 1745. In the 1780s, Roy's British survey, based on the meridian of Greenwich, was aligned with the French map, based on that of Paris. Subsequently, the British organized a survey of Bengal that was later extended to produce a map of the whole of India, while the French applied the method to Egypt during Napoleon's expedition to that country

Planisphère terrestre . . . par M. Cassini (Paris, 1696). Original in William M. Clements Library, University of Michigan; reprint (1941) in British Library, Maps 920.(386).

in 1798–1801. In these projects, geometry was enrolled as an ancillary of imperial rule, as the overseas territories Europeans were trying to subdue were brought under the regime of accurate mapping.

Meanwhile, geodesy (the question of the overall shape of the Earth) had also been preoccupying mathematicians and astronomers. In the 1730s, arguments flared in the Académie des Sciences about whether Cartesian predictions of an elongation of the Earth along its axis would be borne out, or alternatively whether Newton was correct in claiming it was slightly flattened at the poles. Two expeditions were dispatched to make the relevant measurements. Maupertuis was sent to measure a one-degree length of a meridian crossing the Arctic Circle in Lapland, while a team led by Charles-Marie de La Condamine went to perform the same task on the equator in Peru. Maupertuis did better out of the experience, returning to Paris within a few years with results that he was able to show vindicated Newton's claim. La Condamine took nine years to return, having lost many of his measurements on the way; and the botanist on his team, Joseph de Jussieu, only made it back to France after thirty-six years wandering in the Amazon rainforest. These experiences testify to the extraordinary suffering and endurance sometimes demanded of enlightened savants as they struggled to encompass the world within their systems of knowledge.

Crises and Revolutions

The last quarter of the eighteenth century witnessed revolutionary uprisings against the ruling orders in several European countries and a rupture of the ties that had linked Britain, France, and Spain to their colonial possessions in the Americas. The American Revolution (1775–1783) was followed by that in Haiti (1791–1802) and later by wars of independence in Latin American countries. In Europe itself, the French Revolution (1789–1799) set off revolts in the Low Countries, Poland, and Ireland, and provoked international warfare that embroiled the whole continent for two decades. The era of revolutions spelled the end of the Enlightenment as such, as European societies underwent internal and external conflicts, from which they emerged with transformed political and cultural institutions and new relations with the world at large.

The revolutionary period yielded a series of crises in Enlightenment science. American independence opened the way for the development of new scientific institutions in the United States. The American Philosophical Society (founded in Philadelphia in 1768) was joined by other learned societies, including the American Academy of Arts and Sciences (founded in Boston in 1780), and medical societies in many of the major cities. American scientific publications were established, and the colleges of the colonial era adjusted to the task of educating the new nation's elite. The transition was a traumatic one at King's College in New York: loyalist members of the faculty decamped to Nova Scotia, while those who stayed behind renamed the institution Columbia College (later Columbia University). Another American loyalist, the physicist Benjamin Thompson, fled to Europe, where he was awarded the title Count

Rumford by the Elector of Bavaria and played an important part in founding the Royal Institution of Great Britain.

In France, the outbreak of political revolution was preceded by the 'revolution' in chemistry proclaimed by Antoine Laurent Lavoisier. In 1774, Lavoisier, following a hint by Priestley, produced a new kind of air by heating the substance known as red calx of mercury. As Priestley had noted, the air was better for breathing than normal atmospheric air. Lavoisier first called it 'vital air', and then, deciding it was responsible for the acidic properties of the compounds in which it occurred, renamed it 'oxygen' (from the Greek for 'acid-maker'). Within a few years, he was launching an onslaught against the concept of phlogiston, traditionally recognized by chemists as the principle of inflammability in metals and other combustible substances. Lavoisier proposed that combustion could be understood, not as a release of phlogiston, but as a combination of the combustible substance with atmospheric oxygen. Chemical theory could be comprehensively reconstructed on this basis, provided water—regarded as an element since ancient times—was recognized as a compound of oxygen with another gas, hydrogen. Lavoisier's campaign on behalf of the new theory of chemistry gathered pace in the early 1780s. In 1785, he and his allies staged a large-scale demonstration in Paris of the decomposition and synthesis of water, making meticulous measurements of the quantities involved in both reactions, and claiming to show with a high degree of precision that water was indeed composed of hydrogen and oxygen. Two further stages of the campaign followed. In 1787, Lavoisier and three colleagues published a new system of chemical nomenclature, in which the oxygen theory was embedded in the very names to be used for chemical elements and compounds. In 1789, Lavoisier's *Traité élémentaire de chimie* presented the new system in the form of a basic textbook, capturing the allegiance of the next generation of chemists by co-opting the pedagogical methods by which the science was taught.

The new theory gradually won over the majority of European chemists. Early allies were recruited in Germany, the Netherlands, and Scotland. Many British chemists were reluctant to follow the French lead, but most of them were persuaded by the early 1790s. Priestley, however, resisted until the end of his life. Driven into exile in Pennsylvania, after his home in Birmingham was attacked by a loyalist mob incensed by his support for the French Revolution, he kept up a steady stream of objections and purported experimental refutations. After Lavoisier was executed by the revolutionary regime in 1794, Priestley rather tactlessly suggested that French chemists were using the techniques of the Terror to pressurize others to accept their system. The irony of a political radical (Priestley) who resisted the revolution in chemistry, while the chemical revolutionary (Lavoisier) was destroyed by a political upheaval he did not welcome, has often been noted. But Priestley was objecting against Lavoisier's whole approach to science, not just to a few details of his theory. He understood that the French chemist was introducing more sophisticated and expensive apparatus, which would limit the possibilities for replication of his experiments. He also resented the new nomenclature, which he saw as an imposition of a theoretical framework on what should be the neutral descriptive language of empirical science. In these ways,

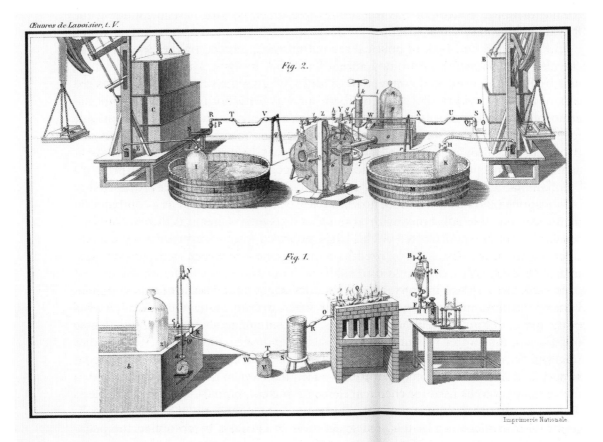

SUR LA DÉCOMPOSITION ET LA RECOMPOSITION DE L'EAU.

Lavoisier's apparatus for the analysis and synthesis of water (1785). From *Oeuvres de Lavoisier* (Paris, 1864–1893), vol. 2, plate 5.

Priestley viewed the new chemistry as undermining the public status of scientific knowledge, which (as we have seen) he associated with the progress of liberty and general enlightenment.

Priestley's critique acutely identified ways in which Lavoisier's chemistry pointed toward the shape of things to come. The instruments deployed by the French chemist set the pattern for the concentration of resources in the new scientific institutions that soon arose. The revolution saw the creation of the 'grandes écoles' in Paris, elite institutions of higher education, including those devoted to scientific and technical training: the École Polytechnique and the Conservatoire National des Arts et Métiers. In the pedagogical programmes of these institutions, Lavoisier's textbook, nomenclature, and instrumentation found their home. The Académie des Sciences was also reformed in the mid-1790s, and taken under the wing of the new Institut de France. Research there could also draw upon substantial instrumental resources, especially when the discovery of voltaic electricity opened up the new field of electrochemistry, which required the construction of large electrical batteries. The French reforms inspired similar developments in other countries. The University of Berlin (founded in 1810) created the model for a research-based university, subsequently adopted worldwide. In London, the Royal Institution (founded in 1799) mobilized private patronage and the enthusiasm of public audiences to support research in its laboratory. Humphry Davy, who worked there in the first decade of the nineteenth century, showed that he could command the resources to match or surpass French successes in electrochemistry. As cross-Channel rivalry arose between resource-intensive national institutions, Priestley would have been right to suspect that the age of enlightened public science was over. The modest apparatus with which he had conducted his experiments, including common household objects, was no longer adequate to the task. And the widespread public replication, which he had regarded as a stimulus to general enlightenment, had been displaced by private or government patronage of privileged institutional locations.

If Lavoisier's 'Chemical Revolution' led the way to the use of more substantial instrumental resources in the sciences, another simultaneous crisis pointed toward quite different features of the end of the Enlightenment. In 1778, the Viennese physician Franz Anton Mesmer arrived in Paris and started offering a new kind of medical treatment. With the aid of such props as metal rods and wooden tubs, or simply by passing his hands over his patients' bodies, Mesmer claimed to be able to direct the invisible fluid of 'animal magnetism' to cure various ailments. His therapies gained popularity among the social elite; they could be rationalized in terms of the prevailing notions of sensibility and the role of imponderable fluids in the nerves. But, in 1784, a commission of the Académie des Sciences, led by Franklin and Lavoisier, decided that Mesmer was a fraud, and his purported cures were due to nothing but the patients' powers of imagination. Mesmer retired from practice in disgrace, but mesmerism as such survived and indeed flourished for decades. Far from dying out, the belief in animal magnetism was bolstered by Luigi Galvani's discovery, published in 1791, of impulses in frogs' nerves that were apparently electrical in nature. The role of electricity in vital processes was to remain a focus of physiological inquiry, part of a renewed fascination with the powers of life that arose in the last years of the century. In addition, by directing attention at the potency of the imagination, the commission had elevated the importance of that mental faculty. In the ensuing years, the imagination was both feared and celebrated—feared when it was manifested as the hysteria of the revolutionary mob, but celebrated as the crucial attribute of poets, artists, and indeed scientific geniuses. In these ways, the crisis over mesmerism foreshadowed trends in the sciences that were to emerge later as part of the movement of Romanticism.

By the beginning of the nineteenth century, these new trends were beginning to become clear. In Germany, defeat and invasion by French military forces prompted the appearance of a newly assertive national culture, especially in literature and

Mesmeric Therapy (1778/1784). Iconographic Collection, Wellcome Library, London, ref. 17651.

philosophy. Johann Wolfgang von Goethe, soon recognized as the leading poet of the German language, made important contributions to natural philosophy in the 1790s and 1800s. His studies of the comparative anatomy of plants and animals suggested the existence of an original fundamental plan or '*Ur*-form', underlying all the manifold varieties of living things. Goethe's criticism of Newton's theory of colours also resonated with the idealist strain of German philosophy, which was increasingly emphasizing the creative role of the individual mind in perception and understanding of the natural world. Around the turn of the century, the Jena philosophy professor Friedrich Schelling emerged as the leading proponent of *Naturphilosophie*, which argued that the fundamental unity of the forces of nature could be perceived intuitively by the human mind. Schelling's philosophy informed subsequent inquiries into the unification of the forces of chemical affinity, electricity, and magnetism.

Naturphilosophie also introduced the idea of dialectical conflict as the cause of transformations in nature, supplying an important motif for the emerging fascination with evolution and historical change.

Two new sciences of this era encapsulated the departure from ways of thinking that had been characteristic of the eighteenth-century Enlightenment. *Biology*, first used as the name of a science in the years around 1800, was devoted to the study of vitality as such, the properties that made living things alive. Though it drew upon aspects of eighteenth-century materialism, it departed from its precursor in firmly demarcating living beings from non-living matter. The way was therefore opened for the development of ideas about evolution as a specifically biological process, rooted in organisms' capacities for reproduction and adaptation to the environment. *Geology* (although the word had been used earlier) came into general use around the same time for the historical study of the Earth. The different strata of rocks were identified with different stages of the Earth's history. This allowed geologists to reconstruct how present-day landscapes had been formed by the action of earthquakes, volcanoes, and erosion over very long periods of time. By the middle of the nineteenth century, the vastly expanded history of the Earth, which geology had revealed, provided the background against which the story of biological evolution could play out. This would have been inconceivable in the previous century, when evolution was the playful conceit of a few freethinkers, and most people believed the history of the Earth was confined within the limits set by Holy Scripture.

Conclusion: The Legacy of Enlightenment Science

Looking back from the standpoint of the nineteenth century, much of the science of the Enlightenment seemed crude and backward. Astronomy was largely confined to the Solar System. Natural history viewed the living world as a static and unchanging system of order. The chemical elements were mostly unknown, and even the concept was poorly defined. Electricity and magnetism were studied in isolation from one another, with no notion of their deep relationship. The idea of energy, which was to unify the physical sciences in the following century, remained beyond the intellectual horizon. The social profile of eighteenth-century science also seemed alien to subsequent practitioners. The men of science who claimed a professional identity in the nineteenth century, and who were eventually to be called 'scientists', looked back upon their precursors as simple amateurs. They did not recognize as their peers the public lecturers, craftsmen, navigators, and naturalists who had populated Enlightenment science. In fact, the whole idea of science as a public enterprise seemed—from a retrospective view—to fail to make the necessary separation between serious intellectual work and mere popularization.

But the nineteenth-century judgement was an unfair one, and it should not place limits on historical understanding. The era of the Enlightenment left an important legacy for the subsequent history of science, even if the prevailing ideas of the time have been superseded in many respects. We may be more inclined now to demarcate the realm of scientific experts from that of the populace at large, but the two remain interdependent. Scientific practitioners still have to communicate to the general public to gain support for their work, and we can all think of individuals who have achieved prominence in that role. Scientific knowledge is still a subject of widespread interest and discussion in the media of our day; and so it should be, given its economic and social importance. We should recall that it was during the eighteenth century that science entered the public sphere, from which it has never departed, notwithstanding the enormous growth of specialist expertise and professional institutions.

Another legacy of Enlightenment science is the connected world within which ideas circulate. Eighteenth-century mariners, explorers, and navigators established many of the ties that continue to bind us together, such as the North Atlantic routes along which communications, trade, and human traffic still flow. They mapped the globe with unprecedented accuracy and extended worldwide the system by which plants and animals were classified. Enlightened investigators tried to subject the whole planet to a kind of grid of uniformity. They assumed that everywhere could be located in terms of latitude and longitude, that every living thing could be placed within the order of classification, and that every human group could be situated at some point on a scale of historical progress from savagery to civilization. These expectations were often thwarted in practice. Navigational and surveying instruments proved hard to handle in Lapland or the upper reaches of the Amazon. Natural history specimens were often lost in transit to collecting institutions, or arrived rotted or unlabelled. Printed materials did not circulate with perfect faithfulness; they were often overwritten, edited, and pirated. And, in trying to subject human beings to the uniformitarian grid, Enlightenment thinkers were forced to confront the irreducible diversity of human cultures, which would not be easily confined within a linear scale.

Those who shared the age's confidence in reason—a confidence that has often been perceived as a kind of arrogance—set themselves to overcome these obstacles. But they were given many occasions to realize that this would never be entirely possible. Individuals who thought of themselves as enlightened inevitably encountered many wonders and anomalies; and they lived in close proximity to people they regarded as primitive or unenlightened. In view of this, we should not be surprised that a degree of uncertainty about how far rationality could be extended was characteristic of the era. This was what Kant was pointing to with his remark that his contemporaries were living in an age of enlightenment, but not yet in an enlightened age. The same thing, however, could be said of every subsequent period. The aims of universal enlightenment have never been accomplished, and a consciousness of the limits of rationality is a condition of modern life. In this respect, while the era of Enlightenment science ended at the close of the eighteenth century, the process of cultural transformation—with its inherent incompleteness—continues to our own day.

PART II
DOING SCIENCE

7 Experimental Cultures

IWAN RHYS MORUS

WHEN we think about modern science, how it is done, and where is takes place, we usually think about experiments in the laboratory. Laboratories are the iconic spaces of experimental science, where white-coated experts interrogate nature and force it to give up its secrets. It is an image familiar to us from countless B movies from the 1950s onwards, or from television science fiction series: think Jon Pertwee's incarnation of the third Doctor Who at UNIT headquarters, for example. The laboratory is so deeply ingrained as the place of experiment in our perception of what science is and how it should be practised that it can be difficult to appreciate that such places have a history and that, in fact, the laboratory as a unique space for experiment is a rather recent historical phenomenon. The kind of activity we would now describe as experiment took place in the past in a variety of different places—in kitchens, outhouses, and workshops. The idea that experiment needed its own particular place only really emerged during the nineteenth century. The laboratory, as a distinctive space for experiment, rather than a space in which experiments happened to take place, appeared in the context of debate and disagreement about just what experiments were, and how—and by whom—they should be conducted.

The Scottish physicist James Clerk Maxwell suggested in the 1870s that there were two different kinds of experimental culture. On the one hand was the culture of 'sparks and shocks which are seen and felt', on the other was the culture of 'currents and resistances to be measured and calculated'. Maxwell himself was certainly an enthusiastic exponent of measurement and calculation in experiment, though some of his activities suggest a continuing attachment to seeing and feeling in experiment as well. The contrast underlines Maxwell's own recognition that there was more than one way of doing experiments during the nineteenth century, and more than one view about what should take place in a laboratory. Laboratories changed during the course of the nineteenth and twentieth centuries from being the workplaces of individual experimenters to being large centres of collaborative research—and teaching. As they changed, so did the kind of experimental knowledge they produced.

This chapter will survey the contrasting and occasionally competing experimental cultures of the modern age. It will explore aspects of the new and transformative knowledge that has emerged from laboratories over the last two centuries and the practices and procedures that lay behind that knowledge. It will emphasize what went on outside as much as inside laboratories. To understand the culture of experiment fully we need to think about the origins of the skills and materials that were needed for successful experimental performances. The experiments that emerged from nineteenth-century laboratories were often visually striking. They were designed to please the eye as much as satisfy the mind. This is not surprising for an age that typically regarded seeing as the paradigm of knowing. But it will become clear in this chapter that the appeal to sensation in experiment was not confined to the nineteenth century. It has been an integral part of experimental cultures throughout the modern period.

Demonstrating Nature

When Humphry Davy arrived in London in 1801 to become a chemical lecturer at the recently established Royal Institution he was joining a very new kind of institution. The Royal Institution's promoters wanted to devote the new body to 'the speedy and general diffusion' of knowledge, 'teaching the application of scientific discoveries', and 'the increase of domestic comfort and convenience'. Ironically, given its later prominence under its first directors, the institution's laboratory received little mention in the early plans. Where it did appear was as an adjunct to the lecture theatre. To promote 'teaching the application of science to the useful purposes of life', the promoters promised that, 'a lecture-room will be fitted up for philosophical lectures and experiments, and a complete laboratory and philosophical apparatus, with the necessary instruments for making chemical and philosophical experiments'. This was, in fact, exactly where one might expect to find a laboratory at the beginning of the nineteenth century. It was the nearest thing the laboratory had to an institutional setting. Laboratories were rooms where experiments were prepared for public demonstration. The Royal Institution's laboratory was imagined by the institution's founders as a place where 'men of the first eminence in science' would practise their experimental performances before putting them on show in the scientific space that really mattered to them—the lecture theatre.

In many ways, that was exactly what Humphry Davy did. Davy made a name for himself at the Royal Institution as a performer of flamboyant and spectacular electrical experiments. Just a year before Davy arrived there, the Italian natural philosopher Alessandro Volta had demonstrated the instrument that would come to be known as the voltaic, or galvanic, pile. The pile was Volta's ultimate response to his dispute with fellow Italian Luigi Galvani. Galvani had argued that his experiments with frogs demonstrated the existence of what he called animal electricity. Volta developed the pile to show that the same electrical effects could be produced with no animal tissue, so disproving Galvani's theory. In Davy's hands the pile would do far more than that, though. He turned it into an instrument that could be used to astound his audience at the Royal Institution with his mastery of nature and to make equally astounding new discoveries in chemistry. Where Volta had argued that the electricity in the pile originated with the contact of different metals, Davy insisted it was chemical in origin. Using the great battery at the Royal Institution, Davy succeeded in identifying half a dozen new elements as well as establishing that chlorine and iodine were elements too, despite claims to the contrary by French chemists defending their late compatriot Lavoisier's new chemistry. With discoveries like these, Davy made the Royal Institution and its laboratory into an important power-house within European natural philosophy.

Davy's successor, Michael Faraday, carried on in the tradition that Davy established of discovery and display. In 1820, the Danish natural philosopher Hans Christian Oersted demonstrated that a magnetized needle suspended near a wire was deflected when a current of electricity passed through it. Oersted speculated that some kind of force acting in a circle around the wire might be responsible. Faraday took up the suggestion and soon demonstrated how a current-carrying wire could be made to rotate around a magnet, the electric and magnetic powers seeming to interact to produce mechanical motion. Just over a decade later in 1832, Faraday discovered electro-magnetic induction, showing that a moving magnet could produce electricity. From then until the 1850s, Faraday kept up a steady stream of experimental researches flowing from the Royal Institution's laboratory. Like Davy, Faraday was a consummate performer as well as an experimenter. In fact the two things often went hand in hand throughout the nineteenth century. Audiences flocked to the Royal Institution's lecture theatre to watch as Faraday 'expiated on the beauties of Nature, and when he lifted the veil from her deep mysteries'. But those spectacular performances depended on careful, meticulous, and disciplined preparation in the laboratory. Faraday's reputation as an experimenter depended on both these things.

Faraday's experiments demonstrating electro-magnetic induction effectively tied the final knot in the relationship between electricity and magnetism. He had already established that a current-carrying wire would rotate around a magnet. Now, using remarkably simple apparatus, Faraday showed how, if two wires were wrapped around an iron ring in two separate coils, when one coil was connected to a battery a current of electricity was briefly induced in the other coil. When the first coil was disconnected, again a current was briefly induced in the other. He also demonstrated how, if a bar magnet was inserted along the axis of a wire coil, a current was induced in that coil by the movement of the magnet. Having established the principle in his laboratory, Faraday moved quickly on to develop a form of the apparatus that would allow him to take the new phenomenon upstairs to the public arena of the Royal Institution's lecture theatre. This was still an important part of the process of making an experiment public. It had to be something that could be seen by an audience. Others were quick to develop their own ways of showing off the phenomenon too, turning the fleeting and transient flicker of electricity that Faraday had produced in the laboratory into a robust and easily demonstrated spectacle.

Early Victorian experimental culture was to a large degree organized around technologies of display like this. In the hands of some of Faraday's contemporaries and rivals it was explicitly directed at finding ways of reproducing the spectacle of nature. 'Nature's laboratory is well stored with apparatus of this kind', said William Sturgeon, 'and the insignificance of our puny contrivances to mimic nature's operations, must be

Michael Faraday working in his laboratory at the Royal Institution. This was where experiments were prepared before being put on show in the institution's lecture theatre. The scene offers a glimpse of the kind of apparatus an early nineteenth-century experimenter might need—such as the furnace in the centre of the picture.

amply apparent when compared to the magnificent apparatus of the earth.' Where Faraday lectured in London's fashionable Albemarle Street, Sturgeon and his like plied their experimental trade in establishments such as the Adelaide Gallery near the Strand, or the Royal Polytechnic Institution on Regent Street. At emporia of scientific spectacle like these, customers paid a shilling at the door for the opportunity to witness wonder:

Clever professors were there . . . teaching elaborate sciences in lectures of twenty minutes. . . . There were artful snares laid for giving galvanic shocks to the unwary; steam-guns that turned bullets into bad sixpences against the target; and dark microscopic rooms for shaking the principles of teetotalers, by showing the wriggling abominations in a drop of the water they were supposed daily to gulp down by pints.

Performers in such places were experimental entrepreneurs who had as much in common with P. T. Barnum (who also put on shows at the Adelaide Gallery) as they did with Faraday.

These galleries of practical science were very much a metropolitan phenomenon, though there was at least one attempt to establish one outside London when William

The Royal Polytechnic Institution (Great Hall.)

Printed & Published at the Polytechnic Institution, 309, Regent St.

The Royal Polytechnic Institution on Regent Street was a major attraction for Londoners interested in experiment and spectacular inventions. The diving bell that was one of the Polytechnic's main attractions can be seen in the background.

Sturgeon was invited to Manchester as superintendent of the Royal Victoria for the Encouragement of Practical Science. One of the first to lecture at the new institution was James Prescott Joule. Joule shared Sturgeon's fascination with making machines that imitated natural operations. His early papers were in the form of letters to Sturgeon's *Annals of Electricity*, discussing Sturgeon's own electro-magnetic engine and similar efforts in making electricity useful. This was the train of experimental inquiry that eventually led a few years later to Joule's paddle wheel experiment and his discovery of the mechanical equivalent of heat. A wealthy Manchester brewer's son, Joule was less interested in spectacle than in efficient performance. Making better engines was the concern that led to the mechanical equivalent of heat. The concern with visibility was still important to Joule nevertheless. As well as suggesting ways for improving engines, the paddle wheel experiment provided a way for making God's ordering of nature visible. It was a way of making tangible how 'the phenomena of nature, whether mechanical, chemical, or vital, consists almost entirely in a continual conversion of attraction through space, of living force and heat into one another.'

This sort of concern with experiments as ways of displaying the hidden mechanism of nature to new audiences was widely shared across Europe and North America during the first half of the nineteenth century. In Britain, the concern with making God's powers in nature visible to their audiences had been an important feature of the public activities of Newtonian natural philosophers from the early eighteenth century onwards. Similarly, a natural philosopher such as Jean-Antoine Nollet in France was adept at turning experiment into a crowd (and monarch) pleasing entertainment, with his chains of electrified monks and royal guardsmen. The German performer Georg Matthias Bose certainly knew how to attract popular attention with the electrical Venus Kiss and his beatification experiment. Experiments like these, or Benjamin Franklin's exploits cooking a turkey with electricity on the banks of the Skuylkill for Thanksgiving Day 1748, were designed with sensation (in all its senses) in mind. Even experiments such as Antoine Laurent de Lavoisier's investigations during the 1770s and 1780s into combustion and respiration and the tradition of metrology in experiment they helped inaugurate still partook of this culture of experiment as a means of showing nature.

During the first two decades of the nineteenth century, French experimental culture was dominated by protégés of the natural philosopher Pierre Simon Laplace and the chemist Claude Louis Berthollet. The style of experiment favoured by the Society of Arcueil (so named after the location of the neighbouring estates owned by the two patrons) emphasized careful measurement and quantification. Jean Baptiste Biot described this new breed of French experimenter as 'physiciens géometres', with the ability to combine calculation and experiment to produce the 'ultimate in precision'. His own work on the polarization of light epitomized this Laplacian approach to experiment. The laboratories of Arcueil might be private domains, but the French state had an interest in experiment too. In 1843 the French Ministry of Public Works commissioned Henri Victor Regnault, professor of physics at the Collège de France, to carry out extensive experiments on the properties of steam, with the aim of compiling data that could be used to improve steam engines. As well as being useful to the French state, such experiments were significant to industrial concerns and to the broader culture of sensation as well. Instruments such as Biot's own polarimeter offered new ways of analysing colours that would be taken up by artists and by industrial dyers.

In the German lands, by the middle of the nineteenth century the ethos of experiment was changing too. Earlier in the century the dominant view was that of *Naturphilosophie* and the aim of experiment was to investigate the transcendental unity of nature. Oersted's investigations into the relationship between electricity and magnetism was informed by this perspective, for example. Another proponent of this kind of view was Johann Wilhelm Ritter, who took his task to be discovering the 'genuine world-soul of nature'. Electricity was the main focus of Ritter's researches, though he also investigated the phenomena of light, discovering the existence of ultra-violet rays beyond the spectrum of visible light in 1801. Another German experimenter fascinated by the possibilities of *Naturphilosophie* was the Berliner Thomas Johann Seebeck, whose investigations led to the discovery of how to produce electricity from heat. By the 1840s, German experimenters were anxious to disassociate themselves from these kinds of speculations and embraced the French enthusiasm for measurement instead. Hermann von Helmholtz was one of the first exponents of this new approach to experiment and his 1847 essay *Über die Erhaltung der Kraft* (On the Conservation of Force) one of the early outcomes. A few years earlier, Helmholtz, along with like-minded friends such as Emil du Bois Reymond and Ernst Brücke (both, like Helmholtz, originally trained as physiologists) had helped establish the Berlin Physical Society to foster this new experimental culture.

Physiology and physics shared a close connection in early nineteenth-century Italian experimental culture too. Luigi Galvani's experiments on animal electricity at the end of the eighteenth century had been an important resource for German *Naturphilosophie* and Italian experimenters such as Giovanni Aldini (Galvani's nephew) and others continued that tradition. Aldini famously toured France and Britain to defend his uncle's claims, performing electrical experiments on the recently executed body of George Forster when he visited London in 1803. Other Italian experimenters were also interested in the relationship between electricity and the other powers of nature. Macedonio Melloni, professor of physics at the University of Parma, and Leopoldo Nobili carried out experiments on the relationship between light and heat in 1831. A year later, Nobili and Vincenzo Antinori, director of the museum of physics and natural history in Florence, found themselves in hot water with Michael Faraday when they repeated and published his experiments on electro-magnetic induction before he had published them himself. Throughout the 1840s, Carlo Matteucci, professor of physics at the University of Pisa, made a name for himself with a series of experiments investigating the relationship of electricity and the nervous force. Like experimenters in France and, to a lesser extent, the German lands, Italian experimenters were often associated with institutions that were funded by the state, even if the state did not usually fund (or expect them to carry out) experimental research.

Early nineteenth-century experimental culture in the United States resembled more closely that of Britain. Slowly, a network of small, privately endowed colleges was spreading across the east coast. Natural philosophy was on the curriculum at such places, and whilst professors were not expected (or paid) to do anything beyond preparing experiments for teaching and demonstration, they sometimes did. Joseph Henry, professor of natural philosophy at New Jersey College (later Princeton) was one of the more prominent. Following William Sturgeon's invention of the electromagnet as an instrument that magnified magnetic effects for display, Henry made a name for himself as a maker of ever more powerful versions. His experiments on the relationship between electricity and magnetism came close to pre-empting Faraday's discovery of electro-magnetic induction. Outside large cities there were limited resources for experimentation. Henry recorded tearing up his wife's silk petticoats to provide insulation for his electro-magnets. Both Boston and Philadelphia supported small but thriving communities of experimenters, with scientific instrument-makers and institutions where the results of experiment could be put on display. There were close links with British experimenters. Henry visited London during the 1830s to purchase instruments and met both Faraday and Sturgeon, noting in a letter home, with some understatement, that they were 'not on good terms'. In a double-edged compliment he described Sturgeon as being 'at the head of the second-rate philosophers of London'.

These growing experimental cultures across Europe and North America depended for their success on developing communities of skilled workers and the availability of material resources. Henry's European visits in search of experimental apparatus are a good illustration of the importance of such things. Such traffic across the Atlantic was not all one way. The Philadelphian instrument-maker Joseph Saxton spent much of the 1830s in London, building apparatus for the Adelaide Gallery (itself founded by the American Jacob Perkins). London, as well as other major European cities such as Paris and Berlin, had developed as significant centres for the instrument-making trade by the end of the eighteenth century. These cities supported a substantial industry in

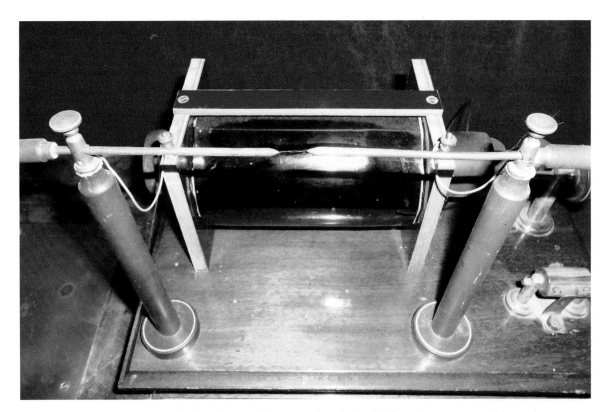

An induction coil of the kind improved by Heinrich Ruhmkorff. These instruments could be used to generate spectacular sparks for demonstration and experimental investigation.

scientific instruments, with close links to other trades such as clock-making, spectacle-making and metal-working. Instrument-makers were quick to pick up on the latest scientific developments and turn them into useful apparatus. The Parisian instrument-maker Hippolyte Pixii transformed Faraday's discovery of electro-magnetic induction into a magneto-electric machine that generated a steady flow of electricity. Joseph Saxton soon produced a similar instrument for the Adelaide Gallery as well. Following Nicholas Callan's invention of the induction coil (another instrument that exploited Faraday's discovery), instrument-makers were soon advertising their own versions of the new instrument in their trade catalogues.

Callan's invention of the induction coil and the improvements made by other instrument-makers to the original design produced what was to be the nineteenth century's most significant piece of scientific apparatus. During the early 1850s the Paris-based German instrument-maker Heinrich Rühmkorff made significant improvements to the coil that made the device much more powerful. With this new source of intense electricity, experimenters could produce and investigate a variety of spectacular phenomena. In 1854, the English experimenter John Peter Gassiot used a Rühmkorff coil borrowed from his friend William Robert Grove to produce the effect known as Gassiot's cascade. The cascade—in which the appearance of flowing liquid fire was created by passing electricity through a glass cup inside the vacuum of an air-pump—was an example of a discharge phenomenon, produced by passing high-tension electricity through an attenuated gas in a sealed glass tube or similar container. It was widely hailed as one of the most beautiful of experiments. The German instrument-maker Heinrich Geissler specialized in the production of intricately designed glass tubes in which different gases glowed in various colours with the passage of electricity. Gassiot, and others during the 1850s such as Grove and the German experimenter Julius Plücker, used Rühmkorff coils to try and discover the properties of these curious discharges.

William Crookes was one of the most energetic and prolific researchers into discharge phenomena during the 1860s and 1870s. His experiments provide illuminating evidence of the complex interplay between spectacle and systematic investigation. Crookes wanted to use discharges in order to investigate the properties of what he called the fourth state of matter. He produced some spectacular experiments, prompting the mathematical physicist George Gabriel Stokes to marvel to a correspondent that, 'I know of nothing like what Crookes has been doing for some years. ... I wish you could see some of the work in his laboratory.' In one experiment, Crookes propelled a tiny glass locomotive down a track inside an evacuated tube by using the power of the 'radiant matter' flowing between the electrodes. Others, such as Plücker's student Wilhelm Hittorf in Germany, or Warren de la Rue and Hugo Muller in England, worked on discharge phenomena too. Experiments like these took place at the boundary of the real in Victorian science. They were at the edges of what experiments could do. Crookes, for one, thought that they offered a good model of how experimenters might approach the study of other liminal phenomena such as those produced in spiritualist séances.

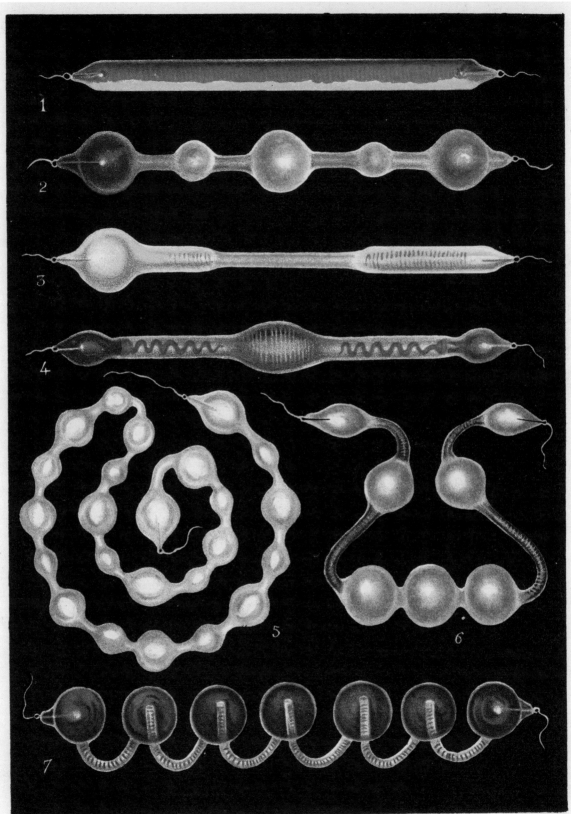

ELECTRIC DISCHARGES IN RAREFIED GASES.

1, Vacuum Tube showing Fluorescence of Sulphuret of Calcium ; 4, Vacuum Tube showing Nitrogen Vacuum (Spirals of Uranium Glass) ; 7, Vacuum Tube showing Hydrogen ; 2, 3, 5, 6, Vacuum Tube showing Geissler Tubes.

Spectacular performances and displays at places such as the Royal Polytechnic Institution in London played important roles in bringing the products of this mid-century experimental culture to a wider audience. Just as important was the proliferation of popular books on experiment. John Henry Pepper's *Boy's Playbook of Science* gave detailed accounts and descriptions of how to perform a wide variety of experiments. Pepper, who managed the Polytechnic from the early 1850s onwards, offered his readers 'a series of philosophical experiments detailed in such a manner that any young person may perform them with the greatest facility, and ... will afford him a delightful and profitable recreation when satiated with mere *play*, or imprisoned by bad weather, or gloomy with the unamused tediousness of a long winter's evening.' Magazines aimed at young boys, such as the *Boy's Own Paper*, published by the Religious Tract Society with the aim of instilling the values of muscular Christianity in a new generation of schoolboys, offered instructions for experiment along with the morally improving tales of derring-do. Scientific instrument-makers manufactured philosophical and optical toys of all kinds aimed at this kind of market. From the 1830s onwards, chemistry sets were produced and sold as philosophical amusements for children.

One of the most important changes that took place during the course of the nineteenth century in this context was in the pedagogical role of experiment. At the beginning of the century, when universities possessed laboratories at all, they were regarded as spaces for the preparation of lecture demonstrations. When experiments were performed in such settings, they were carried out as demonstrations, rather than as activities that students might carry out themselves; experiments were performances to be witnessed rather than practised by students. During the course of the century, however, teaching laboratories designed to make students themselves into experimenters became common in universities. The trend started in continental Europe and was later adopted in the British Isles, with William Thomson establishing the first teaching laboratory at the University of Glasgow in the 1850s. Even as late as the 1880s, university laboratories were still relatively rare. Oliver Lodge reported in his autobiography that 'laboratories were rather novelties in those days', and that he still felt it necessary to 'make a tour of the Continental laboratories and gain experience that way'. Proponents sometimes faced an uphill struggle persuading university authorities that experimental culture, tainted with showmanship and manual labour, had a place in their hallowed halls.

Experimental Institutions

By the end of the nineteenth century laboratories seemed well established as their own particular kind of places. Being in possession of well-stocked and managed teaching

---

Opposite: An array of Geissler tubes. Invented by the German instrument-maker Heinrich Geissler, apparatus like this, in which electricity passed through different gases producing brightly coloured discharges, offer a vivid example of the importance of spectacle for Victorian science.

and research laboratories was recognized as an essential requirement for any European or North American university with pretensions of any kind to being real factories of knowledge. To varying degrees, states were learning the lesson that science mattered for national prosperity and that providing space for experiments was not just a national duty, but a prerequisite of continued economic progress. A rhetoric of discipline was an important feature of the experimental culture that developed in these late nineteenth-century laboratories. Experiment, according to the men who managed these laboratories, was not just a way of finding out about nature, it offered a means of disciplining the minds and bodies of its practitioners too. Successful experiment meant attention to detail, patience, and a high degree of manual dexterity. Not just in Europe and North America were laboratories valued. As the Meiji restoration took place in Japan during the final decades of the century, the new regime placed the reorganization and advancement of knowledge as one of its core objectives. Authorities in laboratory science such as William Thomson's student, William Ayrton, were imported to teach Japanese students the values of experiment.

In his furious broadside against the corruption and nepotism of English scientific culture in 1830, *Reflexions on the Decline of Science in England*, Charles Babbage called on the English state to emulate its French counterpart and make science a state-funded activity. The place of experimental culture in France during the first half of the nineteenth century was, however, rather less secure than Babbage imagined. Scientific and technical education had certainly become a prized commodity in the decades following the revolution. Scientific institutions as well as the state were re-ordered, and men of science such as Laplace or Berthollet held positions of significant political power. The laboratories where such men and their disciples experimented often remained private fiefdoms nevertheless, and French savants complained often about the lack of facilities at their disposal. The state patronage accorded Regnault for his experiments on heat were the exception rather than the rule. Whilst Babbage gazed enviously at French experimental culture, the French themselves gazed just as enviously at the German states—and Prussia in particular. French science—again like the French state—was highly centralized and a career in Paris represented the pinnacle of scientific success. In Paris itself, an ability to turn experiment into spectacular public demonstrations, as Claude Pouillet did with his electrical researches at the Sorbonne during the 1830s, often helped secure such a career.

Under the Second Empire, a new generation of big beasts emerged to represent and defend experimental culture in France. Claude Bernard at the Collège de France set out to turn physiology into physics by putting experiment at the centre of his science. His *Introduction a l'étude de la médicine expérimentale* aimed to place experiment at the heart of medicine. Endowed with a new professorship of physiology at the Sorbonne, he succeeded in persuading the Emperor to provide him with a laboratory as well. The Sorbonne's professor of physics, Jules Jamin, awarded the Rumford Medal in 1858 for his experimental researches on the properties of light, continued in his predecessor's tradition. As Pouillet had done before him, he drew enthusiastic crowds to his spectacular experimental demonstrations. The biggest beast of the new generation was the chemist Louis Pasteur. Like Jamin, Pasteur knew how to use experiment to spectacular effect, attracting audiences and patronage for his work. His comprehensive defeat of his opponent, Félix-Archimède Pouchet, in a public debate on spontaneous generation at the Sorbonne in 1864 was an object lesson in public experimental performance. Experimenting with sealed flasks on stage, Pasteur convinced his audience that spontaneous generation was a chimera that only seemed to take place in circumstances when life was already present.

By the late 1860s, discontent at the lack of state funding for laboratories in France reached a climax. Pasteur wrote that laboratories were 'temples of the future'. He contrasted the miserable state of affairs in France with the flourishing support for experiment that existed in Germany. Pasteur's prestige was high enough to gain him the Emperor's ear, and a visit to Pasteur's chemical laboratory at the École Normale Supérieure in 1868 convinced Napoleon III that he should act on the matter. Laboratories and the experimental cultures they housed were sold to the French state, as they would be to others during the final decades of the century, as vital accompaniments to national progress, both economic and military. The result was a proliferation of state-endowed laboratories across the emerging scientific disciplines, though too late, of course, to save France from the humiliation of losing the Franco-Prussian war a few years later. Physicists at the Sorbonne ended up with two laboratories to accommodate their experiments. Paul Desains presided over the Laboratoire d'Enseignment de Physique, primarily devoted to teaching, whilst Jules Jamin ruled the smaller Laboratoire des Recherches Physiques. Both men were keen exponents of the cult of precision measurement as the cornerstone of experimental work. Pasteur, of course, ended up with an entire institution of his own, with the Institut Pasteur founded in 1887.

At the beginning of the nineteenth century it was clear that it was France's scientific institutions that seemed to offer the best model for how science should be organized, and Napoleon's march across Europe led to the export of French institutional models too. By the end of the century, however, the ideal scientific state was without question the newly united Germany. It was to the new Reich that European experimenters looked for inspiration. Admiring Hermann von Helmholtz's experimental facilities from afar John Tyndall noted that, 'you will find in the Berlin Laboratory the very things which my American and British friends and I would like to see in operation in all college and university laboratories in America and the British Empire'. German superiority in laboratory provision certainly helps to explain how Heinrich Hertz at the University of Karlsruhe succeeded in beating his British counterparts to the discovery of electro-magnetic waves in 1888. The laboratory facilities he would have at Karlsruhe were an important factor in Hertz's decision to accept a professorship there in 1885. There he had access to the resources that allowed him first to produce and detect electro-magnetic oscillations in the ether, and then show that they could be diffracted, polarized, reflected, and refracted just like ordinary light.

More generally, however, German experimenters themselves were often less happy about the state of their laboratories and continued to press for better resources, arguing that 'our most dangerous enemies in the struggle for survival' were busily creating their own research institutions. The outcome of all this lobbying in 1887 was the establishment of the Physikalisch-Technische Reichsanstalt in Berlin. This was to be an institution entirely devoted to experiment and its potential industrial applications. Much of the funding for the new institution was provided by the electrical engineer Werner von Siemens. Siemens, who had made his fortune in telegraphy and in the new electrical industries that started to flourish during the second half of the nineteenth century, was a powerful advocate for state laboratory funding, arguing that experimental research needed to become 'a professional activity within the state structure'. He regarded himself as being a physicist as much as he was an industrialist, and argued that experiment and industry were inextricably linked. Even with Siemens's money on the table, the Reich bureaucracy needed some persuasion. Siemens worried that the powerful Otto von Bismark thought that experiment was 'a type of sport without practical meaning', rather than an activity that was essential to the future prosperity of the Reich.

The institution's first director was to be Hermann von Helmholtz, widely celebrated as the most eminent of German experimenters. He had made his name as an early pioneer of the conservation of energy and was already the director of the Berlin Physics Institute, running the laboratory of which Tyndall was so envious. The scientific section under his leadership was divided into three separate laboratories carrying out experimental work in thermodynamics, electricity, and optics respectively. The aim across the board was to establish new standards of precision measurement that could be put to industrial use. The thermodynamics laboratory worked on improving thermometry, for example, whilst the optical laboratory developed a more reliable photometer for comparing the intensity of light from different sources. The institution's building was built with the requirements of experiment in mind. The entire edifice was built on a foundation of a thick thousand-square-metre concrete slab to ensure maximum stability and the external walls were carefully shielded from the Sun to help ensure an even temperature. The location of the various laboratories within the building was arranged with the specific requirements of different kinds of experimental work in mind. The thermodynamic laboratory was on the ground floor, for example, since that was where it would be easiest to control temperature.

Following Helmholtz's death in 1894, the directorship of the Reichsanstalt was given to Friedrich Kohlrausch, author of the *Leitfaden der praktischen Physik*, one of the most widely used of German textbooks on experiment. He was a master of precision measurement, having spent his experimental apprenticeship with the electrical experimenter Wilhelm Weber at the University of Göttingen during the 1860s before going on to join his former master as co-director of the Göttingen Physics Institute. Even more than Helmholtz, Kohlrausch was determined to establish the Reichsanstalt as an indispensible arm of the German state. By 1903, the Reichsanstalt had more than doubled in size and had become a powerhouse of scientific testing for German industry. The culture of experimentation that both Helmholz and Kohlrausch encouraged was one that emphasized discipline and meticulousness. Some of his

Charlottenburg   Technische Reichsanstalt

The Physikalische Technische Reichsanstalt in Berlin. This was a highly visible statement of the new German state's commitment to physics and experiment. It was a building purpose built for experimental physics.

colleagues thought that Kohlrausch quite literally embodied those sorts of values. The Swedish chemist and physicist Svante Arrenhius, who had collaborated with Kohlrausch whilst he was a professor at the University of Würtzburg, remarked that he 'always lived as orderly as a chronometer'. The experimental world of the Physikalische-Technische Reichsanstalt was meant to place physics at the service of the Reich, but it was also and as surely meant to stand for the virtues of self-discipline and orderliness that Germany's rulers wanted to develop in its citizens.

Exponents of laboratory experiment in Britain made similar claims about the values of experiment. There, the pre-eminent institution for experiment by the end of the nineteenth century was the Cavendish laboratory in Cambridge, established in 1874. It took some persuasion to convince a conservative university establishment that a laboratory of experimental physics was really something that would sit comfortably in an institution devoted to liberal education. As with Siemens's offer in Germany, it was the willingness of William Cavendish, the seventh Duke of Devonshire, to fund the new laboratory that finally convinced them. The first Cavendish professor was the Scottish physicist James Clerk Maxwell. Maxwell understood that making the kind of laborious experimental work that would be carried out at the Cavendish attractive to students and university authorities might be difficult. To some, experimentation seemed contaminated by the worlds of the workshop or of showmanship. Maxwell did not want to 'bring the whole university and all the parents about our ears' by giving the wrong impression of what a laboratory was for. Instead, he emphasized the moral aspects of experimental work. The process of precision measurement promoted self-discipline and helped demonstrate the divine plan, Maxwell argued. A Cambridge education was meant to produce disciplined minds and Maxwell needed to convince the sceptics that learning how to perform experiments offered a means to that end.

Precision measurement was at the heart of the experimental regime at the Cavendish laboratory too. In particular, Maxwell brought with him to Cambridge the work that he had been doing as part of the British Association for the Advancement of Science's efforts to establish a standard unit of electrical resistance. Electro-magnetism was the science in which Maxwell had made his name and his *Treatise on Electricity*

A practical class being taught how to experiment at the Cavendish Laboratory in Cambridge about 1900. Students developed their experimental skills by working their way through a set of standard experiments.

and Magnetism was published in 1873, shortly after he was appointed to his professorship. Developing reliable experimental techniques for establishing standard electrical units was of potentially significant benefit to the country's telegraph industry as well as a means of developing skills in precision measurement. The British Association's magnetometer was moved from Kew to the Cavendish and Maxwell set all his students the task of using it to measure the Earth's magnetic field as their first experimental exercise. Lord Rayleigh, Maxwell's successor, continued Maxwell's emphasis on precision measurement. More than Maxwell, however, Rayleigh emphasized system and large-scale collaboration. Arthur Schuster recalled how Rayleigh was anxious to 'identify the laboratory with some research planned on an extensive scale so that a common interest might unite a number of men sharing in the work'.

When Rayleigh retired and returned to his own private laboratory at his country estate in Terling, his replacement as Cavendish professor was J. J. Thomson. Thomson had grown up at the laboratory under Rayleigh's direction and was fully imbued with its ethos of precision measurement. His vision of what the Cavendish was for was as

J. J. Thomson at the Cavendish Laboratory demonstrating the apparatus that he used to discover the electron. Note the induction coil in the background.

'a place to which men who had taken the Mathematical Tripos could come, and, after a short training in making accurate measurements, begin a piece of original research'. The laboratory expanded significantly under Thomson's direction. In 1895 it started allowing graduates from outside Cambridge to enter as research students and a year later a new wing was opened to accommodate the growing numbers. As his own 'piece of original research', Thomson during the 1880s turned his attentions to the discharge tube phenomena that that Grove, Gassiot, Crookes, and other British experimenters had investigated earlier in the century. Thomson brought the experimental standards of precision measurement to bear on discharge phenomena. His research was provided with fresh impetus by Wilhelm Röntgen's discovery of a new kind of radiation (what we now call X-rays) in 1895. The result of all this careful experimental work was Thomson's announcement in 1897 that the cathode rays developed in discharge tubes under certain conditions were streams of corpuscles carrying a negative electrical charge—later named electrons.

University and state-funded laboratories were not the only new spaces for experiment that emerged during the nineteenth century. The importance of experimental science for industrial progress was a constant refrain throughout the period. In Britain, commentators harked back to the role Joseph Black's experiments on heat had played in the development of the steam engine. Arguments regarding the relationship between experiment and industry played an important part in discussions about state funding for laboratory science in France and Germany, as we have seen. Laboratory work of various kinds did play an important role in the new electrical and chemical industries. Laboratory-trained electrical engineers were needed to maintain the telegraph networks and institutions such as Finsbury College in London established laboratories to provide that training. Telegraph companies established their own experimental and testing workshops as well. There was a shared culture of measurement between these industrial establishments and academic research laboratories, although telegraph engineers were often more concerned with developing robust and reliable measuring apparatus rather than precision. The Cavendish laboratory's involvement in the establishment of universal electrical standards was an example of this shared culture. The Society of Telegraph Engineers (later the Institution of Electrical Engineers) was another shared space. Electrical engineers such as William Henry Preece also clashed fiercely with academics, however, over the question of authority and whether practical men like him were better equipped than laboratory experimenters to understand the apparatus they operated.

In the United States, Thomas Alva Edison's Menlo Park laboratory provided a model for a new kind of industrial laboratory. The dominant image of the experimenter in nineteenth-century America was of the independent, often self-taught, practical inventor. This was the image cultivated by the telegraph inventor Samuel Morse, for example (despite the fact that his early experiments were largely funded by the United States Congress). Thomas Edison was the most successful promoter of this individualist view of the practical experimenter. He was celebrated in the press as the Wizard of Menlo Park. Behind the scenes, of course, the experimental research that took place there required large-scale collaboration. Menlo Park was an industrial laboratory in all senses. Edison's laboratory was a production line of patent applications. His rivals such as George Westinghouse also recognized that the large-scale industrial exploitation of electrical technology needed systematic and organized experimental research. This was what lay behind the myth of the heroic individual inventor. Edison might present himself as a self-made electrical autodidact, but many of his laboratory researchers—such as Nikola Tesla—were trained in the discipline of experiment. Increasingly, by the beginning of the twentieth century industrial rather than academic laboratories were where most of the products of university teaching laboratories would pursue their careers.

The chemical industry that developed during the second half of the nineteenth century formed another focus for laboratory experiment. Throughout the first half of the century, consulting chemists offered their services as analysts to industrialists and others who wanted to identify particular substances, for example. Such individuals also developed new chemical processes for industrial and other purposes of various kinds. Michael Faraday's work at the Royal Institution on improving the quality of optical glass at the behest of the Board of Longitude was an example of this kind of chemical laboratory work, although it was in all probability on a larger scale and involved more resources than would be available to most consulting chemists. The potential of chemical experimentation for industry was made clear in 1856, when William Perkin, a young student at the Royal College of Chemistry in London, invented the first synthetic chemical dye—aniline purple, or mauve. The dye was a derivative of coal tar, a common substance in Victorian Britain. Significantly, Perkin was not actually attempting to develop a new dye. At the suggestion of his teacher at the College, the German chemist August Wilhelm Hoffman, he was trying to find a way of synthesizing the anti-malarial drug, quinine. Perkin abandoned his chemical studies, acquired a patent and set about exploiting his invention.

In 1858 Perkin established a chemical factory in Greenford, near London, to produce his artificial dye on an industrial scale. Over the next decade or so he went on to develop a number of other artificial colours as well. By the late 1860s German chemical laboratories were providing strong competition. The German chemist Heinrich Caro had gained experience of the chemical dye industry whilst working in England before returning to work in Robert Bunsen's Heidelberg laboratory. He was soon working for the industrial chemical company Chemische Fabrik Dyckerhoff, Clemm, and then the Badische Anilin und Soda Fabrik (BASF) developing new industrial dyes. There he was instrumental in developing new kinds of artificial dyes from coal tar in their laboratories. By the end of the 1860s Perkin's company and BASF had to all purposes divided the European market in artificial dyes between them. In Britain, Germany, and France there were close links between academic and industrial laboratory chemistry. Academic laboratories produced chemists and industrial laboratories employed them. Individuals such as Caro moved easily back and forth between the two. German dominance in organic chemistry by the end of the nineteenth century was mirrored by its dominance of the chemical industry.

By the end of the nineteenth century, then, institutionalized laboratories both industrial and academic were the places of experiment. These were places where discipline mattered. The scientists who worked in them had been through rigorous regimes of training that had been designed to instil in them all the necessary skills and mental attitudes that were essential for the painstaking business of precision experimentation. Discipline mattered in these laboratories because that was regarded as the best way of achieving the goal of making the meticulous order of nature both visible and useful. James Clerk Maxwell expressed it best: 'those aspirations after accuracy in measurement, truth in statement and justice in action, which we reckon amongst our noblest attributes as men, are ours because they are essential constituents of the image of Him who in the beginning created, not only the heaven and the earth, but the materials of which heaven and earth consist'. Discipline in the laboratory mirrored the discipline of Creation. The fundamental material components of the universe had the character of 'manufactured articles', according to Maxwell. It was no surprise from such a perspective that the laboratory discipline needed to understand and demonstrate that manufactured character turned out to be useful for industrial processes too.

Big Science

By the beginning of the twentieth century, laboratory science had come to occupy a dominant cultural position. A new kind of laboratory culture had appeared that intersected the worlds of university, state bureaucracy, and industry. The disciplines and practices of precision measurement that modern laboratories needed and inculcated in their researchers were widely understood as being essential prerequisites of scientific, political, and commercial progress. Laboratory personnel moved back and forth between universities, state-funded laboratories, and industrial workplaces. The physicist and philosopher Pierre Duhem complained at the end of the nineteenth century about the ways in which British physicists, in particular (he was reviewing Oliver Lodge's *Modern Views of Electricity*, published in 1889), seemed to view the universe as a factory, rather than as a 'tranquil and neatly ordered abode of reason'. But it was not just in Britain that this link between laboratory culture and industry was forged. French and German laboratory experimenters were certainly just as committed to sustaining the relationship. The culture of experiment at the dawn of the twentieth century revolved around visibility and utility. Researchers were increasingly mobile, moving between laboratories, institutions, and even countries, carrying their skills and their visions of how nature was organized—and therefore of how laboratories ought to be organized—with them.

Experimenters at the dawn of the new century had every reason to suppose that they were on the brink of a new era of discovery that would transform their

---

Opposite: A portrait of an aged William Perkin, offering a roll of cloth dyed with his artificial mauve—the first synthetic dye to emerge from a chemist's laboratory. Synthetic dyes were cheaper and more durable, helping to make fashion accessible to the Victorian middle classes.

understanding of the universe. Certainly, many experimenters thought that the fundamental laws governing the operations of nature were well understood. Oliver Lodge, for example, suggested that the existence of the luminiferous ether was as well established as the existence of matter. 'Persons who are occupied with other branches of science or philosophy, or with literature, and who have therefore not kept quite abreast of physical science', he suggested, 'may possibly be surprised to see the intimate way in which the ether is now spoken of by physicists, and the assuredness with which it is experimented on.' Nevertheless, the same experimenters thought that their experiments so far had only scraped the surface of reality and that a whole legion of spectacular phenomena remained to be uncovered. Experiments such as the ones William Crookes carried out with discharge tubes, or Heinrich Hertz's discovery of electro-magnetic waves, seemed to probe at the boundaries of the real. There was much speculation about what other secrets the ether might reveal. New apparatus and new skills, developed in part in response to the requirements of industry, provided a powerful set of tools for further probing.

The German physicist Wilhelm Roentgen's discovery of X-rays in 1895 is often celebrated as an example of serendipitous discovery. In fact, there was little that was serendipitous about it. The kind of apparatus that Roentgen used for his discovery—induction coils, discharge tubes, and a variety of equipment to detect and manipulate any strange emanations that emerged from them—were common currency in European laboratories by the 1890s. Roentgen was also quite deliberately looking for strange and unusual phenomena in the space surrounding his discharge tube. If what he found was not quite what he was searching for, he was certainly searching for something. Nevertheless, having discovered these strange new rays and noted their capacity to pass through solid objects, and even taken a photograph of his wife's skeletal hand with them, Roentgen clearly understood that he had found something new and significant. He rushed into print just before Christmas and sent copies of the publication to leading physicists across Europe to make sure his claim to the discovery was recognized. The speed with which others succeeded in repeating his experiments is further testament to how widely distributed the apparatus and the ability to operate them skilfully were by the end of the nineteenth century.

Roentgen's amazing discovery spurred other physicists not just to repeat his experiment but to investigate the origins of the strange phenomenon. The French physicist Antoine Henri Becquerel suspected that the mysterious rays discovered by Roentgen might be related in some way to the phosphorescence produced in the discharge tube by the passage of electricity. He set out at his laboratory at the École Polytechnique in Paris to investigate whether these X-rays could be found emanating from other sources of phosphorescence when exposed to sunlight. In this he was continuing a family tradition, since his father, Alexander Edmond Becquerel, who had preceded him as professor at the École Polytechnique, had also been interested in the relationship between phosphorescence and solar radiation. Becquerel had also inherited a supply of uranium salts from his father and used these for his experiments. He soon discovered that the salts seemed to produce some kind of radiation that caused a

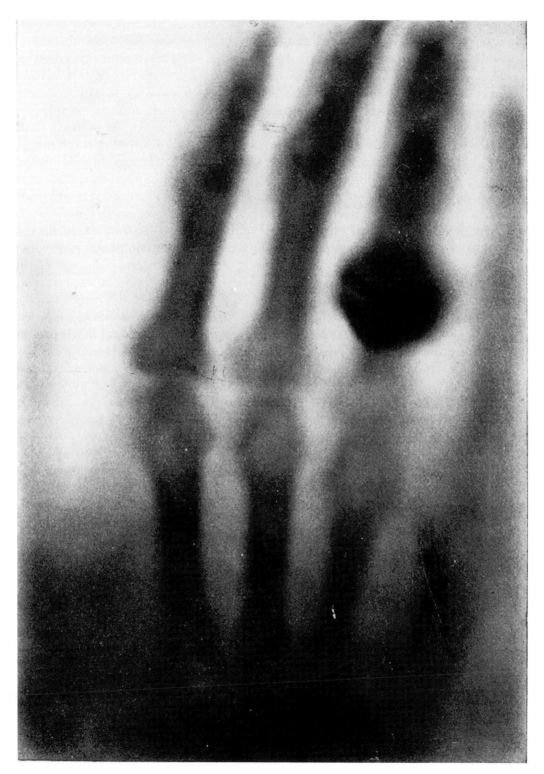

Wilhelm Roentgen's X-ray photograph of his wife's hand. Images like this were a tangible reminder of what experimental physics could achieve. Physics could reveal the interior of the human body.

photographic plate to become foggy and that the radiation persisted even in the absence of sunlight. Becquerel eventually concluded that the uranium salts must themselves be a source of radiation. He found that these mysterious rays could ionize gases and that, unlike Roentgen's X-rays, they were deflected by both electrical and magnetic fields.

Bequerel's researches on this strange new form of radiation were extended a few years later by Maria Sklodlowska (now better known under her married name of Marie Curie). Marie Curie set out in the first instance to study the ionizing effect of the new radiation, using the sensitive measuring apparatus developed by her husband, Pierre. Her research required significant amounts of uranium which she acquired in the form of pitchblende. It became clear that the pitchblende produced more radiation than would be expected from the quantity of uranium it contained. Following much dirty labour processing the huge volume of pitchblende needed for the task, the Curies announced the discovery of two new radioactive elements—polonium and radium—in 1898. Their work involved the processing of pitchblende on an industrial scale to produce even the tiniest samples of the new elements. The Curies' research on radioactivity provides another illustration of the symbiosis between precision physics and industrial production at the beginning of the twentieth century. It depended on the highly sensitive electrical measuring apparatus that Pierre had developed along with his brother, Jacques Curie. It also required the application of industrial methods for refining and separating the sources of radioactivity. Radium, in particular, rapidly became a rare and valuable commodity and laboratory directors such as Marie Curie were fiercely protective of their supplies.

Only experimenters whose laboratories had access to the right kind of resources could investigate the phenomena of radioactivity. In this respect, the Cambridge-trained New Zealander Ernest Rutherford, newly appointed in 1898 as professor of physics at McGill University in Montreal, Canada, was particularly lucky. A few years before Rutherford's appointment, the tobacco baron William MacDonald had donated money to fund and equip a physics laboratory at McGill and it was this funding that was to make Rutherford's ground-breaking experiments on radioactivity possible. At McGill's well-endowed laboratory, Rutherford and his assistant Frederick Soddy carried out experiments on the apparent origins of radioactivity inside the atom, developing the concept of radioactive half-life and suggesting that elements were transformed into other elements as they gave off their radiation. Using the kinds of techniques he had learned whilst working with J. J. Thomson in Cambridge, Rutherford succeeded in establishing that there was more than one kind of radiation: alpha rays that acted like streams of positively charged helium atoms and beta rays that seemed to be streams of Thomson's own recently discovered electrons. When Rutherford moved to Manchester in 1907 he took his radioactive researches with him. At Manchester, helped by a donation of radium from Vienna, Rutherford concentrated his attention on alpha rays, hoping to use them as probes to investigate the properties of the atom.

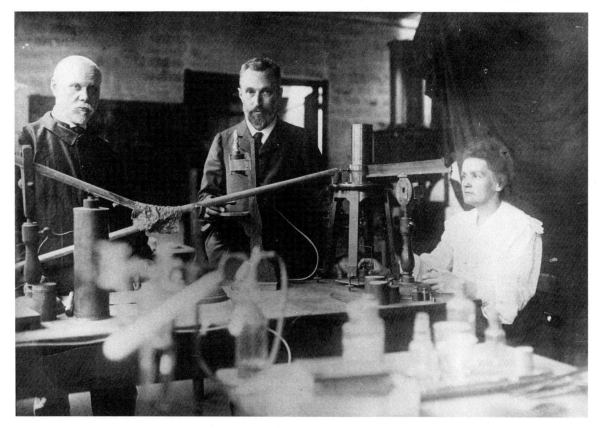

The Curies working in their makeshift laboratory in Paris. Purifying even tiny amounts of the new elements radium and polonium required processing huge amounts of pitchblende.

Rutherford's and others' experiments on the properties of radioactivity drew on a tradition of skilled practice in making the liminal and evanescent phenomena produced in discharge tubes visible and tangible that stretched back half a century and more to experiments such as Gassiot's cascade during the 1850s. Finding ways of making radiation visible remained at the core of early twentieth-century experiments. One of Rutherford's fellow experimenters at Manchester was Hans Geiger. Geiger worked with him in developing various ways of detecting and recording the presence of radioactivity. One of the electrical instruments that Geiger devised worked by registering the ionization caused by individual alpha particles with an electrometer. This later developed into what is now known as a Geiger counter. More useful for Rutherford's experiments was the technique of counting the scintillations that individual alpha particles created on a fluorescent screen. Counting scintillations was a tedious task that required a high degree of discipline. These were the technologies that led Rutherford to the observation that alpha particles fired at a thin foil were sometimes deflected from their path by as much as 90° and resulted eventually in his new model of the structure of the atom as a dense nucleus around which electrons

238 Iwan Rhys Morus

orbited. Other experimenters devised different techniques and instruments to make radiation visible. Charles Wilson, working with Thomson at the Cavendish, developed the cloud chamber, which made the paths of individual particles of radiation visible through the tracks of condensation they created by ionizing moist air.

Research on radioactivity, like research on X-rays, needed the bringing together in a single laboratory of a wide range of experimental skills and disciplines, as well as access to the right materials. One of the key factors in the success of Rutherford's Manchester laboratory was his ability to marshal all these resources and bring them together in one place. Work in the laboratory depended on the glass-blowing skills of Otto Baumbach, the university's glass-blower, who had emigrated from Germany to Manchester a few years earlier, and the experimental experience of the spectroscopist Thomas Royds. Rutherford's reputation as an experimenter attracted researchers from across Europe and North America to come and work with him at Manchester, developing their own experimental skills and taking their new-found expertise back with them to their own laboratories. Geiger, for example, returned to Germany in 1912 to work at the Physikalisch-Technische Reichsanstalt in Berlin. Royds also worked in Germany for a few years from 1909 to 1911, carrying out spectroscopic work in Tübingen and Berlin. He then moved to India to become assistant director of the Kodaikanal Solar Physics Observatory. This, often, was how the experimental skills and expertise that were needed for the new physics moved around. They moved with the people who possessed them.

Just as experimental work with radioactivity and X-rays depended on bringing together a number of different disciplines, it also had an impact beyond experimental physics. In 1912, for example, the German physicist Arnold Sommerfeld carried out some interesting experiments in which X-rays passing through crystals produced a regular pattern of spots on a photographic plate. These were interpreted as diffraction patterns and further evidence that X-rays were indeed a form of electro-magnetic radiation, like light. William Bragg, recently appointed as Cavendish professor of physics at Leeds, along with his son, William Lawrence Bragg, recognized that this meant that X-rays could become a powerful tool to investigate crystal structure. Some decades later, when Bragg junior was himself Cavendish professor at Cambridge, the techniques and instruments that they had played a key part in developing played a vital role in the discovery of the helical structure of DNA by Francis Crick and James Watson. Their discovery depended on the careful and meticulous experiments in X-ray crystallography that had been carried out by Rosalind Franklin at King's College London. The crystallographer J. D. Bernal, to whose laboratory at Birkbeck College Franklin moved a few years later, described her photographs of DNA's helical structure as 'among the most beautiful X-ray photographs of any substance ever taken'.

The success of these pioneers in experimental radioactivity depended to a large extent on their ability to move back and forth between the worlds of laboratory and industry—to recruit allies and secure resources. The same flexibility was crucial to the success of other experimental enterprises as well, as experimenters worked to take their experiments out into the world and turn them into reliable technologies. When Gugliemo Marconi arrived in Britain in 1896 he knew that demonstrating his wireless telegraph under laboratory conditions would not be enough to convince potential backers. They would need practical demonstration before they were prepared to open their wallets. Marconi arrived in the country with letters of introduction to the influential William Henry Preece, the Post Office's chief engineer and superintendent of telegraphs. With Preece's patronage he organized a series of spectacular demonstrations of wireless telegraphy, culminating in the spring of 1897 with a successful transmission between Lavernock Point, near Cardiff, and the island of Flat Holm in the Bristol channel. The Flat Holm experiment almost failed. It was only when the apparatus was carried from the cliff-top down to the beach, adding twenty metres to the antenna in the process, that the signal was detected. As another wireless pioneer, J. J. Fahie, put it: 'Result, magic! The instruments which for two days failed to record anything intelligible, now rang out the signals clear and unmistakable, and all by the addition of a few yards of wire!'

Success for Marconi at Flat Holm depended on being able to adapt laboratory apparatus to a world outside the laboratory that was difficult to control and manipulate in the same way as the space inside. The same sort of challenge faced chemists in developing chemical weapons during the First World War. By the time war broke out in 1914, chemists such as the German Fritz Haber had already forged successful careers moving back and forth between academic and industrial laboratories. Haber had made his name as the inventor of a new process for synthesizing ammonia which required exactly the kind of scaling up from a laboratory to an industrial regime that would prove essential in using chemistry for warfare. By the beginning of the war, Haber was head of the Kaiser Wilhelm Institute for Physical Chemistry and Electrochemistry. There he was well placed to promote the research that led to the first deployment of chlorine gas on the western front in April 1915. Chemists in Britain, France, and the United States were similarly heavily involved in the production of chemical weaponry for themselves. Scientists such as the Scottish physiologist John Scott Haldane, who had carried out his own experimental work on the effects of breathing different gases on human physiology, developed techniques and technologies such as gas masks to alleviate the effects of chemical weapons. Chemical weaponry was often promoted by its supporters as a cleaner, more efficient and scientific way of waging war.

Throughout the First World War, chemists and other experimenters played an important role in developing and implementing new technologies for warfare. In doing so, they cultivated new kinds of connections between the experimental culture of the academic laboratory and the worlds of industry and the military. Similar connections were forged during the Second World War. The invention of radar offers a good example of these interactions. The Committee for the Scientific Study of Air Defence brought together experimenters from different disciplines as well as air force officers. They adapted apparatus from the growing wireless industry to carry out large-scale field experiments to assess the possibility of detecting aircraft by detecting radio waves reflected from it. In February 1935 the Scottish meteorologist Robert Watson-Watt carried out a key experiment using a BBC radio transmitter which succeeded in detecting aircraft flying eight miles away. By 1937, experimental radio detection arrays could detect aircraft as far as one hundred miles away. Turning these experimental arrays into robust and reliable systems required a huge mobilization of personnel and resources. Laboratories such as the Telecommunications Research Establishment had a workforce of thousands. Crucial to their success was the ability to orchestrate large numbers of experimenters working together in teams to achieve specific research targets.

A similar mass mobilization of resources proved essential to the discovery and deployment of penicillin as well, despite the common perception that this was the result of individual genius and serendipity. Alexander Fleming's observation of mould in a petri dish in August 1828 was only a small part of the story. Ten years passed before two experimenters at the Dunn School of Pathology at Oxford University— Ernst Chain and Howard Florey—took up Fleming's researches and set about trying to find ways of producing the penicillium mould on a sufficiently large scale for it to be therapeutically useful. It is an indication of how difficult this was to achieve that the first patient to be treated with penicillin died because the supply that Chain and Florey had available ran out before treatment could be completed. It was as part of the war effort that resources were mobilized on a sufficiently large scale to produce penicillin on an industrial basis. Most of the production took place in the United States since British pharmaceutical companies were already fully engaged in other wartime activities. Just as with wartime innovations like radar, bringing penicillin from the laboratory bench to practice in the field was a complex process that depended on a disciplined division of experimental labour and industrial resources.

The best-known wartime mass mobilization of laboratory workers was, of course, the Manhattan Project to build a nuclear bomb. The project had its origins in a letter sent in August 1939 by Albert Einstein to the then United States president Franklin D. Roosevelt, warning him of the possibility not only that an atomic bomb was theoretically and technically feasible, but that the Germans were already in the process of producing one. Though signed by Einstein, the letter was largely written by Leo Szilard, a Hungarian emigré physicist based at Columbia University in New York. Szilard, along with his colleague at Columbia, the Italian physicist Enrico Fermi, had played an important role in investigating the possibility of producing a nuclear chain reaction that could release huge amounts of energy from the atom. The theoretical possibility of nuclear fission (the splitting of an atom releasing energy during the process) had been explored by the Austrian physicist Lise Meitner and her collaborator and nephew Otto Frisch. By 1939 a number of experimenters had produced fission in their laboratories. Experimenters like Szilard and Fermi quickly realized that there were conditions under which fission would continue indefinitely. If such a process were allowed to continue uncontrolled the result would be an explosion on a hitherto unimaginable scale. This was what had prompted the Einstein letter.

Einstein's celebrity ensured that the letter was taken seriously. The result was a mobilization and organization of experimental resources, in terms of people, spaces, and materials, entirely without precedent. In 1942 Roosevelt placed US army general Leslie Groves, who had been acting as deputy chief of construction for the Corps of Engineers, in charge of the innocuously named Manhattan Engineer District. As an engineer with extensive experience of running large projects Groves combined the sorts of managerial and technical skills the project needed. Under his supervision the atom bomb project's disparate elements starting pulling together. By the end of the year, Enrico Fermi and his team of experimenters at the University of Chicago had succeeded in achieving a self-sustaining nuclear chain reaction. A few months later industrial separation of the required uranium isotope (^{235}U) began at Oak Ridge in Tennessee. Groves also centralized the highly sensitive experimental work on developing fast fission in a single laboratory under the direction of physicist Robert Oppenheimer. The laboratory was based at Los Alamos in New Mexico. There, Oppenheimer was soon in charge of coordinating the experimental work of a hundred or so researchers as they laboured to transform the experimental pile built at Chicago into a practical weapon. This was laboratory physics on a truly massive scale.

The main tasks at Los Alamos were to establish the critical masses of plutonium and uranium needed to produce a reliable and fast chain reaction, and to find ways of bringing that critical mass of material together quickly enough so that it resulted in a reliable explosion rather than a damp squib. These turned out to be immensely complex matters. There were difficulties, for example, with the purity of the plutonium being produced at Oak Ridge. It appeared to have different properties from the small quantities produced in the laboratory and it was eventually established that it contained two different isotopes of the metal. This meant that early schemes for successful detonation had to be abandoned. Turning all the work carried out at Los Alamos during the early 1940s into a practical bomb rather than a complex piece of experimental apparatus was another challenge. Even following the awesome spectacle of the successful detonation of the first atomic bomb at the Trinity test site on 16 July 1945 the experimental work was not over ('I am become Death, the shatterer of worlds', Oppenheimer later recalled thinking). Oppenheimer and his team had treated the detonation itself as one huge experiment and now proceeded to interpret their results to see if the detonation confirmed their expectations. In some ways, it did not, producing more radiation and a larger explosion than they had predicted. But as well as delivering the atom bomb, the Manhattan Project also delivered a new model for laboratory collaboration on a huge scale and demonstrated what industrial-scale experimental physics could achieve.

The transformation that had taken place in the sheer scale of experimental culture becomes particularly clear in the case of atomic physics. The apparatus that J. J. Thomson had used in the experiments that led to the identification of the electron could fit comfortably onto a table top. The same could even be said of the apparatus that the Cambridge physicist James Chadwick used for the experiments that led to the

A nuclear explosion. The characteristic mushroom cloud produced by the atomic bomb came to symbolize the dangers of experimental physics during the Cold War.

discovery of the neutron in 1932. In fact the key piece of apparatus in Chadwick's experiment was only six inches long. But even in 1932, some experiments in atomic physics needed rather more room. The particle accelerator that John Cockcroft and Ernest Walton used at the Cavendish laboratory to split the atom that same year was the size of a small room. When Enrico Fermi in 1942 succeeded in producing a controlled nuclear chain reaction for the first time he used the squash court in the basement beneath the University of Chicago's football stadium as a laboratory. The synchrocyclotron that Fermi had built as chief of Chicago's Institute of Nuclear Studies in 1951 was the size of a small building. Carrying out an experiment in nuclear physics was no longer something that could be performed by a single individual, or a handful of individuals. It needed the collaborative work of dozens if not hundreds of experimenters.

Experimentation on this kind of scale was, clearly, hugely expensive. This was an experimental culture that required the resources of a state—or even a number of states—for its continued existence. In 1952 the Atomic Energy Commission (successor to the Manhattan Project) funded the construction of the Cosmotron at Brookhaven Laboratory in New York. Not to be outdone, the Soviet Union commissioned the building of a synchrotron in Dubna, near Moscow, completed in 1956 and capable of achieving energies of up to 10 GeV, compared to Brookhaven's 3.3 GeV. In 1952, the Conseil Européen pour la Recherche Nucléaire (CERN) was established with twelve collaborating member states (Belgium, Denmark, the Federal Republic of Germany, France, Greece, Italy, the Netherlands, Norway, Sweden, Switzerland, the United Kingdom, and Yugoslavia). Its first particle accelerator was built near Geneva in 1954. By the 1960s and 1970s, the size of these kinds of experimental facilities were to be measured in kilometres and the energies they produced in hundreds of GeVs. When the particle accelerator at the Fermi National Accelerator Laboratory (FermiLab) near Chicago was completed in 1972 it generated energies of up to 200 GeV and by the mid-1970s routinely operated at energies of 400 GeV.

The huge investment in personnel and material resources represented by CERN or FermiLab was at the extreme end of the spectrum of experimental culture during the final decades of the twentieth century. This does not mean that these places were entirely unrepresentative, nevertheless. Experiments at these gargantuan laboratories were both highly specialized and multidisciplinary. Certainly no individual and often no single discipline or sub-discipline now possessed the range of skills and know-how needed to bring experiments to a successful conclusion. Developing state-of-the art experimental facilities was a matter of intense international competition (as in the example of the race to build ever more powerful particle accelerators between America and the Soviet Union), often driven by military and state funding. At the same time, large experimental projects came to require far more international collaboration and the movement of experimenters from laboratory to laboratory. As we saw with early twentieth-century experiments in radioactivity, this was how skill and know-how travelled between individuals and institutions too. The best way to learn how to make a new kind of laser, or a superconductor, was to spend time at the institution that had

CERN from the air. The laboratory itself is underground but the superimposed outline gives a sense of its size. This was Big Science on a massive scale.

originally perfected the technology, or with someone else who had already done so. To the skill of manipulating complex instrumentation, the successful experimenter needed to add the skill of manipulating people and institutions.

It is absolutely clear that experimental culture has been transformed profoundly over the past two centuries. It is difficult to imagine what, if anything, an experimenter such as Michael Faraday would find familiar if he were transported by a convenient Tardis into a twenty-first-century laboratory. It is quite likely that he would not even recognize the strange space in which he found himself as a laboratory at all. It would certainly not contain any of the experimental paraphernalia that someone like Faraday would associate with laboratory life. Despite this transformation, we can still identify important continuities in experiment across the period. Much experimental practice now, as it was for Faraday, is devoted to finding new ways of making nature visible. Experimenters continue to search for novel ways of making the invisible, the intangible, the evanescent, apparent and reliable. In many ways, experimentation is still geared towards generating spectacle. Experimenters want to see—and want their audiences to see—nature. Watching a modern experimenter at work, Faraday might not immediately recognize what they were doing, but he would surely appreciate that he was observing a highly skilled manipulator. Now, just as they did two centuries ago, experimenters need to be meticulous and painstaking. They also need to cultivate an intimate understanding of how their apparatus works and its idiosyncrasies.

Laboratories are only one part of experimental culture—though often the most visible and significant part. The Royal Institution's laboratory at the beginning of the nineteenth century depended for its success on a wider network of workshops and instrument-makers. Davy and Faraday might be its most visible experimental performers but they depended entirely on the hidden labour of a host of others who made glassware, coiled wire, or manufactured acids for the batteries of galvanic cells. Experiments have always been multi-authored, even when there has only been one name on the title page. This is even truer now than it was two centuries ago. Despite this, our view of the authority of experiment (and of science more generally) remains as stubbornly individualistic as it has ever been. It may have taken many hundreds, if not thousands, of experimenters and technicians to do the work of identifying the Higgs Boson at CERN in 2013, for example, but the elusive particle will almost certainly remain associated with only one name—that of Peter Higgs who first hypothesized the particle's existence in 1964. The authority of experiment as a way of reliably discovering and displaying nature seems inextricably linked to this notion of individual authorship.

It is clear, nevertheless, that the position experiment occupies in modern culture is a collective achievement. The experimental life requires far more than a white coat (a very recent innovation—Faraday probably wore a frock coat in the lab) and a laboratory bench. The running of a modern laboratory requires a staff of administrators and technicians as well as the experimenters themselves. Budgets need to be managed and equipment made and repaired. These backroom operators usually remain invisible, but without them there would be no experimental culture to speak of. Beyond the confines of laboratories themselves, experimental cultures depend on the wider networks of modern industrial culture as well. Like factories producing commodities, laboratories need raw materials, except that the commodity that laboratories produce is knowledge itself. The laboratory, as a unique space for making knowledge, is a very modern invention and depends on hidden armies of toilers for its successes. We may think of experiment as a solo performance, but history shows that it bears more resemblance to a variety act.

8 Exploring Nature

AMANDA REES

W‎HEN people tell the stories of the Scientific Revolution, they often use the language of travel and exploration. We describe the voyages of discovery that figures such as Galileo and Harvey took as they encountered novelties in inner and outer space, we speak of the quest to understand planetary movement begun by Copernicus and completed triumphantly by Newton, and we discuss the difficulties encountered on the journeys away from superstition and the pursuit of a reasoned way of defining and exploring the natural world. These descriptions could be dismissed simply as colourful metaphors—but they could, of course, also be considered literally. The era commonly associated with the period of the Scientific Revolution is also the dawn of the European age of exploration and discovery: from the mid-sixteenth century to the end of the eighteenth century, nascent nation-states and houses of commerce began and intensified programmes of internal and external exploration. Voyagers brought back specimens of unfamiliar flora and fauna, and drawings and descriptions of strange peoples and landscapes, all of which needed to be accounted for and to find their places in the European world-view. Travellers surveyed (sometimes even stranger) peoples, practices, and landscapes that were to be found not so far from home, and strove to tie them tighter to urban centres and institutions as part of the processes of nation-building. And by the end of the eighteenth century, the sciences that were to become known as geology, botany, zoology, ecology, ethnology, archaeology, meteorology, cartography, hydrology, oceanography, ichthyology, amongst others, were beginning to emerge from the catch-all discipline of natural history.

But at the same time, the idea of the laboratory was fast becoming the iconic image of science, and the experiment its defining practice. Even today, when most people think about science, the pictures that come to mind are ones of test-tubes, lab tables, white coats, and (possibly) caged rats, all located within the enclosed, indoor space of the laboratory. When surveys of scientific literacy ask how one should do science, correct answers tend to include reference to hypotheses, variables, and experimental trials. For field sciences, the dominance of this particular kind of manipulative practice can be a problem—because the field sciences, by their nature, are very different. Most

Opposite: Frontispiece of Edward Lhuyd's *Archaeologia Britannica*, vol 1. (1707).

Archæologia Britannica,

GIVING SOME ACCOUNT

Additional to what has been hitherto Publifh'd,

OF THE

LANGUAGES, HISTORIES and CUSTOMS

Of the Original Inhabitants

OF

GREAT BRITAIN:

From Collections and Obfervations in Travels through
Wales, Cornwal, Bas-Bretagne, Ireland and *Scotland*.

By EDWARD LHUYD M.A. of *Jefus College,*
Keeper of the ASHMOLEAN MUSEUM in OXFORD.

VOL. I.
GLOSSOGRAPHY.

OXFORD,
Printed at the THEATER for the Author, MDCCVII.

And Sold by Mr. *Bateman* in *Pater-Nofter-Row,* London: and *Jeremiah Pepyat*
Bookfeller at *Dublin.*

obviously, they were and are dependent on work carried on outdoors, in the open air, often far from the urban centres where the laboratories and other institutions of science are usually found. Frequently, these places were and are hard to get to (either physically or politically) and make strenuous corporeal demands (both emotionally and materially) on scientists working there. In almost every respect, the characteristics of these places could not be more different from those associated with those of the laboratory.

Consider these points: where laboratories were designed to be spaces for the experimental manipulation of the natural world, more often than not it was the techniques of observational recording and measurement that tended to be privileged in field sites and spaces. Where laboratories were valued for their uniformity and similarity to each other, field sites were valued for their individual variability and unpredictability. While laboratory workers were trying to translate some aspect of the natural world into a form that could be scrutinized, displayed, and made to run through its paces at human convenience, fieldworkers were concerned to look at 'nature' under 'natural' conditions, seeking to understand the interrelations between organism and environment, to find ways of reading and recording the landscape, and to establish means of tying information from this incredible variety of sites and scenes together into authoritative laws of nature. The differences between the two sites should not be over-stressed—but undoubtedly, doing science successfully in either space required the researcher to develop and deploy a very different set of skills. The problem lay in the fact that the laboratory was privileged as the iconic scientific space. This meant that results from the field were often treated with suspicion: what warrant was there for believing that they represented an accurate account of some aspect of the natural world? Additionally, this was exacerbated by the fact that—again in contrast to the laboratory—fieldworkers were not in complete control of the place in which they were trying to do research.

At a very basic level, the laboratory is an indoor space, meaning that access to the lab can be restricted by something as simple as a locked door or cabinet; its population and conditions are regulated by its researchers operating under commonly held scientific conventions. The field, on the other hand, does not have walls, and this means that the population of the site is far less susceptible to scientific governance. Not only might the phenomena under investigation decline to manifest or choose to migrate, but attempts to study it/them might be actively disrupted by the presence at the site of other (non-scientific) humans. All fieldworkers have to deal with the presence at 'their' site of other residents: game wardens, or national park managers, hunters, farmers, fishermen, loggers, tourists, agents of local or national governments, soldiers, sailors, or eco-warriors. These represent a bewildering variety of different tribes, each with their own agenda for the use of the different places in which scientists want to work. On the positive side, however, they also bring with them an equally wide variety of practices for accessing and extracting the resources of the land, sea, and air, a pool of potential skill on which scientists have eagerly drawn. Fieldworkers have adopted and adapted the strategies and technologies originally developed for

Mobilizing the laboratory? William J. Burchell *Inside of My African Waggon* (1820).

divers, mountaineers, balloonists, film-makers, surveyors, arctic explorers, and have themselves perfected techniques that fed back into the habits of other professions or pursuits.

As a result, looking at the history of field science can require orientation to a rather different set of concerns than those normally associated with the study of science indoors. Historians of science have long been interested in the places where science gets done—but studying field science means putting the variability and volatility of these places, and the demands that working in them makes on human bodies, at the forefront of understanding the process of knowledge production there. And the story of field science does not stop at knowledge production. Understanding field science also requires an abiding attention to the circulation of knowledge, and the uses to which knowledge is being put. Not only are field scientists not in control of events at their sites and dependent for access to those sites on successful negotiations with potential funding bodies and the owners and inhabitants of these spaces, but they are also not in control of the way in which field-based knowledge is put to work. This is particularly problematic when most people's understanding of 'how science works' is based on a particular—and even then, not entirely accurate—picture of laboratory science, where experiments can be done and hypotheses can be directly tested. Field science—being concerned with observation, rather than manipulation—is not susceptible to such apparently clear-cut checks, and as a result, can appear far more uncertain, and sometimes, far more vulnerable to political or ideological interpretation, than its practitioners would wish. When dealing with field science, as this chapter will show, one is dealing not just with the history or the geography of science, but also with the profound political, economic, and social consequences of such work.

Why Do Field Science?

In the first place then, if the field is often so hard to get to and work in, why go there at all? If field sites are such uncertain places, replete with variability and unpredictability, then how were scientists to make authoritative knowledge claims based on their experiences in such ambiguous locations? The answers to these questions depended on the nature of the work that needed to be done. Scientists, travellers, and natural philosophers all went to the field because it enabled them to do work under conditions that either could not be obtained, or could not be accurately replicated, in the laboratory. There are many different sites at which fieldwork was and is carried out: on mountains, moors, or the decks of ships; in forests, mines, or space; at the seaside or specially constructed field stations—the list can be as long as that of the different environments which exist on, in, and under Earth. And despite the clear differences between field and lab, it should be remembered that they exist as categories at either end of a continuum of scientific spaces, between which scientists themselves travel. Researchers who spend the summer vacation working on the savannah will return in the autumn to teach undergraduates and subject their gathered samples to laboratory analysis. But it is useful to consider how different kinds of spaces are linked with, or perhaps produce, different kinds of approaches to the study of the natural world—and different kinds of responses to the knowledge produced there.

Categorizing field science

Very broadly, historians have identified four different categories of field science: survey science, historical/observational science, salvage science, and, for want of a better word, 'extreme' science. In the first part of this chapter, we will look at how these have developed over the course of the last two centuries.

Survey science—the kind of work first associated with the exploratory voyages and expeditions of the late eighteenth and early nineteenth centuries, and with which we might now link cartography, meteorology, or biogeography—was the Big Science of its day. It was most closely linked with the work of the German explorer-naturalist

Humboldtian science: engraving after Humboldt *Geographie der Pflanzen in den Tropen-Ländern* (1807).

Alexander von Humboldt (1769–1859), and is, in fact, often described by historians as 'Humboldtian science'. It was characterized by absolute attention to meticulous measurement, as travellers and explorers crossed sea and land, accompanied by multiple versions of scientific instruments such as thermometers, magnetic compasses, scales, and barometers. In Britain alone, dedicated voyages of scientific discovery, such as that of *HMS Beagle* (1831–1836) or *HMS Challenger* (1872–1876), and overland expeditions such as those sponsored by the African Association, the Geographical Society of London, or the Royal Society were sent out: all recorded information *in situ* so as to apply it on a global scale. There was nothing new about surveys, of course—people like John Ray (1627–1705) or Edward Llwyd (1660–1709) had travelled through northwest Europe in the late seventeenth century in order to record botanical, geological, and linguistic details about the different lands and their peoples. What made the situation different a hundred years later was the turn to instrumentation and the drive for accuracy in measurement. The acceptability of observations as authoritative depended, amongst other things, on their repetition, often using many different versions of the same instrument, so as to minimize error and maximize range. It was no longer sufficient simply to describe what was seen on a journey: instead, travels had to be quantitatively recorded. What's interesting here is that notions of measurement and standardization could be held to apply to the field just as they did to the laboratory—as long, that is, as the person doing the work could be trusted. After Humboldt, the notion of the 'scientific traveller' had been created, constantly on the move, harvesting information as 'he' went, in order to establish universal laws of nature through the combination and culmination of observations made in particular places.

In contrast, the observational/historical field sciences such as botany, ethology, geology, or palaeontology involved and involve spending more time in the study of particular places. The influence of the laboratory can be seen here in the *selection* of these sites—one might, for example, choose to work in a place where fewer (potentially) confounding variables might be present, as a result of altitude, isolation, or the ways in which local people used the area. Alternatively, one might seek to work at a range of sites, chosen so as to complement each other, with variations in one site being accounted for by constancy elsewhere. In this way, one might find in comparisons between these sites the functional equivalent of a series of lab-based experiments. As with survey work, the need for exactitude in sampling, observing, and recording was a key theme, along with the desire to be able to extrapolate from the close examination of the constellation of characteristics in specific locations to other areas of the natural world. Geology, in particular, relied on this, as people like Adam Sedgwick (1785–1873), Henry de la Beche (1796–1855), Roderick Murchison (1792–1871), as well as the young Charles Darwin (1809–1882), used intensive study of particular places to predict the sequence of rock formations elsewhere, while Abraham Werner (1749–1817) taught students how to identify valuable deposits from external physical characteristics—a reminder that mining was as significant as theology to geology's history. Being able to identify the location of coal-bearing strata or ground-water reserves, for example, was to prove central to the institutional independence of geology—and depended on extended fieldwork in the areas in question, whether these were the Welsh Marches or the Ural Mountains.

For the botanical and zoological field sciences, increasingly the impetus for field-work came from the need to examine organisms within their natural environment. By the early nineteenth century, for example, debates on the nature of evolution between figures such as the Comte de Buffon (1707–1788), Jean-Baptiste Lamark (1744–1829), and Georges Cuvier (1769–1832) which turned on the relationship between function, form, and environment encouraged naturalists to examine animals and plants *in situ*, whether that be in gardens, woods, moors, or seashore. Later on, the work of Humboldt, Darwin, and Joseph Hooker (1817–1911) on the reasons for and barriers to the geographical distribution of organisms was also emphasizing the importance of context and the tangle of relationships in which individual entities existed. Studying these kinds of questions could only be done in the field. By the twentieth century, such work was both benefiting from, and sometimes coming to require, a long-term commitment to the study of a particular place. In these cases, doing fieldwork successfully meant making a virtue of the singularity, the uniqueness of that specific space, in a way that almost turned the field site itself into a scientific tool.

For lab science, the context in which scientific work was done was relatively irrelevant: the uniformity of lab space across space and time was meant to minimize the potential impact on that work of uncontrolled variables. For the field, context was essential to understanding the meaning of the work done there. In some cases, it was the detailed description of the particularities of a given field site that could be used to lend authority to the work that was being carried out there—and by the twentieth century, this was especially the case in relation to ecological and ethological studies. At the most basic level, the inclusion of such detail provided reassurance to the reader that one had actually been to the place in question. But far more importantly, and especially in relation to sites where work had been carried out for a long time, the accumulation of quantities of detail meant that there could be a sea-change in both the quality of the data and the understanding of phenomena. Once a site had a recorded history—where variables such as rainfall, plant cover, species counts, and so on had been recorded for an appreciable period of time—that data then provided the constant backdrop against which the significance of current events could be judged. For ethology/zoology in particular, prolonged work in a specific place was important for two reasons. First, long exposure meant the animals under observation would become used to human presence, and would therefore behave more naturally—and it was natural behaviour that the observers had come to see. But it also meant that scientists could identify and recognize individuals, and thus come to record the animals' own histories. This was important, because knowing, for example, that what one was seeing was not an embrace between two sexually mature monkeys, but between a mother and her adult son, could make a major difference to the interpretation and significance of their behaviour. In both cases, the longer studies had been carried out in a particular place, the more valuable and authoritative such observations became.

Yet another category of field science, however, relates to observations that are valuable because of their fragility, rather than their longevity, usually because of the activities of other humans. Although many examples of botanical, ecological, and even geological field science can fall into this area, perhaps the disciplines of archaeology and anthropology present the best examples. This is due, of course, to the extent of global political and economic development during the course of the nineteenth and twentieth centuries, which wrought tremendous changes in human societies and their relationship with each other and with the land. Anthropologists, for example, found themselves in a position where they were studying ways of life that were in the process of deliberate or inadvertent eradication; archaeologists were confronted with a situation in which the physical integrity—such as it was—of their subjects of study were threatened by agricultural and industrial development. In both cases, information had to be salvaged before human activities damaged or altered it beyond recognition, although—ironically—it was often these potentially destructive human activities that permitted the scientists to gain access to the information in the first place. The same networks of ships, railways, and political institutions that were marginalizing traditional ways of life were the routes that anthropologists used to access the field. Archaeologists found that mines and quarries, for example, could become very useful places to search for fossils and artefacts, as long as the companies involved were willing to accommodate scholars; indeed, some quarry owners discovered that archaeological interest in their diggings meant that another extractive industry could be tapped, as tourists came to see their results. As such, salvage science is a good— but by no means singular—example of the ways in which the history and practice of the field sciences were thoroughly embrangled with commercial and industrial development.

A final category of field science could be characterized as 'extreme' science: fieldwork done under conditions of great hardship, or in settings exceptionally inimical to human life—but which were also, ironically enough, often linked with popular leisure-time activities, sometimes making it hard to see if scientists were at work or play in these areas. Such environments included those of mountain, marine, polar, and space science, where altitude, element, or temperature made it hard for human beings to survive without technological assistance, much less conduct scientific research. Mountains, for example, became places to do science from the mid-eighteenth century on. Sometimes described as 'laboratories of nature', they were places where the influence of altitude and pressure on physical variables and biological variation could be studied, astronomical observations made, and the impact of the mountain itself on both global and local climate—as well as the body of the observer— could be considered. Additionally, they were increasingly fashionable places to see

J. B. Noel kinematographing the ascent of Mt Everest (1922).

and be seen, as upper- and middle-class citizens discovered the pleasures of mountain climbing and sketching. In a similar manner, the oceans had increasingly become the subject of scrutiny by the beginning of the nineteenth century. A number of economic and cultural factors combined to account for this, including such developments as the expansion in the size and range of national fishing fleets supported by improvements to ship and harbour design in the late eighteenth century, as well as the laying of the trans-Atlantic cable lines in the nineteenth century. Additional encouragement, as with the mountains, came with the discovery by the middle classes of Europe and America that the seaside was a pleasant, uplifting, and potentially stylish place to be. By the twentieth century, undersea exploration had taken a central place in marine science, especially after Jacques Cousteau and Emile Gagnan developed the aqualung in the 1940s—which in turn was taken up as an increasingly popular leisure activity. In contrast, neither the poles nor space were available for exploitation by the casual visitor, but as with the survey science of the nineteenth century, work in these areas was facilitated by the interests of nation-states and commercial operators, and supported by the avid and abiding public interest in consuming the accounts of the triumphs and disasters of the polar and space expeditions.

Embodying field science

What the cases of both salvage and extreme science make clear, however, is the extent to which doing science in the field was dependent on the body of the fieldworker. When people think of science, they often think of it as an overwhelmingly intellectual activity; in contrast, looking at field science shows how important the body and its physical and emotional resilience or adaptability was to investigating nature. This was the case on a number of levels, ranging from managing to remain physically active in a stressful environment to training one's eyes and hands to record accurately what one saw.

Before they could go about the business of studying what they had come to see, fieldworkers had to habituate themselves to a novel, and usually unfamiliar environment. In the first place, this related to the fact that fieldwork required movement through space, meaning bodily strength and mastery was a key element in successful fieldwork. Secondarily—but just as importantly—it also connected to the use of the body itself as a scientific tool. Fieldworkers often had to be extremely creative in developing technologies to extend their physical access to different areas of the natural world, in a process that wasn't just physical, but often involved both imagination and emotion as crucial resources. In the case of mountain science, for example, the eighteenth century 'discovery' of the Alps for science also involved their aesthetic discovery: climbing became a literally uplifting activity, both morally and physically. Various expeditions to Mount Everest, for example, cited not only the geological and cartographical advantages that an attempt at ascending the mountain would accrue (not to mention the adventure, or the kudos, that awaited the first team to reach the summit), but also the spiritual benefits of such an effort. In the case of the 1922 expedition, it may be that this element was deliberately over-emphasized as part of a strategy to gain the approval of the lama of Rongbuk monastery; however, it is certainly the case that members of later expeditions described their efforts as a pilgrimage. Imagination and embodiment also had a role to play in the case of anthropology. Although trained as an experimental psychologist, W. H. R. Rivers (1864–1922) argued, based on his experiences on the Cambridge Torres Strait Expedition (1898), that one must share the life of anthropological subjects, doing as they did in order to understand from the inside out how they lived their lives, before one could go about analysing their social structures.

More generally, fieldworkers had to train themselves to observe and record in the field: not just what to record, but how to observe it in the first place. Geological students like the young Darwin had to learn to walk the land and to distinguish meaningful observations from accidental deposits; it was necessary to learn, for example, that the discovery of a tropical shell in a Shrewsbury gravel pit did not mean that all previous interpretations of the geological history of the area needed to be jettisoned. The significance of this process of learning to see can be judged from the emphasis that was consistently placed by learned societies and individuals on the necessity of teaching people how to do it. From the *Directions for Sea-men, Bound for Far Voyages*, published by the Royal Society in the 1660s, through the various editions of the *Hints to Travellers* (first published by the Royal Geographical Society from 1854), and the manuals of inquiry put out by various branches of the British Association for the Advancement of Science, scholars regularly tried to advise and instruct travellers on what their eyes should look for and their hands should record. The log- and sketch-book were key tools of knowledge, and for naval surveyors like John Roe (1797–1878), their most precious possessions often included treatments for eyestrain brought on by the effort to record as accurately as possibly what they saw.

Some kinds of fieldwork were, of course, more stressful than others, both physically (in terms of acclimatizing to new environments) and emotionally (in terms of getting equipment and research strategies to work in that environment). For those maritime observers new to the sea, learning how to cope with sea-sickness and accustoming oneself to the physical challenges of life aboard ship were fundamental to social and scientific success on the voyage. Altitude sickness could be an issue for those conducting mountain-based research (as could simply taking photographs in the era of collodion technology), malaria and other tropical diseases were constant threats to those working at lower altitudes and closer to the equator, and just doing science at sea, under water, and in the air usually required more extensive technological interventions and support. But all fieldworkers faced less tangible threats to their survival as scientists, threats that emerged from the particular nature of field-based research, and which centred around the question of scientific *identity*. Their very status as scientists was a matter that was often rendered equivocal, usually as a direct result of the hybrid and ambiguous nature of the places in which they worked, and the social groups with which they shared that space.

We have seen the ways in which various political, economic, and national interests made it possible for field scientists to access the subjects they wanted to study, and the role of culture in making certain aspects of field science fashionable and popular. We will now go on to look rather more closely at the political economy of field research and at the different cultures and societies that helped to structure the spaces in which scientific work was carried out.

The Political Economy of Field Science

Doing field science with a global reach was expensive: what this meant was that those interested in studying such matters had to find ways of enlisting the support of other people and institutions in the funding of this research. To put it crudely, the political and economic benefits that would or could accrue from scientific research had to be made clear. But the trouble was that the interest that other groups (politicians, civil servants, soldiers, sailors, entrepreneurs, financiers) had in these matters might not be wholly scientific—and this had consequences for when and what kind of research was done, as well as who carried it out. We can see this if we look at the kind of science that was done at sea.

Magnetic measurements and the maritime economy

At the beginning of the nineteenth century, for the military, the seas and oceans were largely just hostile environments over which they travelled and on which they fought. By the mid-twentieth century, however, changes in the nature of international warfare had turned the sea floor into a theatre of conflict, and focused military attention on the constitution and structure of the oceans themselves. Science and scientists had key roles to play in this shift—but this was not a simple process of mutual symbiosis. Those with an interest in pursuing scientific problems—establishing the shape of the Earth, for example, mapping the ocean floor or searching for evidence of the giant squid's existence—had to enlist the support of others in their endeavours, to seek the political or economic backing that would lead to financial support for their studies. But especially during the nineteenth century, the period during which science was becoming a professional activity, it was not always clear who could best be relied on to identify and solve the scientific questions that seemed to present themselves.

For example, interest in pendulum experiments which could determine the shape of the Earth had intensified in the years following the Congress of Vienna (1815). Captain Henry Kater (1777–1835) had published a series of papers on pendulum design and observations in the Royal Society's *Philosophical Transactions*, which closed by calling for a series of globally comparable pendulum measurements that needed to be taken from geographically distant locations. This was a project that required not just national, but international support and cooperation—but fortunately, it was also a problem that could help solve immediate questions of national prestige and military practicality. As more and more ships were made of iron, magnetic studies and means of improving navigation became more and more interesting to the world's navies. Additionally, in the post-Napoleonic era, voyages of survey and exploration would give at least some of the suddenly under-employed Navy something to do: no longer required to fight the French, being assigned to such a journey was one way of keeping one's job and even successfully pursuing promotion.

Commander John Ross's (1777–1856) initial voyage in search of the northwest passage (1818–1819) was the first to carry out extensive pendulum work, primarily conducted by Edward Sabine (1788–1883). Sabine, educated at the Royal Military

---

Opposite: Doing science on the deck of HMS *Challenger*.

Academy and an officer in the Royal Artillery, was the expedition's astronomer, having been personally trained by Kater in operating the new kind of pendulum. In the years that followed, observations from the ships searching for the northwest passage—a potentially highly lucrative trade route, and for which discovery the Admiralty offered prizes—were an important, although not sufficient, source for the accumulation of pendulum observations. By the 1830s, then, Sabine was at the forefront of a campaign to establish a global network of geomagnetic observatories under British leadership. The Second Secretary to the Admiralty, John Barrow (1764–1848), who had been a key figure in promoting both the search for the northwest passage and in African expeditionary ventures, had also sponsored and facilitated several more voyages of Arctic exploration. For both men, however, the issue of who was qualified to make accurate observations about the natural world remained a problem, and it was one that was tied to the broader relationships between science, exploration, and commerce in the period.

Sabine became the personal target of criticism from Charles Babbage, inventor of the Difference Engine, who was then in the midst of a quarrel with the Royal Society, which had failed to appoint him as junior secretary. In his *Reflections on the Decline of Science in England* (1830), Babbage argued that most of the Society's Fellows were elected for their social, rather than scientific, qualifications—and attacked, in particular, Sabine's appointment as scientific advisor to the Admiralty. His condemnation was based on his contention that Sabine was 'an officer of artillery on leave of absence from his regiment': a military man, not a scientist, whose observations were not to be trusted. Babbage had looked closely at Sabine's recorded measurements and results from the Ross voyage, and concluded that they were uncannily and suspiciously in agreement. They must, he argued, be the product of an amateur observer who was not even competent to fake observations convincingly—indeed, an entire section of the polemic was devoted to the 'Frauds of Observers' who wish to be considered as accurate witnesses for science. This use of Sabine's military identity to pre-empt his scientific credibility was part of Babbage's wider concern with the social and institutional status of science—part of his wider attack on the state of the Royal Society and its relationship with the Admiralty. But the issue of identity and the proper qualifications for observation was also one that troubled the Admiralty Second Secretary, John Barrow.

In Barrow's case, the question was whether anyone *other* than a member of the Royal Navy was qualified to carry out observations and measurements on the ocean. He felt very strongly that exploration and survey work was the prerogative of the Navy, despite the fact that it was whaling ships, in search of economic profit, that were returning with vitally important news and observations from the far north. In fact, it was the reports of whaling ships that had first suggested that Arctic ice might be sufficiently fragmented to permit the resumption of the search for the northwest passage. Barrow was particularly hostile to one William Scoresby Jr (1789–1857), the son of a Whitby whaler. Scoresby, a correspondent of Joseph Banks, had made extensive charts of Greenland's coasts, as well as many observations of polar ice, sea temperatures, and marine life as it related to the northern whale fisheries—but his request for government finance for a voyage of discovery under his command was rejected at Barrow's behest. Arctic observations and explorations were not to be trusted to a commercial operator; nothing daunted, Scoresby borrowed against the voyage's profits and went ahead anyway. The venture was a success, with Scoresby reaching further north than any previous sailor—but Barrow's hostility to the 'mere whaler' meant that ships were soon despatched to sail a few degrees further along Greenland's coast, potentially overshadowing the geographic and biological discoveries that Scoresby had made.

What both these examples show is the extent to which economic, military, and scientific identities and interests were entangled during this period of survey and exploration, when the sciences themselves were in the process of disciplinary formation and professionalization. Over a hundred years later, the story of sub-surface exploration illustrates the continuing close relationship between science, national interest, the military, and commercial operations.

Science and the sub-marine

By the twentieth century, war was being carried on in, not just on, the water. The increasing use of submarines both during and after the First World War meant that it became progressively more important to understand the properties of the sea itself, as a medium that things travelled through, as well as over. Naval officers needed to be able to draw on expertise in order to understand the behaviour of waves at the sea-shore, to appreciate the military applications of thermoclines and the correct operation of bathythermographs, to understand underwater acoustics whether produced by submarines or cetaceans, as well as (in peacetime) to be able to identify the species of fish that fishing boats were permitted to harvest. But this interest in, and support for, oceanography and maritime studies did not come without potential costs both at the individual and the institutional level, particularly in relation to the Navy's approach to basic, as opposed to applied, research and the role of national security.

In the United States, for example, two key oceanographic institutions had been established on either coast by the early 1930s—the Scripps Institution of Oceanography, founded in 1903 in La Jolla, California, and Woods Hole Oceanographic Institution (WHOI) in Massachusetts, founded in 1930. It was Athelstan Sphilhaus of WHOI that invented the bathythermograph in 1938. This instrument was a means by which the reliability of sonar could be estimated, since the movement of sound through water was affected by temperature and depth, and they could be fitted to the exterior of submarines to aid them in both attack and defensive manoeuvrings. This development inaugurated a productive and reliable relationship, based on the exchange of data, instruments, and personnel, between the US Navy and WHOI that persisted throughout the Second World War and beyond. It was the presence of oceanographers on Navy ships and Navy funding for research that facilitated the

study of a range of geodetic and oceanographic data, from the shape of the ocean floor to the behaviour of waves at the seashore—key scientific information and observations that also had clear military applications when it came to marine landings and sub-surface nuclear warfare.

But the relationship between basic and applied research, and the Navy's interest in supporting the former, was not always clear, as the development of manned deep-sea submersibles illustrates. The idea of building such a vehicle, a means by which humans could reach hitherto impossible depths and pressures, was in its modern form that of Allyn Vine (1914–1994); its realization was the product of extended negotiations on finance, ownership, and design between WHOI, the Office of Naval Research, and Reynolds Metal Corporation. DSV *Alvin* was commissioned in June 1964, but spent its first few years on specific tasks for the Navy—carrying out practical jobs like inspecting hydrophone arrays, and looking for a lost hydrogen bomb. It was not initially used to fulfil the research proposals that had been put forward in order to address significant questions of marine biology and geophysics. By the early 1970s, there were serious conflicts over funding between the Navy and the other interested parties, which meant that the *Alvin* came near being axed. It did not help that there had been many sub-surface accidents and losses in the late 1960s—including the sinking of *Alvin* itself in 1968—although the discovery of a still-edible bologna sandwich on board when *Alvin* was recovered almost a year later did stimulate microbiologists to expand their work on decay in the deep sea, and eventually, to challenging the notion that the ocean trenches might serve as receptacles for human waste: a somewhat inadvertent contribution to blue-sky (-sea?) research. But once the Navy commissioned its own dedicated submersibles, they cancelled *Alvin*'s funding, forcing (ex-Naval) researchers like Robert Ballard (born 1942) to seek other supporters. In this, they were remarkably successful: beginning in 1974, *Alvin* was used in a collaborative project between academic and military institutions and personnel, which not only discovered the deep sea hydrothermal springs, but established that there was life there.

Here, the various interested parties and institutions had managed to create a basis for cooperation and collaboration, resulting in an event with implications that ranged from the geological through the biological to the philosophical sciences, not to mention the consequences for space exploration. Other examples of deep-sea work had less fortunate results. In particular, issues of security and bureaucracy caused problems, especially during the Cold War. Some scientists found their nationality or personal political histories restricted the projects with which they could become involved. In other cases, permission to publish data or results was withheld. For example, by the 1940s, the US Navy was restricting the nature and kind of information that could be published about the oceans—in particular, information about the ocean floor, which could potentially be used to disrupt both submarine traffic and long-distance communication and monitoring. Much of the data-points from which

Overleaf: The rescue of the sub *Alvin*, 1969.

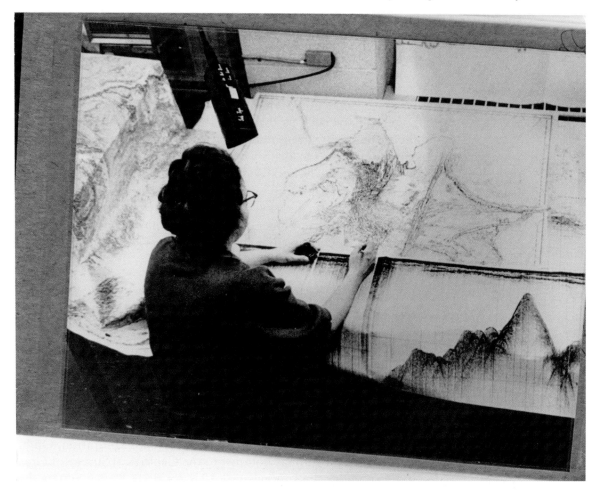

Marie Tharp at the drafting table.

maps of the sea-bed could be built came from the automatic depth-recorders attached to naval ships—but these specific bathymetric details of depth contours and soundings represented knowledge that had to be restricted in its circulation. As a result, for example, the famous Heezen-Tharp map of the ocean floor was eventually published in a less specific, physiographic form. As it happened, this made the image popular with, and familiar to, a much wider audience, especially after an artist's version of it appeared in *National Geographic*—but this does not detract from the fact that scientific work had effectively been censored. Many scientists had related concerns about the sources of the funding for their research and the potential impact that this might have on their academic independence.

But despite these examples of tension between exploration, politics, science, and commerce in making observations of the natural world, what is clear is the extent to which the various groups interested in studying the sea were involved in establishing a shared culture of collaborative maritime science. At the outset, the emphasis on accurate and repeated measurements, for example, was an element already present in naval culture, officers and scientists alike kept journals, and the popularity of natural history as a middle-class pastime provided mutual ground on which they could meet. While social relationships aboard ships were spatially mirrored, with access to certain areas dependent on rank, on at least some voyages—not just dedicated scientific expeditions such as that of HMS *Challenger*—space was made for scientists and their equipment. The question of recompense and financial support was always problematic: after all, the identity of the 'natural philosopher' was closely tied to that of 'gentleman' precisely because of the assumption that financial independence meant freedom to speak the truth. But by the mid-nineteenth century, a class of 'scientific servicemen' had emerged, drawing not just on naval officers and military men, but also on colonial officials and other government employees who had a remit to collect information and specimens from the various posts to which they found themselves sent.

Science in the state's service

In fact, a key element of the professionalization of science in the nineteenth century was the emergence of a number of institutions, often state-supported (in Britain at least), that were dependent on and would employ field-based agents, such as hydrographic offices, national geological surveys, ordnance surveys, census bureaus, and so on. Again, these operated both at home and abroad, and key figures in the development of the field sciences often moved between them. Joseph Hooker is one such example. In his youth, he had participated in the 1839–1843 voyages of Arctic exploration. On his return, in 1846, he took up a position as botanist to the Geological Survey of Great Britain, founded in 1835 under the supervision of Henry de la Beche, and at that point entering a period of intense activity, since an act of parliament in the previous year had called for the production of a complete geological map of the country. Hooker spent the next year studying fossil plants in the coal beds of Wales and the South West, before being sent by his father (who was soon to become the director of the Royal Botanic Gardens at Kew) on a plant collecting expedition to India and the Himalayas. His travel to India was facilitated by the long-established military and commercial links between Britain and the sub-continent: he sailed on Navy ships and lodged with officials of the East India Company. In recompense, he managed to cause what could have been a serious international incident—and possibly the punitive execution of a border guard—when in his search for novelties he crossed into the forbidden territory of Tibet without permission. On his return to Britain, he was appointed as assistant director at Kew in 1855, succeeding his father as director in 1865. In these positions, and from the connections he had forged

---

Opposite: Joseph Dalton Hooker, *Snow Beds at 13,000 feet 1854*.

SNOW BEDS AT 13,000 FT. IN THE TH'LONOK VALLEY, WITH RHODODENDRONS
KINCHINJUNGA IN THE DISTANCE.

through all these voyages at home and abroad, Hooker established a network of correspondents and representatives, formal and informal, who sent information and specimens back to London.

This work had both a scientific and an economic element: indeed, it was often very hard to distinguish between the two. Kew and the network of colonial botanic gardens were central, not just to the collection and breeding of new plants, but to their improvement with an eye to potential exploitation. Specimens came from designated expeditions, from civil servants stationed in remote outposts, from customs officials at ports and borders, from travellers and locals, and even from street markets. They could be given, traded, bought, or stolen—or even sabotaged, as disgruntled gardeners who had failed to germinate Chinese seeds liked to believe. The successful naturalization of a valuable plant or animal could lead to immense profit. It was in Paris that the first *Sociétié d'Acclimatation* was established (in 1854 by Isidore Geoffroy Saint-Hilaire), but the economic and business significance of acclimatizing colonial plants and animals—and of transplanting to the colonies productive domestic flora and fauna—was quickly recognized elsewhere. By 1900, there were more than fifty acclimatization societies, mostly located in the colonies of Europe.

Trade in spectacular exotic animals and plants had always flourished alongside travellers, but these societies were interested in the possibility of deliberately using science and expertise to improve (make more profitable) indigenous ecosystems by means of species transplantation. Some advocates, such as Richard Owen or Francis Trevalyan Buckland, were keen on the prospect of introducing new food animals to European tables, but the greatest successes came in the colonies with the establishment of tea and coffee plantations, and the export of salmon, trout, and sheep (and rabbits) to Australia and New Zealand. Here, it was not just the scientists that moved from place to place: the products of fieldwork and practices also travelled in support of the wider economy. This relationship between science and economic improvement—of which Scoresby's interest in improving the Arctic whale fisheries was another example—was a significant factor in the development of the field sciences. As noted, surveys and expeditions, both at home and abroad, had been sent out not just to make a record of land- and seascapes, but as part of this process to identify potential sources of botanical, mineral, and animal wealth. This process was not restricted to the early days of exploration, but was a key part of the imperial and colonial process: the African Research Survey (1929–1939), for example, sought to identify and mobilize the natural resources of British colonial Africa. Nor was it always only carried out by travellers: by the early twentieth century, European colonies were actively seeking to attract settlers with technical expertise in forestry, medicine, game, and agriculture amongst other specialities, to manage and maximize local production.

Early contributors to what was becoming the science of ecology, such as Reginald Stapledon (1882–1960), were firmly rooted in this background of improving agricultural science, moving between domestic and colonial sites of study, and making theoretical contributions as they struggled with practical problems of production. One of Staple-don's concerns, for example, was with the problem of grassland, which had long been an issue for British farmers—specifically, with the problem of how to turn arable land into pasture, since once land had been ploughed, returning it to a state in which it produced appropriate fodder for grazing animals was very difficult. In 1912, Staple-don began a comprehensive survey of the vegetation of Mid and North Wales, and in 1919, he became the first director of the Welsh Plant Breeding Station, a position of great significance given the importance of productive grassland to the policies of imperial Britain. His vegetation maps, together with the experimental work being done at places such as the Royal Agricultural College, made it possible to show how the quality of grassland (measured in yields of meat or milk) could be incrementally improved. But it was his travels in Australia and New Zealand that brought home to him one of the key insights that was to underpin ecological fieldwork—the need to consider relationships and contexts, and in this context, the recognition that when it came to productive pasturage, the actions of livestock animals were just as important, if not more so, than deliberate efforts at human management.

It was one thing to improve on nature, quite another for humans to produce it from scratch. The experience of scientists like Lauren Donaldson (1906–1994) in the American Northwest showed how impossible it was to experimentally recreate natural relationships. Donaldson believed that it was possible to manipulate nature in the field in order to create an economically productive ecology—specifically, to create a salmon run in the Pacific Northwest. By the 1920s, concern was growing at the damage done to fisheries by industrialization and over-fishing. Donaldson tried seeding a nutrient- and salmon-deficient lake with minerals and fish ova, alongside the eradication or margin-alization of undesirable animals and plants. Ultimately, he failed. Salmon grew, and even bred in the lake, but they did not migrate; beavers could be shot, but they kept coming back and damming streams. Donaldson could create a fish farm, not a salmon run. For both Stapleton and Donaldson, ecological fieldwork had demonstrated the limitations that natural complexity placed on human intervention.

Untangling the convoluted and complex relationship between field science and economic and commercial demands over the past two hundred years is beyond the scope of this chapter. But it is worth noting that just as the identification and *extraction* of biological and mineral resources has been a central plank in this relationship (from the explorers of empire to Shell's sponsorship of both the Royal Geographic Society and the National Geographic Society), so has the *defence* of these resources against development, in the name of conservation, biodiversity, and demo-cratic freedoms. Again, this is a problem that is closely related to the question of how scientific identity is established and maintained in the context of field science—how, when, and why might a scientific researcher become better described as a scientific activist? What influences do, or can, field scientists have over economic and political development? And—perhaps most importantly—what consequences might the attempt at doing so have for their ability to speak with authority about the natural world? The next section of this chapter will consider these questions.

Activism, Authority, and Field Science

Doing fieldwork usually means working on territory, whether land or sea, that either belongs to, or is used by other human groups. This can cause both practical and epistemological problems. In terms of research methodology, for example, many researchers, especially those involved in observational/historical field sciences, had committed themselves to observing 'natural' behaviour or events. They went to the field in the first instance because they wanted to see how organisms related to each other within a 'natural' context, since recreating the complexities of natural relationships in an artificial context was impossible. Particularly valuable, then, were pristine sites that had remained relatively unaffected by human development, but since these sites were, almost by definition, impossible either to find or to access, most twentieth-century workers contented themselves with approximations. Nevertheless, the claim that a rival was doing work on a site where conditions had been materially affected by human alteration remained a rhetorical resource in controversial debates. Turning from epistemology to practicality, doing fieldwork virtually always meant getting permission from those groups that had a previous claim—whether legal or moral—on the space in which the researcher wished to work.

Getting access to the field

In other words, field sites are not just scientific spaces, but working landscapes. They are occupied or used by other human beings, and put to economic, practical, even spiritual uses. Depending on the nature of the site, this has a number of consequences for the scientist and for the research carried out there. First, there is the need to get permission to be at the place, and the related issue of what, if anything, the scientist must or should offer in return for that permission. Second, there is the issue of managing relationships with the other people at that site, and in particular, managing any contribution that they make to the conduct of the research there. In the third place, there is the problem of presenting the results of that research to the people back home, whether they be academic supervisors/colleagues, politicians, civil servants, business leaders, or members of the general public. All three of these matters can have a significant impact on the field scientist's ability to do research and the reception which that research receives.

It should be evident by now that fieldwork is carried out in an incredibly varied range of places—from ship deck to mountain top to rainforest—and the political and bureaucratic process of getting permission to work in the field can be incredibly complex. In the later twentieth and early twenty-first century, for example, permits to carry out work in national parks need to be obtained, research visas stamped to allow entry into foreign countries, licences to import technology or export samples acquired. Once in the field, and perhaps more informally, scientists have often had to offer inducements to ease their entrance to their actual site. These might range from offering credit, whether academic or monetary, to helping with a legal problem or supporting an economic claim, and their nature has changed over time. Anthropologists, for example, in the days of empire were allowed access to tribal peoples not least because their work might help colonial officials manage the 'natives'; in the twenty-first century, in contrast, elders might agree to cooperate with anthropological research in return for help in courtroom battles to reclaim land from incoming settlers.

Archaeologists and paleontologists, especially in the American West, found themselves in a position where the most intellectually interesting sites were either already private property, or within Indian reservations, or were in the process of becoming so—as the director of the Carnegie Museum, William Holland (1848–1932) discovered in the early twentieth century. Extremely rich fossil beds had been found around a ranch in Nebraska owned by the Cook family who, as it happened, were very willing to accept and support visiting scientists, although they were not keen on the Carnegie's attempt to insist on privileged access. But it was when Carnegie researchers refused to give the Cooks credit for discovering the fossil beds in the first instance that the relationship between the family and the institution became badly damaged. It finally broke when Holland tried to obtain legal title to the land containing the fossil beds that was outside Cook's property line, with the intent of excluding other scientists. Harold Cook promptly homesteaded the area, and was able to parlay his practical support for research into a scientific career for himself with institutions like the University of Nebraska and the American Museum—who were willing to be more accommodating to the contributions of amateurs than was the Carnegie at this point.

Locals, laypeople, and locating expertise

But the contributions of people who did not primarily identify themselves as scientists, whether properly credited or not, had always had a major role to play in the conduct of field science. Earlier in this chapter, the achievements of Edward Sabine and William Scoresby were discussed, but even after science had largely grown professionalized, amateurs continued to be central to science in the field. At home or abroad, field scientists found themselves drawing on a range of local talent in the collection of data and specimens, whether these were sailors aboard ship, local farmers, or schoolchildren. Perhaps one of the most famous amateurs in the history of science—certainly in the history of geology—was Mary Anning (1799–1847), although the use of the term amateur in her case is ironic. Fossil collecting was for her, above all, a business: her discoveries, whether of the iconic marine vertebrate fossils of *Ichthyosaurus* and *Plesiosaurus*, or the more common ammonites and coprolites, were sold to make a living for her and her family. And her approach to that business was utterly professional: she was widely known to be an expert and experienced excavator who was very familiar with the relevant scientific literature. She was recognized by naturalists and geologists like de la Beche and Murchison, but her skill at finding and unearthing fossils was not mirrored in formal credits: when her finds appeared in museums they did so not under her name, but that of her donors.

Mary Anning, sketched at work by Henry de la Beche.

Anning is unusual in that she is so well known—other amateur contributors to field science, whether past or present, are largely forgotten, precisely because they tend to hand their data (specimens or observations) over to scientists for analysis and reporting. In this, they are following a pattern established in the very early days of scientific field research. Originally, the fieldworker's role was simply to collect information, which they would then transmit back to university and museum scholars in the metropolitan centres of knowledge production. This was one key reason why institutions, learned societies, and associations were so keen to produce manuals instructing people on how and what to observe: this was a means of directing people at a distance to the points that these 'armchair scholars' thought were of interest. It was also, and remains, a means of collecting large quantities of data relatively efficiently, as in the case of Charles Elton's (1900–1991) work with the Hudson Bay Company in the early twentieth century: local agents forwarded to him data on the numbers of fur-bearing animals and their prey, which he was then able to use both to advise the Company and to elaborate his own ecological work. But these examples focus only on one aspect of fieldwork—what happens to information, specimens, or data *after* they have been extracted. What precedes that process is equally important.

Amateur, lay, and local contributions to fieldwork went far beyond simply the delivery of information—their skills and practices were, in fact, a crucial factor in the development of the reliable scientific field researcher. As previously noted, when scientists went to the field, they found it already populated by a variety of different social groups—and it was from the culture and practices of these groups that they learnt to navigate the field environment. This chapter has already looked at the relationship between marine science and the military, but ordinary fishermen also provided a resource for researchers. For example, the ecologist Stephen Forbes's (1844–1930) study of the Illinois River drew inspiration and information from the local rivermen: he and his colleagues used their local techniques for working the river, collecting samples, and so on, as well as listening to and profiting from vernacular versions of the river's natural history. Eugenie Clark (1922–), sometimes known as 'The Shark Lady', paid tribute to the knowledge and help she had received from 'the best spear-fisherman' Siakong, the Pacific Islander who tutored her in the technology and psychology of fish collection. This was not a phenomenon confined to the twentieth century: Horace-Bénédict de Saussure's (1740–1799) contributions to mountain science, Alpinism, and geology are recognized with a statue in Chamonix—where he is shown alongside Jacques Balmat (1762–1834), the mountain guide who showed him it was possible to climb Mont Blanc. Expeditions and the survey field sciences benefited hugely from local knowledge and guidance, not just in terms of accomplishing the expedition's aims, but often with regard to simple survival.

When it comes to the observational/historical field sciences, local knowledge and practice is extremely significant—not least because while the success of this work is often dependent on detailed knowledge of specific places, the work is usually carried out by people who are not local to that particular area. For ethology and zoology, for example, locating the animals that are of interest—especially if they are rare—often depends in the first instance on talking to local residents. Ironically, often the most knowledgeable people can be the ones who are involved in making the animals rare in the first place—such as the hunters, the loggers, and the fishermen. Fishermen, for example were employed to bring samples to marine biological stations: scholars would travel to the seashore, request specimens for study and dissection, and find them at their desks the next day. But increasingly through the twentieth and twenty-first century, field researchers found themselves employing local assistants, not just to find the animals or plants in the first instance, or to provide guidance or

protection—but to maintain the kind of long-term record-keeping that was going to prove vital to the ultimate authority and acceptability of the work emerging from that field site. Most field scientists, after all, had commitments at home that they needed to meet—they either needed to return to write up and defend their theses at the end of their projects, or they needed to be back on campus to teach at the start of the academic year. What happened to the site and the study then became an issue, especially in relation to zoological and ornithological studies: was all the work done there in terms of habituating and identifying the animals to be abandoned, or was there a way of managing events so as to maintain continuity? One might pass over a site to an incoming colleague on a new project—but one might also try to ensure that local field assistants could maintain site records and ensure a permanent research presence at the site.

The history of primatology in particular is characterized by such efforts and transitions, although few of them have been unproblematic, and some have been intensely controversial, with questions raised about the qualifications of field assistants to make long-term observations. The Gombe Stream Research Centre, for example, is one of the longest running sites of field research in the world, not least because of the ability of local assistants to maintain observations in the absence of researchers. In 1960, it was a condition of Jane Goodall's (1934–) permission to work at Gombe that she be accompanied while there by local men. Their role became more prominent after 1970 when it was decided that as well as accompanying students, they would collect basic data on chimp social grouping and behaviour. Their role grew with Gombe, as more and more students came to work with Goodall and the chimps—and became utterly essential when Zairean rebels took four students hostage in 1975. Research permission for foreigners to work at Gombe was promptly withdrawn by the Tanzanian government, and for the next years, social and ecological data was collected by the field assistants. Goodall later noted the importance of instituting training programmes for such people, arguing that—in this instance—their ability to consistently and accurately identify the individual chimpanzees was the fundamental qualification for being able to record data reliably.

In many other examples of field science, providing training or employment for local people has become part of the process of acquiring permission to do research—as well as, for many scientists, a moral obligation. But the question of the reliability of these field results can remain, especially when they're assessed by people far from the field site in question.

Pleasure, pain, and popular narratives

Managing the problem of the role played in the research by people without formal qualifications was just one of the issues that researchers had to deal with when it came

---

Opposite: Balmat and Saussure remembered at Chamonix.

to the social relationships and identities of field science. They had to be sociable, in order to—as we saw earlier in relation to marine science—build coalitions that would support their research. They had to convince people—whether scientists or members of the lay public—who lived and worked in places far distant from their research sites that their work represented an authoritative account of some aspect of the natural world—that they were, actually, *doing* science in the field. This could be a structural as well as an academic problem. When scholars went to the field, they usually did so in the 'vacation' from formal teaching. Marine biologists went to the seashore, glaciologists or astronomers to the mountains, botanists to the woods and heaths—and they did so using the railways and roads created for the leisure industry that had emerged by the late nineteenth century. They often travelled with their families, and stayed in the lodging houses and road inns used by tourist and other travellers; they camped, they went hiking, they sketched and took photos, and to a casual glance were indistinguishable from other middle-class professionals on holidays. Marine biology after the development of the aqualung by Cousteau and Gagnan is a particular case in point: scientists belonged to diving clubs, and diving clubs fed information to scientists. Were these people working, or having fun at someone else's expense?

This problem could be exacerbated by the fact that, while no scientific project necessarily produces reliable results, unproductive fieldwork can look particularly suspicious. If a lab project proves inconclusive or produces contradictory effects, then at least the researcher's visible presence in the laboratory shows that they have tried their best to succeed. But how do you rate or reward the long periods of time when a fieldworker can't find the data they're looking for, or gauge, from the perspective of domestic academic institutions, the physical strain of getting to, constructing, and

Jacques Cousteau and team being lowered in a shark cage, 28 July 1954.

defending a research site in the rainforest? Letters between researchers in the field and those at home reveal the tensions that could emerge when the latter tried to exert too much control at a distance—for example, the problems caused for the Geological Survey when de la Beche tried to insist that survey officers account for their daily movements in detail (1850), or the acrimonious correspondence between Harvard astronomers William Pickering and Robert Black (1889–1890). Field researchers felt that their professional autonomy and dignity needed to be respected: administrators and more senior figures worried that they would not see an academic return on their funding. Field science was a much chancier business than was lab work.

In order to succeed, fieldworkers had to possess or develop a range of personal characteristics and interpersonal talents. They had to be sociable—capable of making common cause and communicating with a range of different social groups, from their supervisors at home to the local assistants in the field. They had to be physically and emotionally robust and resilient. As noted earlier, by definition, doing fieldwork required movement through space: not just the ability to work long hours with concentration, but physical fitness and endurance, as well as the capacity to tolerate a range of different environments. As the primatologist Clarence Ray Carpenter (1905–1975) put it, to succeed, a fieldworker needs 'the endurance and patience of a pack mule'. Accounts of fieldwork, from the explorer's stories of the nineteenth century to the blogs of the twenty-first, often reference a kind of muscular science, with the body again as a tool for wresting data from the land- and seascape. Bleeding feet, blistered hands, bouts of disease, monotonous or no rations, cold, heat, and abrupt and uncomfortable encounters with large and decidedly uncharismatic megafauna abound in these narratives. Emotional and psychological stress was also a potential problem, particularly where fieldwork was being conducted solo in unfamiliar places.

But critically, fieldworkers were also able to turn these problems to their advantage, particularly when giving accounts of their work to wider audiences, both academic and lay. In some contexts, they reported that prolonged solitude could be theoretically and intellectually productive: encountering new phenomena under conditions of complete absorption in one's work could generate creative innovation. More often, it was the physical and emotional suffering they had endured in the field that was stressed, especially in works intended to be consumed by the general public. From the arctic and tropical narratives of early nineteenth-century explorers to the twenty-first-century accounts of herpetologists in the Congo, the difficulties and problems encountered and (sometimes) overcome by the fieldworker were foregrounded, whether they sprang from the physical or the social environment. Partly, this arose from the structural demands of writing for a mass audience: in 1752, John Hawkesworth (1715–1773) had argued that, in order to appeal, explorers' stories needed to stress the heroic nature of the venture. They should show the traveller sacrificing self for knowledge, and should preferentially give first-person testimonials of eyewitness accounts of the natural world, giving their audiences the chance to watch over their shoulders as they met and mastered extraordinary challenges.

Despite the fact that the production of popular science became a steadily less respectable pursuit for scientists as the twentieth century progressed, for a number of reasons field scientists remained keen to publish such accounts. First, and most obviously, they were a potentially profitable means of raising their profile. More importantly for most writers, however, they were not just a means of publicizing their work, but of showing the public the importance of their work, thereby making it more likely that financial support for their research would be maintained. John Barrow's eagerness to promote Arctic exploration through the narratives published by Murray and sons showed his early awareness of the need to establish broad public backing for publicly funded research. By the late twentieth century, it was not just support for research that scientists were seeking through popularizing their work, but support for broader projects of conservation. By this point, the field sciences characterized earlier as examples of observational/historical science had begun to show some of the characteristics of salvage science, as researchers hurried to study organisms and environments before they disappeared forever. Particularly with regard to zoological work, what is notable about these accounts is the way in which scientists took Hawkesworth's arguments about the need to dramatize the scientific story a step further. It was not sufficient to portray the scientist's heroic struggle: in these accounts, the subjects of scientific research themselves become personalities, struggling to survive and protect their families in a hostile physical and social environment. As this chapter showed earlier, long-term research meant that it was possible for scientists to know the individual histories of their research subjects, portraying them as characters with back-stories, motivations, and emotions of their own. Such an anthropormorphic strategy did mean that the scientific status of the authors might be called into question; for some, this was a cost worth paying if it meant public support for conservation work.

Conclusion

Public interest in field science has always been intense, both in terms of information and practice. The nineteenth century—or more specifically, the period from 1820 to 1870—has been called the heyday of natural history, as botanizing became a respectable leisure activity and drawing rooms became crowded with collections of living and preserved creatures, from aquaria to pressed flowers. But in terms of the content of scientific knowledge, it was and is the capacity of the field sciences to tell origin stories that was responsible for most of its appeal. From geology to astronomy to primatology, accounts of fieldwork that seem to shed light on where humans came from and where we might be going have been transmitted through magazines, books, radio, film, and blogs, and avidly consumed by the public. Of course, this adds another element of uncertainty to the role and identity of field scientist—as noted earlier, they are not necessarily in control of the information that emanates from their field site, and they are certainly not in control of how that information is deployed in public life by different groups. From Hawkesworth to the present day, accounts of life in the field have been edited and adapted—not with any nefarious purpose, since the construction of any narrative relies on the foregrounding of some elements and the marginalization of others—but with consequences for the way in which the Western public thinks about nature and our relationship with it. Present in, but often treated as largely marginal to these accounts, are the networks of political support, commercial backing, and intellectual obligation that underpin fieldwork. Particularly over recent years, highlighted are the role of local peoples in both helping and hindering the research process.

But as this chapter has repeatedly shown, fieldwork is intimately tied up with issues of economic development, political manoeuvring, and policy-making: just as some fieldworkers went to the field to study relationships between organisms, so field science exists within a wider ecosystem of relationships, both reciprocal and denied, between humans and the rest of the natural world. This is not to argue that other kinds of science don't show such characteristics, only that they are much more obvious in the context of field science, since field science is conducted out in the open, rather than behind laboratory walls. But, of course, the combination of the fact that fieldwork happens within the public domain with the ready availability of the results and accounts of field-based research means that the expertise of field scientists can sometimes become blurred. This is exacerbated when fieldwork is judged and assessed against a fairly simplistic account of the scientific method—one that assumes, for example, the ability to carry out clear-cut experimental tests of hypotheses. There may well be aspects of field science susceptible to such assessments, but most work in the field—done, as we have seen, in specific, unique places, and using methods of observation rather than manipulation to exert control over their environment—is not. And in addition, more than most scientists, fieldworkers are aware of the care that needs to be taken in the interpretation of their observations and measurements, and the need to be reflexively self-critical and cautious when considering the relationships between theory and observations, instrument and data. Whether dealing with global climate measurements or considering an interaction between two juvenile baboons, most modern field scientists show a self-awareness and humility that scholars in the humanities and social sciences would do well to emulate.

Understanding this aspect of field science is particularly important, given the significance that the results of field research have for the human future. As noted, accounts of field science have been consumed for what they can tell us about human origins; it is ironic that understanding more about the history of field science can help us appreciate more the complexity of the political, scientific, and cultural responses to the prospect of human extinction.

9 The Meaning of Life

PETER BOWLER

The Rise of Materialism

In 1818 Mary Shelley's novel *Frankenstein* created the enduring image of a lone, deranged genius using mysterious forces—with disastrous consequences—to create life. But the prospect of revivifying a composite of dead bodies was not, in fact, beyond the limits of the known; indeed, as the book's preface noted, it was a simple extrapolation from the latest scientific insights into the nature of life: 'The event on which this fiction is founded has been supposed, by Dr. Darwin, and some of the physiological writers of Germany, as not of impossible occurrence.' Shelley's vision was a warning of what might come if the latest developments in science were given free rein. If the nature and origins of life were to be seen as subject to natural law, and hence to human manipulation, the traditional framework of Christian thought would be undermined. Animals and people would be seen as nothing more than machines, and the notion of an immortal soul would be lost along with the myth of supernatural creation depicted in Genesis.

These disturbing consequences would indeed emerge and would be driven by a broad range of developments in the life sciences. Most obvious in the context of Shelley's novel was the emergence of a science of physiology which sought to understand the functioning of the body in materialistic terms. For Frankenstein and the more radical physiologists of the period, the body was just a machine, driven by the still mysterious force of electricity (thus borrowing one of the most exciting new discoveries in the physical sciences). Materialists had used this model of life to challenge the religious view of human nature since ancient times, and their philosophy had received a new impetus during the eighteenth-century age of Enlightenment through the writings of Julien Offray de La Mettrie and others (La Mettrie's 1748 book had been called *L'homme machine*). There was no mysterious vital force enlivening the body and hence no point in supposing that human beings had a soul that survived the death of the physical body. In the nineteenth century that model gained ground in science as physiologists probed the body's functions with an increasingly sophisticated range of techniques.

The threat to traditional values became even more real when the new way of thinking was applied to the brain itself. Physiologists showed that the nerves worked by electrical activity, while the anatomists demonstrated that sections of the brain

Portrait of Erasmus Darwin.

were responsible for producing particular mental functions. The public flocked to phrenologists who claimed to be able to read one's personality from the shape of the skull (presumed to reveal the underlying structure of the brain). But the popular fad concealed a deep threat to the Christian vision of a soul existing apart from the body, gifted with free will by its Creator precisely because it was not subject to natural law. While religious thinkers recoiled with horror, a whole generation of psychologists and philosophers were inspired to develop a more scientific theory of how the mind operated.

Frankenstein reveals another disturbing aspect of the new techniques. The experimenter didn't just study the living body—he sought to manipulate and hence to control it. The horrific implications of this control were highlighted by fictional characters from Frankenstein to H. G. Wells's Dr Moreau, although their real-life manifestations would not become fully apparent until later. Of more immediate concern to many ordinary people was the apparent indifference of vivisectionists to the suffering of the living animals used as experimental subjects. To moralists it seemed obvious that in hardening themselves to such activities the scientists were paving the way toward a wider indifference to the consequences of how their research might be applied.

There was, however, another strand of materialism that raised its own sources of concern. Shelley's reference to Dr Erasmus Darwin reminds us that one of the scientific thinkers involved was also the source of an early version of what would later be called evolutionism. He saw the origin of life in the Earth's distant past as a natural process, followed by a steady but inevitable progress by which the first primitive life forms had become more complex over a vast period of time. Evolutionism involved a rejection of the Genesis account of creation, and by linking the human race to the 'brutes that perish' it paralleled the physiologists' assault on the concept of an immortal soul. Religious believers inevitably rejected the idea, but the radical thinkers who followed in Erasmus Darwin's footsteps sought to retain a sense that the cosmos had a moral purpose by stressing that evolution inevitably moved animals toward higher levels of physical and mental organization. This sense of purpose would be challenged by Charles Darwin (Erasmus's grandson), whose *Origin of Species* of 1859 seemed to reduce evolution to a chapter of accidents driven by a brutal struggle for existence. Much late nineteenth-century evolutionism can be seen as a rearguard action fought to retain the element of progress that gave the theory its moral foundation.

By the 1850s it had become impossible to defend a literal reading of Genesis because geologists had demonstrated that the Earth had witnessed a long series of developments antecedent to the appearance of humankind. An outline of the fossil record had been established, confirming that life had first appeared in primitive forms and had then ascended the scale of organization in the course of geological time. Even those who wanted to defend the idea of divine creation had to accept a series of distinct creative episodes, while the evolutionists exploited the record's message of progress for all it was worth. The popular understanding of evolution inevitably focused on the fossil evidence, but new ideas about the development of life also came from a different area of discovery, the geographical exploration of the globe. Scientific travellers such as Alexander von Humboldt revealed a pattern of distribution among the plant and animal species of the various continents that begged for an explanation. Charles Darwin was led to his theory of evolution not by the fossils, but by seeking to understand the geographical distribution of the South American animals he saw on his voyage aboard HMS *Beagle*.

These developments took place against a background of immense change in the status and organization of science. The popular image of the lone (and probably mad) scientist performing bizarre experiments in some remote location was increasingly out of step with the real world. Charles Darwin was a wealthy amateur (the term had no pejorative connotations at the time) but science was increasingly being done in large communal laboratories located in universities, government-sponsored museums and surveys, and industrial plants. The governments of France and Germany led the way in providing support for science, and the physiology that improved our understanding of the body's functions was done in the laboratories they supported. Natural history was transformed by the work of comparative anatomists based in the natural history museums of Paris and other great cities. Darwin's supporter T. H. Huxley played a leading role in the professionalization of the scientific community in Britain—his first job was as a paleontologist at the Royal School of Mines, and he worked tirelessly to establish laboratory-based instruction for science students at a range of other institutions. Another Darwinist, J. D. Hooker, was director of the Royal Botanic Gardens at Kew, centre of a global network of plant collectors.

Hypothetical cross-section through the Earth's crust to illustrate the superposition of strata. From William Buckland, *Geology and Mineralogy Considered with Reference to Natural Theology* (1837).

Charles Darwin in 1840. Portrait by George Richmond at Down House.

All of this came about in part because science was displaying its value to governments and industries as a source of practical information aiding industrial, medical, and agricultural progress. Physiology offered obvious benefits to the medical profession by revealing the factors necessary for a healthy diet, while other scientific discoveries such as the germ theory of disease also offered the prospect of future cures. The medical profession was slow to take up the offer, seeing its practice more as an art than a science, but in many other areas the expectation that scientific discoveries would offer control as well as understanding of nature was welcomed. The public was keen to experience what might be on offer, seeking information through exhibitions, museums, public lectures, and an ever-expanding range of printed literature.

Here we have a very different aspect of the new materialism, building in a more realistic way on the hope of dominating nature caricatured in the image of Frankenstein and his monster. If much of the new technology came from the physical sciences, the life sciences were not far behind, as the medical applications of physiology revealed. Darwin drew inspiration from the animal breeders who revolutionized agriculture by creating more productive varieties of domesticated species. Kew Gardens was the centre for a project to transform the productivity of the British empire by transporting useful species such as the rubber tree from their native homes to other regions. The voyage of the *Beagle* was also a product of British imperialism—the ship's main task was to chart the coast of South America to aid the shipping that carried world trade. Darwin's theory of evolution threatened the Christian view of humanity's relationship to God, but the ability of science to satisfy the practical demands of human life represented an equally challenging threat to the old morality and the old social order.

The New Science of Life

These generalizations are valid enough in outline, but modern historical research has increasingly revealed the complexity of the actual course of events. Conservative forces fought a desperate rearguard action to defend traditional ideas about the nature and origin of life, sometimes modifying those ideas to bring them more into line with the latest scientific developments. In the case of physiology, the emergence of a new science devoted to explaining life in terms of the physical and chemical processes was a protracted affair influenced by a wide range of factors. The application of experimental procedures to living things opened up a cornucopia of new knowledge, allowing physiologists to set themselves up as an independent discipline free from their original ties to the medical profession. The rejection of vitalism (the notion of a special vital force that was not subject to the laws of the physical universe) was an integral part of this process of modernization. But professionalization was driven by the academic environment as well as the growth of knowledge. In Germany the expansion in the number of physiology departments was driven by university rivalries, while Britain's belated catching-up exercise was speeded by the Royal College of Surgeons' decision to require some training in the field for students. Nor was there any simple connection between the rise of experimentalism and the declining belief that living things were animated by a distinct non-physical force.

The use of experiment to reveal the functioning of the living body was certainly a key feature of the new science. In the later decades of the eighteenth century Albrecht von Haller had begun to define the discipline of physiology based on increased use of controlled observation. This was taken up with enthusiasm in the early years of the following century, first in France, then in Germany, and somewhat later in the English-speaking world. The experimentalists certainly hoped to establish law-like regularities governing the various functions of the living body, and for the more radical materialists (sometimes known as reductionists) this meant that all of those functions would eventually be explained in terms of chemistry and physics. It used to be thought that Friedrich Wöhler's synthesis of urea from inorganic chemicals in 1828 played a major role in undermining faith in a distinct vital force. No longer could it be maintained that some compounds could only be produced via the action of a life-force. But historical research has shown that Wöhler's synthesis produced no such sudden transformation of belief. Throughout the nineteenth century the battle between vitalists and reductionists swayed back and forth, with both sides making use of experiment to support their conclusions.

In the first half of the century the key founders of the new science either retained some form of vitalism or refused to come out openly in support of the reductionist position. Xavier Bichat, the pioneering French experimentalist, defined life as the sum

Vivisection experiment.

of the processes resisting the degeneration produced by the operation of physical forces. For Claude Bernard, the living body was a system set up to maintain its internal environment in the face of a changing external world. Without openly appealing to a vital force, he saw the living body as something more than a simple mechanical system. In Germany, Johannes Müller and Justus von Liebig both continued to favour vitalism, but others adopted a position paralleling Bernard's, seeing the body as making use of physical and chemical processes, but doing so in a way that could not be explained solely in physico-chemical terms. This 'teleomechanist' position saw the body as a complex, self-balancing system designed to continue maintaining its integrity. Later in the century the idea that the body is more than the sum of its parts became popular under

---

Opposite: Louis Pasteur.

M. PASTEUR.

the name of holism or organicism. The body's functions could only be explained in biological terms involving levels of organization not reducible to physical laws.

It was the evolutionists who sought to explain how such higher levels of organization could have emerged by natural processes, but their investigations derived from an entirely different research tradition. Thomas Henry Huxley, one of Darwin's leading supporters, was also a proponent of the reductionist position, openly appealing to the old materialist tradition that animals should be treated as mere machines. The new theory that all living things were composed of cells threw emphasis on the 'protoplasm' within the cells, allowing reductionists to believe that by understanding its chemical properties they would reveal the 'physical basis of life'.

Huxley was one of the few figures to straddle the realms of laboratory biology and the scientific natural history of the Darwinists. Most physiologists continued to address the issue in purely experimental terms, and at this level the debate between reductionism and organicism continued for some time. In the early twentieth century John Scott Haldane was still defending the view that the body was a system designed in ways that could never be understood in mechanistic terms. There was even a brief revival of open vitalism inspired by Hans Driesch's notion of an 'entelechy' capable of overriding mechanical principles. Many early twentieth-century physiologists continued to work with concepts that were defined in purely biological terms, in effect continuing the holist view that the body is more than the sum of its parts. The division between reductionists and organicists was often based more on ideological than scientific grounds. Reductionists tended to be materialists in the broader sense, opposed to traditional religious beliefs and conservative social forces. Organicists remained sympathetic to the idea that there might be some higher powers at work in evolution, if not in the everyday world. The strident anti-vivisection movements of the nineteenth century were also driven by suspicion of science's role as a radical modernizing influence, echoing the fears expressed earlier in the Frankenstein story.

Electricity and Life

Most of the early nineteenth-century discoveries focused on the chemical processes sustaining life, an early example being the explanation of 'animal heat' by analogy with combustion. But as Mary Shelley's story reminds us, there was also intense interest in the possibility that the body was vivified by a force related or even identical to electricity. Luigi Galvani's discovery of current electricity was prompted by the observation that a frog's leg was made to twitch by the current. His idea of a distinct form of animal electricity was soon discounted, allowing electricity itself to be seen as a force which activated the nerves controlling the body. Here lies the origin of the Frankenstein story, fuelled by Giovanni Aldini's demonstrations of the contortions produced in the bodies of executed criminals when stimulated by electric shocks.

Although it soon became apparent that human or animal life could not be restored in this way, the possibility that primitive forms of living organization might be produced by electrical activity was widely discussed by radical thinkers. The pioneering French evolutionist Jean-Baptiste Lamarck proposed that all life began from simple forms generated by this means. In the late 1830s Andrew Crosse claimed to have produced living insects by passing an electrical current through certain chemicals. A link to the Frankenstein story inevitably found its way into the resulting debate, in which Crosse frantically tried to defend himself from the accusation that he was a materialist. His discovery was eventually discredited, although not before it was endorsed by the most controversial evolutionary text of the next decade, Robert Chambers's *Vestiges of the Natural History of Creation* of 1844.

Meanwhile, physiologists continued to study how the nerves operated through the transmission of electrical impulses. Hermann von Helmholtz was later able to measure the actual speed with which the impulses travelled along the nerve ganglions. Among popular writers about science, the belief that electricity was indeed the key to understanding life was widely discussed, along with the obvious possibility that control of the processes involved would have medical applications. Alfred Smee's *Elements of Electrobiology* (1849) proclaimed the whole body to be a complex electrical circuit. The brain itself was a series of electrical connections, linked to the body by nerves that operated in the same manner as the telegraph. The medical profession was highly critical of efforts to apply this philosophy of life to the treatment of nervous and even physical disorders, but unlicensed practitioners catered to a significant popular interest in this new vision of life.

Mind and Brain

The electrical view of the brain's operations dovetailed with another popular movement, also widely associated with materialism and decried in professional academic circles. In the late eighteenth century Franz Joseph Gall used cerebral anatomy to build up a theory in which the various mental functions were thought to be produced by activity in particular parts of the brain. Under the name of 'phrenology' this view of human nature was spread around Europe by J. C. Spurzheim. The phrenologists assumed that the individual's personality was determined by the structure of the brain, and believed that one could deduce this structure—and hence the nature of the personality—from the bumps on the skull. Advocates for the technique such as George Combe of Edinburgh proclaimed it as a science that would reform society by allowing people to make the best use of their inborn faculties. Combe's *Constitution of Man* of 1832 was one of the most popular books of the century. He claimed not to be a materialist, but the implication that the mind was created by the physical processes going on in the brain was evident. Phrenology thus helped to throw doubt on the traditional belief in a soul independent of the body.

Phrenology became hugely popular, despite being condemned in academic and medical circles. Conservative thinkers objected to its materialism, but also pointed out that the surface features of the skull do not follow the underlying structure of the brain. Phrenology eventually came to be seen as a pseudoscience, but the basic idea that mental faculties must result from physical activity in the brain remained

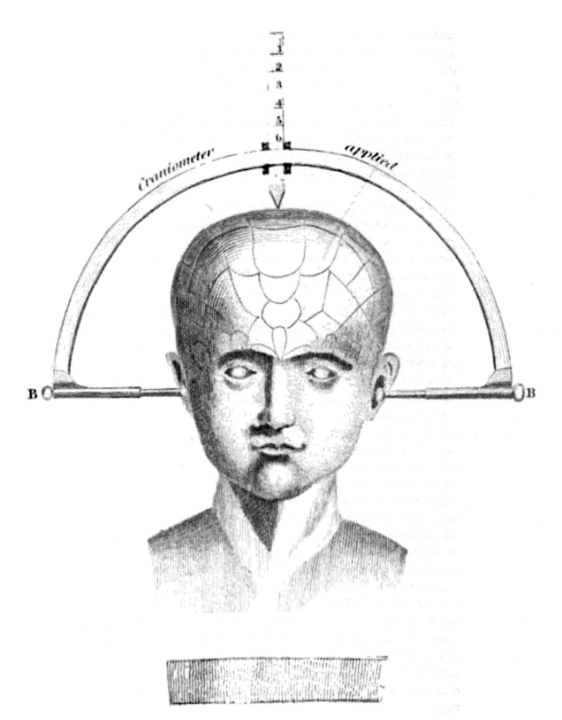

Phrenological Head. From George Combe, *Elements of Phrenology* (1841).

influential, inspiring major thinkers such as Herbert Spencer. Eventually anatomists began to confirm that some faculties are indeed located in certain areas of the brain. In the 1860s Paul Broca identified an area responsible for articulate speech. Physiologists were also investigating the operations of the nervous system and the brain, building on the discovery that the nerves operated by electrical impulses. But neurophysiology and psychology were slow to integrate, mainly because the physiologists were anxious to project an image of their science as based firmly in experiment. Getting involved with issues such as the physical basis of the higher mental faculties looked too much like materialist speculation, so the investigation of mental functions was left to psychologists still working within an older tradition more in tune with the philosophical analysis of the mind. Only in the twentieth century would the link between neurophysiology and psychology become properly established.

Even so, philosophical psychologists such as William James were increasingly attracted to a model of the brain which saw it as a kind of electrical circuit. Given the increased sophistication of the telegraphy and the electrical power industries in the late nineteenth century there were plenty of physical analogies available by which the operations of a complex electrical machine such as the brain could be understood. When Sigmund Freud sought to establish his brand of psychology (psychoanalysis) as an independent discipline in the 1890s, he constructed an image of its origins which downplayed the influence of biological models on his thinking. In fact Freud had begun his career as a neurophysiologist working on the nervous systems of lower animals, and he was well aware of the latest ideas which explained nervous tensions by analogy with electrical discharges.

Freud wanted to cure patients suffering from hysteria, irrational behaviour believed to be caused by emotional stress. He argued that the tension had been generated by traumatic events in early life and could be cured by persuading the patient to recall and confront the trauma. His approach was proclaimed as a pure science of the mind, operating with concepts that could not be explained in physical or biological terms. Mental illness was due to purely psychological tensions, not to physical aberrations in the brain. Along with other, often very different approaches such as behaviourism (based on experimental studies of animal behaviour) psychoanalysis was part of a more general trend by which the early twentieth-century sciences of the mind sought professional independence from their origins in philosophy and biology.

Freud's path to psychoanalysis had in fact been illuminated both by earlier ideas about the functioning of the nervous system and by late nineteenth-century evolutionary models of the mind. Freud's 'unconscious' was really the legacy of our early animal ancestry only imperfectly overlain by the higher mental functions of consciousness that emerged in the last phase of human evolution. Where Freud differed from the evolutionary thinkers of the Darwinian era was in his assumption that the conscious 'ego' was often unable to control the desires of the subconscious 'id' left over from our animal past. An evolutionary model established in the expectation that it would reveal the triumphant progress of the mind in the course of evolution had been turned on its head to reveal the hidden dangers of our bestial instincts. Here was a new form of materialism just as frightening as that created by the Frankenstein story. No wonder Freud's analytical psychology was seen as a potent symbol of the new century's rejection of the Victorians' faith in progress.

The Emergence of Evolutionism

The model of progress challenged by Freud had been supported in large measure by the emergence of evolutionism in biology and related disciplines. Erasmus Darwin had scandalized readers with his evolutionism as well as his ideas on the nature of life. Darwin's *Zoonomia* of 1794 expounded his theory in technical terms, but his radical ideas also found their way into his poetical works, which were quite popular at the time. William Paley's *Natural Theology* of 1802 was written in part to defend the traditional vision of divine creation against Darwin's speculations. Where Darwin saw life beginning from simple organisms formed by spontaneous generation, then rising to ever-higher levels of organization over many generations, Paley argued that the adaptation of species' structures to their environments could only be explained by supposing that they were designed by a wise and benevolent God. Just as the complex structure of a watch proclaims that it was made by an intelligent craftsman, the bodies of living things reveal the power, wisdom, and goodness of God.

For Paley there was only a single creation, but the biblical version of the Earth's origin was already being thrown into doubt by discoveries in geology. Studies of the fossil record by Georges Cuvier in France and followers such as William Buckland in Britain revealed the remains of strange creatures unlike any known today, which had presumably become extinct before the present species had appeared. Perhaps they had been killed off by a vast geological catastrophe, which Buckland tried to identify with Noah's flood. But even he realized that the last catastrophe was only one of many, and by the 1840s an outline of the history of the Earth very similar to that still accepted today had emerged. The outline seemed to confirm a progressive development of life from primitive invertebrates in the most ancient periods, through an age of fishes, an age of reptiles (the term 'dinosaur' was coined in 1841 by Richard Owen), an age of mammals, and finally the modern period in which humankind appeared. To retain the idea of divine creation it would be necessary to imagine a whole series of such events, with many of the products being subsequently wiped out.

Paley's 'argument from design' could be salvaged in the face of these new discoveries, but naturalists were forces to think more carefully about the significance of the changes revealed by the history of life. The alternative view that life developed by natural law rather than divine miracle was certainly in circulation among more radical thinkers, as Erasmus Darwin's speculations reveal. The notion of design was later turned on its head by Erasmus's grandson, Charles, with his theory of natural selection. But Charles Darwin's *Origin of Species* would propound a particular image of how evolution worked that was radical in more ways than one. It explained how species became adapted to their environment as the 'struggle for existence' eliminated all but the most-favoured individuals. But it also insisted that the evolution of life should be seen as a branching tree, the branches diverging as populations migrated around the world, became separated by geographical barriers, and adapted to the environments they encountered. This was by no means how most early evolutionists conceived the process, and to understand the debates of the time we need to think carefully about the different ways in which a theory of evolution (or 'transmutation' as it was called at first) could be articulated.

The alternatives are best revealed in the work of Erasmus Darwin's French contemporary, Jean-Baptiste Lamarck, whose *Philosphie zoologique* of 1809 proposed a similar theory in considerably more detail. Both appreciated that species are adapted to their environments, and suggested that this came about as animals strove to adjust their behaviour to new conditions. If the ancestral giraffes had stretched their necks to reach the leaves of trees, the exercise would have lengthened their necks and the additional length would be inherited by their offspring, who would then repeat the process—leading ultimately to the modern species' long neck. This is the inheritance of acquired characteristics, often termed 'Lamarckism'. Although rejected by modern geneticists this mode of inheritance was accepted by most nineteenth-century naturalists, although few at first saw it as sufficient to generate new species.

Neither Erasmus Darwin nor Lamarck had any notion of natural selection, but nor did they think in terms of ancestral populations becoming separated and diverging in different directions to give a 'tree of life'. Both thought that evolution must also be an inherently progressive process—it must necessarily push living things up the scale of organization until eventually something like the human level of intellect was attained. Lamarck argued that the progressive trend was the primary agent of change, with adaptation to the environment being a purely secondary effect. The model of evolution as a ladder of progress rather than a branching tree was to have a profound effect on nineteenth-century thought, although it was a vision profoundly at variance with the one that Charles Darwin would promote in the *Origin of Species*. Even within particular groups of animals and plants, naturalists thought in terms of rigid trends pushing species in the same predetermined direction. Evolution had to have a goal, and be driven by a built-in trend that would advance toward that goal. It was this goal-directedness of most early evolutionary theories that would eventually make them acceptable to liberal religious thinkers.

Georges Cuvier ridiculed Lamarck's theories, and historians used to think that this signalled their complete rejection by the scientific community. But more recent research shows that there were more radical thinkers, especially among medical anatomists, who were prepared to challenge the traditional professional hierarchy and who saw such materialist ideas as an ideal way to promote their case. At Edinburgh, Robert Grant startled the young Charles Darwin with his support for Lamarckism, although at the time Darwin—then a young medical student—was not impressed. Many of the early supporters of Lamarck, however, focused not on his suggested mechanism of adaptation (which only later became identified as the core of 'Lamarckism') but on the idea of multiple lines of evolution advancing in parallel up the scale of development toward humanity.

The same model of development as a ladder rather than a tree flourished in Germany among the followers of J. F. Blumenbach. These included well-trained biologists such as Johannes Müller who dismissed any notion of divine creation as a satisfactory explanation of the origin of new living forms. They seldom developed well-articulated theories of what we would call evolution, and some still thought that complex living things might be generated spontaneously by purposeful natural forces. This, of course, would relegate evolution to a secondary role, although it was often speculated that each continent might have its own original forms as the starting point for further predetermined developments in the course of time. In its most extreme version this implied that the various human races were independently evolved on separate continents from distinct ancestral forms. By the middle of the nineteenth century the concept of spontaneous generation as applied to complex forms had been abandoned, in part because the new theory that all organisms are composed of cells made it seem implausible. This created more space for an evolutionary perspective, although not necessarily the one that would be promoted in the *Origin of Species*. There was little enthusiasm for divine creation, but as yet no major push to develop the linear model of development into a fully articulated theory of evolution. Instead, students of comparative anatomy focused on the search for 'laws of form' that would establish regularities linking the various species into a coherent pattern.

Some of the early German thinking on these issues had been encouraged by the speculative and mystical tradition of *Naturphilosophie* identified with Lorenz Oken. The search for underlying patterns linking the species was sometimes known as 'transcendental anatomy'. But the followers of Blumenbach were no mere speculators, being more inclined to 'teleomechanism'—the idea that nature is driven by physical forces constrained to operate toward purposeful goals. In France too the search for unifying patterns among living types was promoted from a materialist perspective by Etienne Geoffroy Saint-Hilaire. Like Lamarck, Geoffroy clashed with Cuvier by proposing a theory of transmutation, but his version postulated abrupt saltations or macromutations which produced new species instantaneously through some deviation in the process of embryological development. The influential German thinker J. W. Goethe hailed the clash between Cuvier and Geoffroy in 1830 as the great intellectual debate of the age.

Transcendental anatomy promoted an alternative to the 'ladder' model of development because it stressed the underlying unity of each group of living species. This approach was introduced into Britain by Richard Owen, who wrote in 1848 of the 'vertebrate archetype'—the basic underlying pattern on which all of the living and fossil vertebrates were modelled. The modifications gave each species the characters it needed to adapt to its prevailing environment. Owen presented his archetype theory as a superior form of natural theology, allowing us to see the orderliness of God's creation as well as His benevolence. But the model also pointed toward a branching rather than a linear model of development, and Owen began to see the history of life unfolding in the fossil record more as a tree than a ladder. His evidence would actually be used for a very different purpose by Darwin in the *Origin of Species*. The archetype became the common ancestor from which various more specialized descendants had evolved.

The 'Vestiges' Debate

Continental speculations about evolution were at first dismissed by the conservative scientific establishment in Britain. Owen led the way in attacking Lamarckism during the 1830s, although his opposition became less strident in the course of time. In particular, he refused to criticize the book which introduced the new ideas to a wider public, Robert Chambers's anonymously published *Vestiges of the Natural History of Creation* of 1844. Chambers was an Edinburgh publisher who produced literature aimed at the rising middle class, for whom the idea of progress was central as the foundation for their hopes of social reform. As an amateur naturalist, he realized that one way of promoting the inevitability of social progress was to present it as the continuation of a gradual development which had operated throughout the history of life on Earth. *Vestiges* thus introduced the theory of progressive evolution brought about by natural law rather than divine miracles—although to make the idea more palatable Chambers stressed that the laws were instituted by God to achieve His goal—the production of rational, moral beings—indirectly.

Many of the themes in Chambers's book would have been familiar to any radical thinker. The Earth was created along with the other plants by the condensation of a swirling nebula of gas and dust. The first, very primitive, living things were generated by the action of electricity on matter. Life then gradually ascended the scale of complexity in the course of geological time, driven by a law of development which occasionally tweaked the process of embryological growth to produce the next highest step in the scale by a sudden jump or saltation. Chambers's model of evolution was linear rather than branching (because that fitted better into the general ideology of progress) and it paid little attention to adaptation. The image was of multiple parallel lines of evolution with separate origins each advancing step by step up roughly the same hierarchy of organization. This may have been an evolutionary world-view, but it was very different to the one that Darwin would present fifteen years later.

Chambers made no effort to conceal the theory's most dangerous implication: the idea that humans were the highest products of the animal kingdom. This meant that the human mind's rational and moral faculties had developed gradually from the lesser mental powers of the lower animals. He openly proclaimed that our mental functions, and hence human social relations, are governed by natural law. For traditional Christians this was a direct challenge to the idea that humans were created in the image of God and that our immortal soul would be judged by its Creator. There was a huge public outcry against the book, with Adam Sedgwick, the Cambridge professor of geology who had mentored Darwin, leading the way. Much of the debate was focused on the scientific weaknesses of Chambers's position, especially the discontinuity of the fossil record, which at this point had yet to reveal continuous evolutionary sequences. It was almost as though the wider implications of his theory were too dangerous to discuss openly (especially, Sedgwick hinted, in front of the ladies). Yet the book sold well, and its ideas gradually began to find their way into popular literature. More and more people were becoming dissatisfied with the old world-view based on divine creation and original sin. By the 1850s even the scientific community was beginning to reconsider its position. In 1855 Herbert Spencer published a Lamarckian explanation of the emergence of the human mind. As yet, however, most naturalists were reluctant to promote an openly evolutionary model of the history of life on Earth. But Charles Darwin, who had been working on such a theory since the 1830s, sensed the change in public attitudes and at last began to prepare for publication.

The Development of Darwin's Theory

Charles Darwin was born in 1809, his father (one of Erasmus Darwin's sons) being a wealthy medical doctor. After an unsuccessful episode as a medical student in Edinburgh (where he met Grant, the Lamarckian anatomist) he went up to Cambridge to read for an Arts degree, with the vague intention of becoming a clergyman-naturalist. Here he read Paley's *Natural Theology*, which subsequently focused his attention onto the problem of explaining how species became adapted to their environments. He also began extra-curricular work in geology under Sedgwick and in botany under J. S. Henslow. It was Henslow who gave him the opportunity to travel aboard the survey vessel HMS *Beagle*, which left in 1831 for a five-year voyage to chart the waters of South America. While on the voyage he read Charles Lyell's *Principles of Geology* and was converted to the uniformitarian view that the Earth's surface was shaped not by violent catastrophes but by everyday causes operating over vast periods of time. As he tried to make sense of the discoveries he made on the voyage, this model helped to prime him for an approach based on natural processes rather than miraculous creation.

The most important discoveries were in biogeography and concerned the factors which govern the distribution of species across geological space. He saw that barriers such as great rivers or mountain ranges often separated closely related species of the same group. This point was brought home to him most directly in the Galapagos Islands, where different species of mockingbirds and finches could be found on the various islands. As he pondered these facts after his return to England, Darwin realized that they would make sense if birds derived from original South American parent species had accidentally been transported to the islands, where each local population had then adapted to its environment in a slightly different way. The groupings of species into genera and higher classifications was a consequence of divergent evolution from a common ancestor, driven by some process operating on fragmented populations that could no longer interbreed. Here was the model of a 'tree of life' based on constant branching followed by adaptation to different environments.

In the late 1830s Darwin began a programme of research to discover how the process of adaptation worked. He knew of Lamarck's explanation, and never doubted

HMS Beagle *in Murray Channel*, Tierra del Fuego. Watercolour by Contrad Martens.

that acquired characters are inherited to some extent, but did not believe that this was adequate. In an effort to understand how species vary, he worked with animal breeders and noted that they achieved their ends though artificial selection. They were aware that every individual within a population has its own peculiarities, most of which seem to have no use or purpose. There is a fund of natural variation within the population that is undirected, or (in common language) random. But the breeder can pick out those few individuals who happen to be born with the character he desires, and will select only those to breed. By continuing the process over many generations, all sorts of bizarre characters can be created, such as the various breeds of pigeons or dogs. Darwin wondered if there could be an analogous process operating in nature, but without the conscious directing agency of the breeder.

His clue as to how such a process of 'natural selection' might work came from reading the work of Thomas Malthus on population. Malthus had attacked speculative proposals for social reform by arguing that the human population would always tend to outstrip the food supply, so that poverty was inevitable unless people voluntarily limited their family size. Applying this model to the animal kingdom, where self-restraint does not apply, Darwin realized that there must be a constant 'struggle for

1. Geospiza magnirostris.
3. Geospiza parvula.

2. Geospiza fortis.
4. Certhidea olivasca.

Galapagos Finches showing variation in beak structure. From Charles Darwin, *Journal of Researches*, chapter 17.

existence' caused by shortage of food, and that in this struggle those individuals who were by chance best-adapted to the local environment would survive and breed successfully. The less-fit would be eliminated. Here was a process analogous to artificial selection, but driven by purely natural causes and aimed solely at adaptation to the local environment.

The implications of natural selection were truly shocking to anyone brought up in the traditional world-view based on divine creation. This was evolution by law, but the process was purely one of trial and error, based on relentless struggle and suffering, hardly what one would expect to be instituted by a wise and benevolent God. Its end-product was local adaptation, not progress toward some morally significant goal. Darwin was aware that evolution was progressive in the long-run, but he could see that most branches of the tree of life led only toward increased specialization for some narrowly defined way of life. Extinction was inevitable if the environment changed too rapidly, or if some better-adapted species invaded the territory. Here was a highly materialistic view of life, reinforced by Darwin's growing conviction that the human race would have to be included in the system. Our mental and moral faculties must be extensions of those possessed by the higher animals, brought about because they conferred some advantage to our early ancestors as they separated from the apes.

Partly for fear of the public reaction Darwin decided not to publish his theory. He had now married and taken up the life of a country gentleman, and his own wife was deeply concerned about the implications of his thinking. He continued his species work in private, gathering supporting evidence from other naturalists. He was in any case busy publishing the results of the *Beagle* voyage and, later, on a huge project to describe and classify the barnacles, then a little-known group. These efforts would pay off when he eventually published, because they gave him a solid reputation as a naturalist which would ensure that his theory could not be dismissed as wild speculation. By the mid-1850s he sensed a relaxation in the public anxiety on the issue, and informed a few other scientists of his ideas, including Lyell and the botanist J. D. Hooker. He began work on what would have been a multi-volumed account of his theory for eventual publication.

This work was interrupted in 1858 by the arrival of a paper written by an unknown naturalist, Alfred Russel Wallace, which appeared to contain the essence of his own theory. Wallace was then making a living collecting specimens of exotic species in the islands of what is now Indonesia. In 1855 he had published a paper outlining, in effect, the theory of divergent evolution, but without making the implications explicit. Now, having remembered an earlier reading of Malthus, he recognized that the struggle for existence would drive unfit varieties to extinction and thus cause adaptive evolution. He wrote the idea up in a short paper and sent it to Darwin, who he knew to be interested in these issues, to ask if he could arrange for publication.

Wallace's paper is widely regarded as a classic example of independent discovery. Darwin was certainly panicked by it, fearing he might be scooped, and was advised by Lyell and Hooker to publish Wallace's paper along with a short extract of his own writings that could be verified as having been written earlier. But some historians now suspect that Darwin may have over-reacted, reading too much of his own thinking into Wallace's sometimes opaque language. When the paper is read without Darwin-tinted spectacles, it seems probable that much of the discussion focuses on the extinction of unfit varieties or subspecies, not the elimination of unfit individuals in a single population. Wallace did not invoke the analogy with artificial selection, and remained suspicious of it when he subsequently read the *Origin of Species*. He came from a poor background and did not share Darwin's enthusiasm for a world-view analogous to the competitive spirit of Victorian capitalism. Given these differences and his much later discovery of the idea of selection, it is hard to see Wallace as being in a position to kick-start a major revolution in scientists' thinking.

Darwin, meanwhile, rushed to finish a much shorter account of his theory, which appeared at the end of 1859 as *On the Origin of Species*. The joint Darwin-Wallace papers published in 1858 had very little impact, but now a detailed account of the theory appeared in a format that no one could ignore. The debate over the plausibility of evolution, which had been hanging fire though the previous decade, was now ignited with explosive results.

The Darwinian Revolution

Darwin's book prompted a vigorous controversy both in science and among the general public. So great was the hostility of some conservative thinkers that the outcome hung in the balance through the early 1860s. But by the end of that decade it was becoming clear that the battle had been won by the evolutionists—or 'Darwinists' as they began to call themselves. But this term is misleading, because at the time it denoted anyone who had followed Darwin into the general world-view of evolution-ism. It did not (unlike the modern use of the term) denote acceptance of natural selection as the primary cause of evolution. Even some of Darwin's most vigorous champions were not fully convinced by his explanation of how the process worked.

Late nineteenth-century thought was certainly dominated by the idea of evolution, both in science and in many areas of culture and social thought. It was also dominated by the idea of inevitable progress, with some form of the struggle for existence seen as the driving force. But by the standards of modern Darwinism the influence of the theory of natural selection was often superficial—it turned out that there were other ways of visualizing 'progress through struggle' that did not depend on the selection of random variations within a population. The popular assumption that nature would eliminate the less efficient species is a case in point—this was endorsed by some thinkers who would never have admitted that natural selection could actually produce new species.

Much of the initial controversy was sparked by religious and moral concerns. The clash between Thomas Henry Huxley and Samuel Wilberforce, Bishop of Oxford, at the 1860 meeting of the British Association for the Advancement of Science is often seen as a pointer to how things would turn out, thanks to the perception that Huxley had triumphed in the debate. For traditional Christians there were obvious areas of concern that would be raised by any evolution theory. The status of the human soul was called into question—how could an immortal spiritual entity be evolved from the 'brutes that perish'? Here, however, there were developments in other disciplines that also pointed toward an evolutionary perspective. Archaeologists were uncovering the remains of ancient stone-age cultures which indicated a deep antiquity for primitive humans. Anthropologists studying cultures from around the world began to rank them into a hierarchy ranging from the most primitive—little better than our stone-age ancestors—through to modern industrial civilization. All too often the most primitive (the Australian aborigines were the classic example) were also described by physical anthropologists as having ape-like features and smaller brains than 'modern' humans. In the absence of fossil human ancestors they were made to stand in for the 'missing link' in the chain of progress from the ape. When fossils did turn up later in the century, they too were incorporated into the progressive sequence.

Opposite: Cartoon of Thomas Henry Huxley. From *Vanity Fair* (1871).

"Not a brawler."

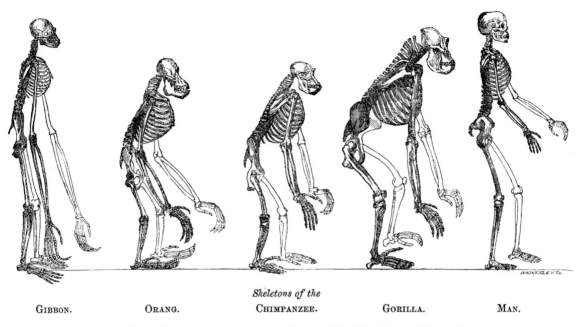

Skeletons of the
GIBBON.   ORANG.   CHIMPANZEE.   GORILLA.   MAN.

Photographically reduced from Diagrams of the natural size (except that of the Gibbon, which was twice as large as nature), drawn by Mr. Waterhouse Hawkins from specimens in the Museum of the Royal College of Surgeons.

Skeletons of the apes and a human compared. Frontispiece to Thomas Henry Huxley, *Man's Place in Nature* (1873). The gibbon (on the left) is shown twice the size of the other skeletons.

Given this barrage of apparent evidence, even liberal Christians became convinced that the human spirit must have evolved in accordance with laws instituted by the Creator. The reverend Charles Kingsley wrote his popular story *The Water Babies* to celebrate how effort and initiative led toward progress, while idleness and sloth marked the path to degeneration followed by the apes. Kingsley supported Darwin and his tale is often portrayed as an expression of Darwinism, but in fact it adopts the Lamarckian view that individuals can improve themselves and hand their advances on to their offspring to promote the advancement of the race. This was the message that was also promoted in the philosophy of Herbert Spencer—and despite coining the iconic term 'survival of the fittest', Spencer was actually a Lamarckian who sang the praises of self-improvement as the foundation of social and biological progress. He advocated a society based on unrestrained free enterprise because he thought that the struggle for existence would encourage everyone to make better efforts to improve themselves and pass the benefits to future generations. In America, clergymen such as Henry Ward Beecher openly proclaimed themselves as followers of Spencer, seeing his philosophy as a modernized version of the Protestant work ethic. These Spencerians

Opposite: Cartoon of Bishop Samuel Wilberforce. From *Vanity Fair* (1869).

PEDIGREE OF MAN.

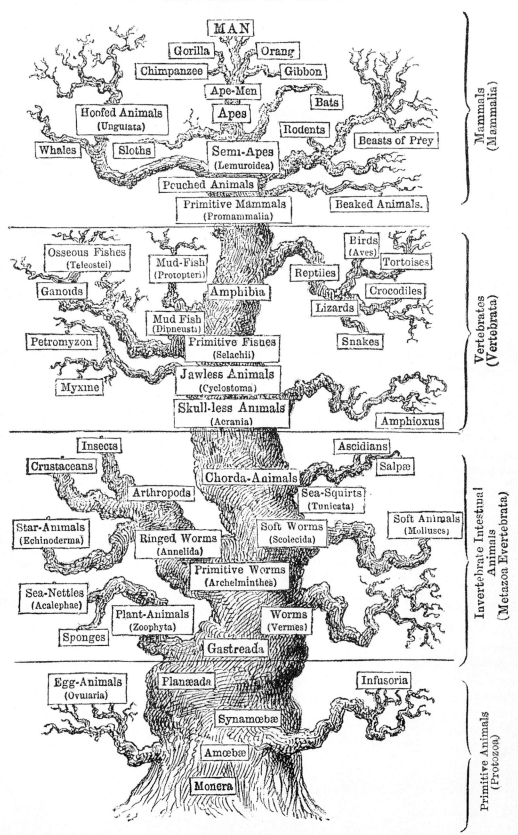

are often described as 'social Darwinists' but their enthusiasm for progress through struggle owed little to the idea of natural selection.

Natural selection in its truly Darwinian form did not look like the kind of process that a benevolent Creator would establish to achieve His ends. By selecting a tiny handful of fitter variants and killing off the rest, it reduced evolution to a process of trial and error, undermining all evidence of design by God. Nor did it necessarily make for inevitable progress, although in his *Descent of Man* Darwin tried to identify the circumstances that had put our ancestors on the road to mental superiority over the apes. He identified our move out of the trees onto the open plains as the key step, a suggestion ignored by most of his contemporaries, who were convinced that there must be an inevitable evolutionary pressure leading to larger brains. The Darwinian model of evolution as an ever-branching tree, with the human twig as just the end-point of one branch among many just didn't fit with the popular enthusiasm for the idea of progress. As far as most Victorians were concerned—including many liberal religious thinkers—evolution had to have a predetermined goal of moral significance. They preferred the model of evolution as a ladder, with themselves at the current top of the scale of perfection.

The Eclipse of Darwinism

These moral concerns were shared by many scientists. Some, like Huxley, were happy to use Darwinism to discredit the Church, seeking to replace it as a source of authority in the modern world. But they were anxious not to be identified as atheists devoid of moral concerns—which is why Huxley coined the term 'agnostic' to denote those who were suspicious of the claims made by orthodox religion. There was a widespread reluctance to see the world as a place governed only by trial and error, and a preference for the more orderly and purposeful schemes of evolution already hinted at earlier in the century. There were certainly technical problems with Darwin's theory, most obviously the weakness of his ideas about variation and heredity. But underlying the scientific objections there was a hunger for something else—either as a supplement to natural selection or an alternative—that would inject an element of order and purpose into the process of evolution.

Significantly Huxley himself was never convinced that natural selection offered a complete explanation. He expressed a preference for evolution by sudden jumps that would create new species instantaneously, something that would help explain the many gaps in the fossil record. He was not convinced that all evolutionary developments conferred adaptive benefits, speculating that there might be laws governing the production of variations that would control the emergence of new characters. In practical terms, this meant that his position was not all that different from his arch-opponents Richard Owen and St George Jackson Mivart. Owen was now a convert to evolutionism, though an open opponent of the selection theory, while Mivart's

Opposite: Evolutionary tree showing the ascent of life toward the human species. From Ernst Haeckel, *The History of Creation* (1876).

Genesis of Species of 1871 offered a cornucopia of anti-Darwinian arguments. Both thought that evolution would be directed along predetermined channels by forces generated within the organism. But they also made an accommodation with traditional religious beliefs by suggesting that the resulting trends were somehow designed by the Creator, and this made their arguments anathema to Huxley.

Huxley certainly accepted a role for natural selection to explain the superficial adaptive characters possessed by all species, which is why he could legitimately promote himself as a Darwinian. He had no time for the Lamarckian theory of the inheritance of acquired characteristics, yet for most of his contemporaries this became the preferred mechanism for explaining adaptive evolution. In the era before the emergence of genetics, most naturalists—Darwin included—believed that such a process was plausible. Herbert Spencer became a leading champion of Lamarckism, and we have seen how it played a vital role in his social philosophy. From a very different perspective the German champion of Darwinism, Ernst Haeckel, also enthused over the progressive character of evolution and invoked a large helping of Lamarckism (his books were also popular in English translations). Huxley, Spencer, and Haeckel represent three figures widely perceived as Darwinians who supplemented natural selection with other mechanisms of evolution to impose a sense of order or progress onto the history of life on Earth. Haeckel's use of the analogy between evolution and the progressive development of the embryo was characteristic of much non-Darwinian thought.

As the century progressed, doubts about the efficacy of natural selection became more rather than less acute. In the mid-twentieth century Julian Huxley, a grandson of Thomas Henry Huxley and a leading architect of the modern selection theory, wrote that the decades around 1900 had witnessed an 'eclipse of Darwinism'. The non-Darwinian mechanisms once invoked as supplements to natural selection were increasingly proclaimed as complete alternatives. In America, paleontologists such as Edward Drinker Cope insisted that only Lamarckism could explain the linear trends they saw in the fossil record, also invoking the analogy with embryological development. Cope was a Quaker, and published a *Theology of Evolution* in 1887 stressing that Lamarckism dispensed with the struggle for existence and allowed an element of purposefulness to be retained in evolution. These moral objections were also articulated outside the scientific community by writers such as Samuel Butler, whose *Evolution Old and New* of 1879 launched a bitter attack on Darwin and his theory. By the end of the century the non-Darwinian explanations of evolution were widely considered to have triumphed, even within the scientific community. In the early twentieth century, enthusiasm for Henri Bergson's philosophy of 'creative evolution' encouraged non-Darwinian thinking among many biologists. A new initiative would be needed to allow a re-emergence of the selection theory.

The Emergence of Genetics

Despite its 'eclipse' there remained a core of support for the theory of natural selection. The theory did, at least, encourage support for the belief that living things must interact with their environment, and Darwin himself had written of the 'tangled bank' of what would later be called ecological relationships. The term 'ecology' was coined by Ernst Haeckel, although evolutionism played a comparatively limited role when a distinct programme of research began to emerge under that name at the turn of the century. The early ecologists were not Darwinians, and their science was by no means a product of growing concerns about humanity's destruction of the natural environment.

Of more direct interest to those who wanted to preserve the theory of natural selection was the problem of heredity. Darwin had assumed that new individual variations were somehow created within the population, and for selection to work these new characters had to be passed on to future generations. But he had no idea what caused the variations, and his theory of inheritance was widely deemed to be unsatisfactory. The very concept of heredity in the modern sense was only just beginning to emerge into scientists' thinking, and it took major conceptual developments in the late nineteenth century for its significance to become apparent. Like most of his contemporaries, Darwin assumed that each organism produces the agents (perhaps material particles) which govern the embryological development of its offspring. This is why he still accepted the Lamarckian view that characters acquired by the parent could be inherited. There was as yet no notion of characters or hereditary units being transmitted through successive generations independently of the parents' bodies. A widely cited attack on the selection theory in 1867 by Fleeming Jenkin asserted that on the traditional model of heredity new characters would be diluted to nothing in the course of reproduction with unchanged individuals.

The Darwinians needed a new concept of heredity, and several of Darwin's followers helped to establish the modern viewpoint. But the detailed structure of the new theory that emerged in the early twentieth century—genetics—also owed much to one of the main alternatives to Darwinism. The new approach was built on the foundation of what is sometimes called 'hard' heredity: the claim that what is transmitted from parent to offspring cannot be affected by changes in the parents' own bodies. ('Soft' heredity is the Lamarckian notion that acquired characters can be passed on.) The principle architect of this concept was August Weismann, who introduced the notion of a material substance, the 'germ plasm', contained within the nucleus of the reproductive cells (egg and sperm) that was responsible for transmitting characters to future generations. Weismann declared that the germ plasm was totally isolated from the rest of the organism's body, so the Lamarckian theory was false. He performed a famous experiment by cutting off the tails of many generations of mice to show that the germ plasm for tail-production was retained despite the mutilation.

Weismann insisted that natural selection was the only conceivable mechanism of adaptive evolution. His followers became known as neo-Darwinians (note that Darwin himself would not have been counted as a neo-Darwinian). They were opposed by the neo-Lamarckians who rejected natural selection altogether. In the closing years of the century it was by no means clear that the neo-Darwinians would triumph, although the subsequent emergence of genetics bolstered Weismann's notion of hard heredity and ensured the eventual elimination of Lamarckism from biology.

The wider consequences of this new view of heredity were explored by Darwin's cousin, Francis Galton. He became convinced that inherited character plays a key role in human affairs. The individual's physical and intellectual capacities are predetermined by what is inherited from the parents—no amount of improvement in education or the environment can have any significant effect. Galton's book *Hereditary Genius* of 1869 insisted that high levels of intelligence always runs in families. He became even more concerned to show that defects of intellect and character are also preserved by inheritance, and called for restrictions on the reproduction of the feeble-minded who were supposedly breeding in the slums of the great cities. This was the 'eugenics' programme, promoted by Galton as a crusade to preserve the human race from degeneration.

In effect, the human species was to be subject to a process of artificial selection. Galton himself thought that natural evolution could only occur through the production of sports or 'saltations'—individuals born with an entirely new hereditary character. His disciple Karl Pearson, however, showed that Galton's view of heredity was consistent with the Darwinian theory of natural selection acting on small individual variations. Along with W. F. R. Weldon he founded the research programme known as 'biometry', which sampled wild populations to measure the range of natural variation and to demostrate the effect of natural selection on that range. One set of observations showed marked changes in the dimensions of crabs in Plymouth harbour during a dredging operation. Pearson was also a strong supporter of eugenics in the human population.

Neither Weismann nor Galton and his disciples thought of hereditary characters as fixed units. Pearson, in effect, showed that Fleeming Jenkin's attack on Darwin's theory had missed the point by assuming that new characters only appear abruptly, although in fact there is a fund of natural variation in any population. This is often depicted graphically as a bell curve for each character, with most individuals grouped around the mean and smaller numbers stretching out to the extremes on either side. This approach made Pearson suspicious of the innovation that led to the founding of modern genetics.

In 1900 biologists who had begun breeding experiments on a range of different species began to focus on a feature often apparent in domesticated breeds, the existence of discrete characters that seemed to breed true as units. Possibly to head off a damaging priority dispute, they hailed the work of the now-deceased Gregor Mendel as the foundation-stone of their new theory. Working on domesticated peas back in the 1860s Mendel had established laws which seemed to govern the transmission of discrete characters such as seed colour through successive generations. When crossed, pure-bred green and yellow peas did not yield yellowish-green offspring—they were pure green, although the yellow character reappeared in one quarter of the next generation if the hybrids were crossed. It was the existence of these discrete hereditary characters that was now being confirmed in a range of breeding experiments, and Mendel's laws of inheritance were posthumously hailed as the basis for a new model of heredity.

Heredity in *Primula Sinensis*.

1. Primrose Queen.   2. Crimson King.   3. F₁ formed by crossing these two types.   4—21. Various F₂ types obtained by self-fertilising F₁.   4, 10, 16. Whites.   5, 11, 17. Various tinged whites. 6, 12, 18. Light magentas.   7, 13, 19. Reds.   8, 14, 20. Magentas.   9, 15, 21. Deeper magentas. 7, 13, 15, 19, 20 have the dark blotches which cannot appear unless the stigma is red.   16—21 are all large eyed, viz. homostyle forms, like 1.

Colour variation in *Primula* to show characters inherited as genetic units. From William Bateson, Mendel's *Principles of Heredity* (1913).

Eventually this new science would revive the fortunes of Darwinism, in part because it did not allow the creation of genes corresponding to acquired modifications. Lamarckism was thus ruled out—yet the origins of genetics lay in another non-Darwinian theory, the belief that new characters are only produced abruptly by saltations or 'sports of nature'—what we today would call macromutations. Several of the founders of genetics began as saltationists, and seem to have been led from the idea that characters are created as discrete units to the possibility that they might be transmitted to future generations as units too. Hugo De Vries, the best-known of the 'rediscoverers' of Mendel's work, proposed a theory of evolution by macromutations. William Bateson, who coined the term 'genetics', had also written in favour of saltations. The leading American geneticist Thomas Hunt Morgan had at first supported De Vries's mutation theory and had criticized the notion that nonadaptive types could be eliminated by natural selection. No wonder that Pearson found the new theory hard to accept: the biometricians saw no sign of discrete characters in their wild populations and dismissed them as artefacts of domestication. Meanwhile the geneticists dismissed the continuous range of individual differences as a transitory product of environmental influence.

The result was a bitter debate which delayed the reconciliation of Darwinism and genetics until the 1920s and 1930s. Eventually a new science of population genetics was created, based on a recognition that sometimes many genes can affect a single character, so the bell curve of normal variation is made up of numerous overlapping genetic effects. R. A. Fisher's *Genetical Theory of Natural Selection* of 1930 was a key text providing mathematical details of how selection by the environment could gradually alter the proportion of genes in the population and thus produce adaptive evolution. Morgan's school had now shown the existence of plentiful small-scale mutations, so these could provide the random variation within the population required by Darwin's theory. This model of selection was linked to field studies and other areas relevant to evolution by figures such as Julian Huxley in Britain and Ernst Mayr in America, establishing a new form of Darwinism that soon came to dominate biology. Huxley's *Evolution: The Modern Synthesis* of 1942 gave the theory its name.

Biology and Modern Materialism

Space forbids any detailed elaboration of biological developments in the twentieth century and beyond, but we may conclude by assessing the impact of all the areas described above on how we think about human nature and our position in the world. Some applications of biological thinking have generated serious social concerns, often because they are based on oversimplifications of the situation popularized by science writers (and all too often condoned by at least some members of the scientific community). When these are pushed to extremes, as the in Nazis' ruthless imposition of eugenic policies to 'purify the race', we can all express our concern—yet the underlying assumption of genetic determinism still drives much of the popular interest in the role played by heredity in human affairs. Meanwhile the image of Darwinism as an agent of atheism and materialism is exploited by religious Fundamentalists who want to challenge the whole programme of science whenever it conflicts with their interpretation of a sacred text.

In the late nineteenth century many religious believers had made an accommodation with evolutionism, especially in its non-Darwinian forms. When religious Fundamentalism began to flourish in America, it soon began to target Darwinism as a symbol of the trends undermining traditional values. Curiously, Darwinism achieved this state despite being still under eclipse at the time. The trial of John Thomas Scopes for teaching evolutionism in Dayton, Tennessee, in 1925 has come to be regarded as the high-point of this first wave of opposition. The re-emergence of the theory of natural selection in the mid-twentieth century re-ignited the fear that science was reducing everything to a product of trial and error, making nonsense of all spiritual values. In fact, some early supporters of the modern Darwinian synthesis such as Julian Huxley hoped that the theory could preserve some sense of progress and purpose in the world. But increasingly the popular image of Darwinism has been dominated by figures such as Richard Dawkins who promote it precisely because it destroys any hope that the world has been designed by a wise and benevolent God. Creationism has moved to the opposite extreme in its Young Earth version, proclaiming the literal truth of the Genesis creation story. The views of moderate religious thinkers find it hard to gain a hearing in the resulting atmosphere of polarization.

The hard-line model of Darwinism has, in fact, been softened in recent decades by the emergence of evolutionary developmental biology (evo-devo) and the study of epigenetics. The old idea that each gene coded for a particular character (which thus determined its fate under selection) has been overlaid by a recognition that gene interactions have complex and sometimes unpredictable results, while the developmental processes by which the genetic information is expressed in the organism also play a role, and can have consequences for evolution such as restricting the range of possible variation. As early as the 1970s biologists such as Stephen Jay Gould were warning that the simplistic gene-centred version of Darwinism would have to be qualified in some respects, including the possibility of rapid episodes of change (the theory of punctuated equilibrium). Although generally rejected by the mainstream scientific community, the emergence of research areas such as evo-devo has spawned another generation of critics who challenge the authority of the theory of natural selection. The old non-Darwinian interests have re-emerged, if not the actual non-Darwinian evolutionary mechanisms. Some even claim that epigenetic inheritance (the transmission of modifications to development by mechanisms other than the genes themselves) allows a kind of Lamarckism.

This point also has consequences for our understanding of the social impact of genetics. Galton's eugenic programme was at first unsuccessful precisely because it denied the positive effects of self-improvement required for the Lamarckian process of social progress to work. But his views gained credence at the end of the nineteenth century as the middle classes became increasingly concerned about the proliferation of 'unfit' individuals in the population. The twentieth-century debate would be polarized between the supporters of nature (biological inheritance) and nurture (environment and education) as the determinants of human character. The eugenics programme flourished around the world in the first half of the century, with several American states, for instance, enacting policies of forced sterilization for those deemed unfit. This movement reached its nadir with the Nazis' policies of racial purification, and the resulting horrors did much to discredit the more extreme applications of the notion of hereditary determinism.

Nevertheless, the explosion of interest in genetics in the late twentieth century has encouraged the survival of less extreme versions of the belief that genes determine the whole character of the individual. When James Watson and Francis Crick identified the structure of DNA in 1953, their work fed into a research programme in molecular biology which hoped to show how genes are encoded to produce characters and how those characters are then developed. The decoding of the human genome, announced with much publicity in 2000, reinforced the popular belief that the genes were all-powerful in determining character. Genetic research generated exaggerated hopes that many illnesses—most notably cancer—would be shown to have a genetic basis, allowing scientists to provide the medical profession with powerful new cures.

Ongoing research has helped to reveal the complexity of the process that translates genes into bodily functions and has undermined the simple notion that each gene codes unambiguously for a single character. It has now become virtually impossible to establish a clear notion of what a gene is—it depends on the particular situation under investigation. Yet the popular view still rests on the assumption that there ought to be a single gene for each character, encouraging the hope that elimination of harmful genes can have an immediate effect on public health. Behind the scenes there are still extremists—even in science—who argue that harmful genes are concentrated in certain social classes or racial groups. The claim that we are all merely puppets of our genes is, of course, grist to the mill of those who see science as a threat to human dignity.

The same issue arises more generally through the huge expansion of the biomedical sciences in the late twentieth and twenty-first centuries. As developments in physiology and biochemistry have uncovered more of the 'secrets of life', the public's expectation that the results will yield cures for diseases has increased. In some cases, cancer again being an obvious example, these hopes have been dashed as the full complexity of the problem has emerged. Genes do play a role, but many genes are involved in cases of what had been thought to be identical cancers, and the genes react with a host of environmental influences to determine the actual outcome of development. The popular expectation that science can produce a 'magic bullet' to target all medical problems thus highlights a dilemma for biomedical research that has, indirectly, been created by the spread of the materialist view of life.

Such concerns are also crucial in the area of mental health, where the reductionist model in which everything depends on brain function threatens to undermine traditional values of moral responsibility. In the later twentieth century the pharmaceutical industry created a range of drugs that were able to reduce the symptoms of many mental illnesses. Freud's psychoanalytical techniques, so fashionable in the middle decades of the century, were discredited. Mental illness, it was now claimed, resulted from physical imbalances in the brain, not from childhood traumas. Yet as a result huge numbers of people became addicted to antidepressants, while mental illness has remained a serious issue in society. The media highlight public concerns about mental illness, but lawyers and social scientists increasingly worry that our definition of moral responsibility has been compromised by the recognition that some individuals may have little control over their actions. One thing, at least, is clear from all of these examples: scientists need to engage ever more closely with the public to manage how the hopes and fears generated by the materialist view of life will be addressed in the future.

10 Mapping the Universe

ROBERT SMITH

STOCKHOLM Concert Hall, 10 December 2011. In the award ceremony for the 2011 Nobel Prize in Physics, astronomers Saul Perlmutter, Adam Reiss, and Brian Schmidt are awarded the prize for the discovery of the accelerating universe, a find that has been widely claimed as *the* scientific discovery of the last quarter of the twentieth century. In a universe they had believed to be dominated by the attractive action of gravity, astronomers had long assumed that the expansion of the universe that began with a 'Big Bang' some thirteen to fourteen billion years ago would eventually slow down, and that perhaps there might even be a reversal of that expansion. What Perlmutter, Reiss, Schmidt, and their colleagues had done was to provide compelling evidence that the opposite was the case, and that the expansion is speeding up. In time beings at our vantage point can expect to look out onto a nearly empty universe as the vast majority of all the galaxies would have faded away except for those in our own 'supercluster' of galaxies.

What is perhaps most striking in this cosmic vision is that astronomers and physicists by the early twenty-first century were comfortable, indeed took it as their central task, to discuss and ultimately explain the entire history of the universe as well as to map all of its contents. Nineteenth-century astronomers as well as most twentieth-century ones would have regarded such confidence as absurd. This chapter is about the changes in the enterprise of astronomy over the last two centuries or so, including the composition of the astronomical workforce, as well as the shifting views of astronomers on the appropriate scope of astronomy and the nature and workings of the physical universe.

A Very Different Sort of Astronomer

The literary figure Fanny Burney wrote in her diary of a visit to William Herschel in 1786:

That great and extraordinary man received us with almost open arms. ...His immense new telescope, the largest ever constructed, will still, I fear, require a year or two more for finishing. ...Already, with that he has now in use, he has discovered fifteen hundred universes [galaxies in modern terminology]! How many more he can find who can conjecture?

In 1786, William Herschel was a famous, but also a most unlikely, astronomer. His life had been transformed on 13 March 1781. On that evening, Herschel, then an organist, composer, conductor, and music teacher in Bath, one of the popular spa resorts of eighteenth-century England, found the first planet to be discovered in recorded history. This was a sensational find. It catapulted Herschel to fame as well as doubled the size of the known Solar System as the new planet lay far beyond Saturn. Royal patronage quickly followed and Herschel readily deserted his musical career to become astronomer to the court of George III at Windsor.

At the end of the eighteenth century professional astronomers focused on the workings of the Solar System. For professional astronomers, the stars provided a background 'grid' against which the positions and motions of planets and comets could be plotted and then explained in terms of Newton's law of universal gravitation. Each planet would be attracted not only by the Sun but also by each of the other planets. The orbit of each planet would therefore not be the sort of simple ellipse that Kepler had calculated early in the seventeenth century. The mutual pulls of the planets on each other meant, Newton had calculated, that the Solar System would inevitably suffer instability and decay within several hundred years. Why, then, had it not perished? God, Newton was sure, directly intervened to maintain the stable order.

In the late eighteenth century the outstanding French mathematicians P. S. Laplace and J. L. Lagrange re-investigated the mutual gravitational pulls of the members of the Solar System with new mathematical tools that went beyond those developed by Newton. Laplace and Lagrange demonstrated to the satisfaction of their colleagues that the Solar System was stable and that stability, Laplace and Lagrange reckoned, flowed from the workings of natural law. There was no need to invoke divine action. For nearly all astronomers and mathematicians in the years around 1800, the Newtonian system was the unassailably true system of the world.

Herschel's interests and approaches to astronomy were radically different from the 'positional astronomy' of the mathematical astronomers. It was as a self-styled natural historian of the heavens that Herschel distinguished himself from his contemporaries. Herschel's grand project and the ultimate object of his observations was the 'Construction of the Heavens'. To this end he strove to determine the layout of our own star system—the Milky Way—as well as the arrangement and development of other star systems (what Fanny Burney had called universes), issues that had generally drawn little interest from mathematicians and professional astronomers. At the same time as some students of the Earth such as James Hutton were advancing ideas of 'deep time', Herschel was fascinated by, and pioneered the study of, deep space, the effort to conceive of the sidereal realm in three dimensions. As a celestial botanist pursuing the 'Construction of the Heavens' Herschel swept the skies with his telescopes to collect astronomical 'specimens', to catalogue them, and to search for and speculate on possible linkages between the various specimens.

Herschel's success as a builder of reflecting telescopes was also fundamental to his astronomical endeavours. Whereas refractors exploit lenses to form images, a primary or main mirror is at the heart of a reflecting telescope. Although he was self-taught in mirror making and polishing, Herschel fashioned the best mirrors and most powerful telescopes ever constructed.

William Herschel's twenty-foot reflecting telescope.

William Herschel's telescope building was also in part a family effort with considerable assistance from his brother Alexander, but most crucially his sister Caroline. An accomplished astronomer in her own right who discovered eight comets, she and William proved to be the most productive collaborative couple in the history of astronomy. At a time when women were generally barred from careers as astronomers, Caroline defied these limits. Family linkages were important for other women too. Jérôme Lalande was one of the leading astronomers of the late eighteenth and early nineteenth centuries and was aided in his researches by several family members including his daughter Amélie.

Caroline Herschel's observations of nebulae—visible as dim smudges of light in the sky—persuaded her brother in the early 1780s to sweep the skies systematically for more. When William and Caroline began their hunt for the nebulae around one hundred were known. By the time they were done they had catalogued over two and a half thousand. But what were the nebulae? William Herschel gave different answers at different times. In 1785, for instance, he had reckoned the Andromeda Nebula's appearance to be the result of the 'united luster of millions of stars', but towards the end of his life, a time when he suggested that many nebulae were single stars or comets in the process of formation, he was unsure of its nature.

The heavens, moreover, exhibited many different 'specimens' of nebulosity. These ranged, Herschel believed, from extremely diffuse nebulosity to nebulosity out of which stars are starting to form to, at the other end, very condensed clusters of stars. These specimens, moreover, were also arranged by age, with the journey from nebulous matter to clusters of stars propelled by the attractive power of gravitation acting over time.

Perhaps there were remarkable events at the end of the sequence, disintegration followed by renewal. 'These clusters may be the *Laboratories* of the universe', Herschel suggested, 'wherein the most salutary remedies for the decay of the whole are prepared'. The light released during the catastrophic demise of star clusters as well as the light emitted more generally by luminous bodies, would, in some places in space, be dense enough to form nebulosity, starting the cycle all over again.

William's son John completed his father's reviews of the heavens by taking a rebuilt version of William's most successful telescope to South Africa to sweep the southern skies. Most likely, John wrote in 1826, the nebulae should be interpreted as 'a self-luminous or phosphorescent material substance in a highly dilated or gaseous state, but gradually subsiding by the mutual gravitation of its molecules into stars and sidereal systems', but it would be best in the present state of our knowledge 'to dismiss hypothesis, and have recourse (perhaps for centuries to come) to observation'.

John's contemporaries were generally not so cautious. Many believed that his father had made a compelling case for truly nebulous matter. The elder Herschel's observations were a crucial resource for P. S. Laplace's theory of the development of our own Solar System and other solar systems, a theory that first appeared as a lengthy note appended to the end of Laplace's *Exposition du système du monde* of 1796.

Why, Laplace asked, did the seven known planets and their fourteen moons all revolve around the Sun in the same plane and in the same direction? The likelihood of this occurring by chance was minute. Nor was it scientific to invoke divine fiat. Instead, Laplace argued, the arrangement and structure of the Solar System sprang from the manner in which the planets and their satellites had condensed out of the shrinking outer, rotating nebulous atmosphere of the Sun.

Opposite: Drawing by William Herschel showing different 'specimens' of nebulae.

Positional Astronomy

William Herschel was a towering figure to his contemporaries, but they preferred to direct their energies to ends that were not Herschel's. At the time of Herschel's death—1822—professional astronomy meant positional astronomy. Founded on Newtonian theory, positional astronomy was also tightly linked to, and in many respects driven by, state interests and there were ready justifications for state support in terms of practical navigation, surveying, and geography. In order to map the Earth, geodesists looked to the heavens. In addition, late eighteenth- and nineteenth-century astronomy was very much a part of the global and imperial projects of Britain, France, and various other nation states. But these rationales do not tell the whole story as the functions of official observatories—those established by national or local governments or by universities—were in part to do with other ends. They were symbolic institutions fashioned to exhibit an enlightened engagement with a highly precise form of science often using state-of-the-art instruments. 'There was', one historian has pointed out, 'generally no clear theoretical aim to the exercise, and the observations themselves often remained unpublished, or if published remained unused.' Such observatories were about prestige and served as tokens of 'stability, integrity, order, permanence'. Constructing a star-catalogue was also a moral project in that astronomers sought star positions as accurate as possible even if the level of accuracy exceeded the demands of the practical purposes to which the catalogues might be put.

The leading exponent of positional astronomy in the first half of the nineteenth century was the German astronomer F. W. Bessel. The relentless emphasis Bessel placed on using excellent instruments and on the rigorous accounting for and reduction of errors was central to his solution of a very long-standing problem: the determination of 'stellar parallax' or the distance to a star by direct measurement.

The basic technique had been well understood for a long time. It relied on using the Earth's orbit about the Sun as a baseline from which to observe a star from different positions. The Earth's motion would be reflected in changes of position of a relatively nearby star compared to the background of distant and essentially fixed stars. By the early nineteenth century, astronomers knew from earlier, failed, attempts that the size of the shifts would be tiny and hard to detect.

After over twenty years of careful observations of 61 Cygni, an obscure star just visible to the naked eye in the constellation of Cygnus the Swan, Bessel announced in 1838 that he had indeed measured a distance, and his answer, of nearly eleven light years, is close to modern determinations. As John Herschel declared when he awarded Bessel the Gold Medal of the British Royal Astronomical Society,

I congratulate you and myself that we have lived...to see the great and hitherto impassable barrier to our excursions into the sidereal universe, that barrier against which we have chafed so long and so vainly...almost simultaneously overlapped at three different points. It is the greatest and most glorious triumph which practical astronomy has ever witnessed.

Herschel's mention of 'three different points' refers to the almost simultaneous measures of parallax by Bessel, Wilhelm Struve at Dorpat—he found a distance for the brilliant star Vega in fact before Bessel detected that for 61 Cygni but he published his result later—and Thomas Henderson at the Cape of Good Hope measured a distance for the bright southern star Alpha Centauri.

The German methods of positional astronomy were adapted and developed by George Biddell Airy in his tenure as the Astronomer Royal and head of the Royal Observatory at Greenwich between 1835 and 1881. Thomas Hardy, in his 1882 novel *Two on a Tower*, could still portray an astronomer's life in a romantic light, the individual grappling with cosmic questions, poised on a watchtower scanning the heavens. At Airy's Greenwich and at many other observatories too, the reality was quite different. While industrialists measured their profit in terms of money, Airy measured Greenwich's 'profit' with respect to public utility and scientific prestige. The Observatory ran in some ways as a sort of accountant's office or small factory where there was a strict hierarchy of staff and a rigid division of labour. Observers, not just astronomical objects, were themselves the objects of scrutiny as Airy endeavoured to mechanize observational practices.

The capabilities of astronomy were underlined by the discovery of Neptune in 1846 as a direct result of the calculations by the French astronomer U. J. J. Le Verrrier of the existence and location of such a planet. The unprecedented find generated high passions and a ferocious battle for credit with a number of British scientists pressing the claims of John Couch Adams who had made similar calculations to Le Verrier but had not published them before the discovery. There was even an angry debate on the name of the new planet before 'Neptune' was settled upon. Benjamin Peirce, an American astronomer, also claimed that the discovery had been a 'happy accident'. In Peirce's opinion, the calculated planet and the observed planet had such dissimilar orbits that the find must have been due to luck.

The outcome of these controversies was that Le Verrier and Adams were generally lauded as the co-mathematical discoverers of the planet and the finding not a happy accident. Neptune was widely hailed as a spectacular triumph for science. Indeed, what more striking evidence could there be of astronomy's perfection than the prediction of the existence and place of a major planet? Astronomy truly seemed to be the Queen of the Sciences.

The Nature of the Nebulae

William Herschel had turned the reflecting telescope into a serious tool of research at the end of the eighteenth century. Others were inspired by his efforts to attempt to fashion even more powerful reflectors. One such was the third Earl of Rosse, an Irish nobleman, who began his experiments at Parsonstown in central Ireland in the 1820s. Rosse's most ambitious telescope had a primary mirror 72 inches in diameter and some four tons in weight. A visitor who witnessed the erection of the telescope was awestruck at the 'gigantic scale' of the effort. This colossal telescope—it came to be

Lord Rosse's giant reflecting telescope, the Leviathan of Parsonstown.

known as the 'Leviathan of Parsonstown'—was completed in 1845. Despite the limited area of sky it could examine and its location under the often cloudy skies of central Ireland, Rosse and his observing colleagues would be important figures in the wide-ranging debates on the nature of the nebulae.

In the 1830s the debates over the existence of nebulous fluid in the universe had become politically, morally, and religiously charged. The Glasgow political economist and astronomer John Pringle Nichol, among others, seized on Laplace's nebular hypothesis and evolutionary view of the Solar System to argue that this presented a general model of universal progress that justified political reform. Robert Chambers was a member of Nichol's circle and his sensational *Vestiges of the Natural History of Creation*, first published anonymously in 1844, offered a grand theory of cosmic development that ran from nebulous matter to human beings.

In contrast, Thomas Romney Robinson, director of the Armagh Observatory and one of the observers to regularly use the Leviathan of Parsonstown, was politically conservative and a robust critic of the nebular hypothesis. He had maintained even before the Leviathan was directed to the skies that it would undermine Laplace's hypothesis (and claims for political reform) by resolving the nebulae into clouds of stars. When he announced a number of nebulae had indeed been resolved into stars by the Leviathan soon after it went into operation, Robinson further contended that it would be 'unphilosophical' not to conclude that given a big enough telescope, any nebula could be resolved into stars.

The extraterrestrial life debate was also woven into disputes about the resolution of nebulae. In the middle of the nineteenth century, most astronomers accepted that life existed beyond the Earth. One figure who took strident issue was the Cambridge polymath and devout Anglican William Whewell. In 1854 he wrote *Of the Plurality of Worlds: An Essay* and, to the consternation of many of his contemporaries, declared that the pluralists' arguments were scientifically defective. Nebulae were not star systems grown milky and nebulous because of their great distance. The accepted ideas on extraterrestrial life were also religiously dangerous. If beings inhabited other planets, how could Christ have revealed Himself to all these different inhabitants in all these different places? For Whewell, anxious to ratchet down the number of possible abodes of life, our own galactic system, the Milky Way, comprises the entire visible universe and the nebulae are not distant star systems that lie beyond its boundaries.

Big telescopes were central to the debates on nebulae in the first half of the nineteenth century. In the 1860s a new tool, the spectroscope, took these debates in strikingly new directions.

Spectroscopy

Many early nineteenth-century astronomers held what would later seem to be a very restricted vision of the proper scope of astronomy. Perhaps the most famous expression of the limits of astronomy was that of the positivist philosopher August Comte who in speaking of stars reckoned that while we 'can conceive of the possibility of determining their shapes, their sizes, and their motions, we shall never be able by any means to study their chemical composition or their mineralogical structure'. Yet in 1861, Warren de la Rue could exclaim that the 'physicist and the chemist have brought before us a means of analysis that...if we were to go to the sun, and to bring away some portions of it and analyze them in our laboratories, we could not examine them more accurately than we can by this new mode of spectrum analysis'.

What had happened between the declarations of Comte and de la Rue? Early in the nineteenth century, through the researches of William Wollaston in Britain and particularly Josef Fraunhofer in Germany, it became widely accepted that after their light was passed through a prism the Sun and some stars exhibit an array of colours crossed by dark lines—these lines became known as Fraunhofer lines—in their spectra. Soon investigators decided that the observed spectral lines are somehow characteristic of the chemical substances that give rise to them and from the 1820s on there were two main kinds of spectral research. The first focused on the ways in which the spectra of different substances changed when they were subjected to various conditions in the laboratory. In the second, researchers searched for links between the solar spectrum and laboratory spectra.

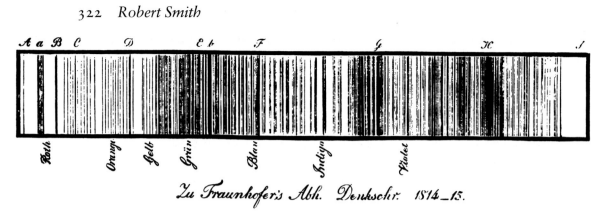

Fraunhofer's depiction of spectral lines in the spectrum of the Sun.

One of Bunsen and Kirchhoff's early spectroscopes.

The telling breakthroughs came in the years around 1860 with the fundamental investigations of the physicist G. R. Kirchhoff and his colleague at Heidelberg in Germany, the chemist R. W. E. Bunsen. Kirchhoff and Bunsen established that a substance capable of emitting a certain spectral line has a strong absorptive power for the same line. Laboratory analysis could reveal which substances emitted or absorbed which lines and so the composition of celestial bodies could now be analysed by detailing their spectra. The interpretive framework hammered out by Kirchhoff and Bunsen handed new and remarkable powers to astronomers. By a comparison of the lines in the solar spectrum with laboratory spectra, Kirchhoff himself quickly identified several terrestrial elements in the Sun. In a manner no one had anticipated, the stars themselves had become laboratories.

For later astronomers, Kirchhoff and Bunsen's researches marked the birth of what would come to be called astrophysics, the study of the physics and chemistry of astronomical bodies. This novel sort of astronomy meant new sorts of institutions came into being that were quite unlike the traditional observatories directed towards positional astronomy. As one pioneer of astrophysics later recalled, then 'it was that an astronomical observatory began, for the first time, to take on the appearance of a laboratory'.

The spectroscope also had a dramatic effect on the debates on the nature of the nebulae. Until 1864 these had been very much shaped by observations made with powerful telescopes and the drawings, sketches, and published reports that sprang from them. In that year, however, William Huggins, an amateur astronomer in London, inspected several nebulae spectroscopically. When he directed his telescope and spectroscope to a so-called planetary nebula (a name given by William Herschel to nebulae distinguished by their circular shape) in the constellation of Draco he glimpsed bright lines. Such lines, Huggins was well aware, are characteristic of a glowing gas and so these sorts of nebulae must be vast clouds of luminous gas or vapour, and certainly not clusters of stars. This was a dramatic and convincing find that underlined the force of the new astrophysics.

Huggins was a new sort of astronomer, an astrophysicist. He also had the wealth to supply much of his own equipment, as well as sufficient leisure, to pursue astrophysics. Like a number of male scientists of the eighteenth and nineteenth centuries, Huggins also had for part of his career the invaluable collaboration of his wife in all manner of observatory tasks. Margaret Huggins, for instance, brought a new sophistication to Huggins's use of photography to record spectra.

The leading roles played by amateurs like William and Margaret Huggins in astrophysics was not to last long, however. A number of nations quickly established astrophysical observatories, as well as incorporated astrophysics into the activities of their existing astronomical institutions. The Royal Observatory at Greenwich in London, a bastion of traditional astronomy, for example, also began to undertake astrophysical investigations in the 1870s. The first observatory specifically established by a state for the pursuit of astrophysics was built in 1876 at Potsdam in Germany. Others soon followed at Meudon in France and South Kensington, London. With new institutions, such as Potsdam, and the changes in the existing ones, such as the Vatican Observatory and Greenwich Observatory, and with growing interest in astrophysics in universities, it even became possible for someone to earn their living as an astrophysicist. By the end of the nineteenth century, professional astrophysicists had very largely supplanted the amateurs.

During the late 1870s and 1880s a new and relatively easy-to-use photographic plate—the dry plate—became widely available. It was the major technical factor in ushering in the era of astronomical photography as the exposure times of the dry

William Pickering and some of the women astronomers of the Harvard College Observatory.

plates could be extended almost indefinitely and features never before noted brought to light. In the late nineteenth century, both positional astronomers and astrophysicists added photography to their armoury and were now able to permanently record the light from sources and review the photographic plates at their leisure.

Photography enabled astrophysicists to pursue large-scale observing programs on stellar spectra. The great exponent of data collecting of this sort was E. C. Pickering at the Harvard College Observatory in Cambridge, Massachusetts. Airy at Greenwich had employed a strict division of labour and a rigid hierarchy for the ends of positional astronomy. Pickering did the same but now for the collection and analysis of the light from many thousands of stars. In the mid-1880s Pickering embarked on two huge projects on the stars, one on spectra and the other on brightnesses. Pickering's scheme of organization relied on poorly paid, but often highly skilled women. Perhaps the most impressive tribute to Pickering's organizational skills and the talents of the Harvard staff was *The Henry Draper Catalogue*, published in nine volumes between 1918 and 1924, with spectral classifications of a remarkable 225,300 stars performed by Annie Jump Cannon.

An outpouring of philanthropic support for American observatories in the years around 1900—including Pickering's Harvard College Observatory—helped make astrophysics in the United States the best funded anywhere. And the most richly supported of all was the Mount Wilson Observatory in California, founded in 1904 as a part of the Carnegie Institution that was brought into being by the munificence of

---

Opposite: William Huggins in his observatory at Tulse Hill.

the steel magnate Andrew Carnegie. The rise of the US as a leading economic power was also mirrored by its growing importance in the manufacture and use of giant instruments. In 1919, the most powerful telescope in the world, a reflector with primary mirror 100 inches in diameter, went into service on Mount Wilson with very major consequences, as we will see in the next section.

Beyond the Galaxy?

In 1905, the eminent historian of astronomy and science writer Agnes Clerke argued that: 'No competent thinker, with the whole of the available evidence before him, can now, it is safe to say, maintain any single nebula to be a star-system of coordinate rank with the Milky Way.' The great majority of astronomers echoed her opinion. They generally pictured the universe as divided into two parts. One was visible and confined to the Milky Way. The other was infinite and was believed to be beyond observation. Galaxies might exist in this region, but could not be observed in even gigantic telescopes and professional astronomers were little interested in this unseen universe.

Photographs of the Andromeda Nebula provided one body of influential evidence against the existence of distant star systems. The Oxford astronomer H. H. Turner recalled in 1911:

Many of us remember the occasion [in 1888] when the Andromeda nebula in its true shape was first put upon the screen at the Royal Astronomical Society by Dr Isaac Roberts, but the shape [a spiral] is so familiar now that only those whose memory does go back to that date (end of 1888) can realize the revelation that seemed to come with the picture. One heard ejaculations of 'Saturn', 'the Nebular Hypothesis made visible', and so on.

For the astronomers in 1888, the Andromeda Nebula was a solar system in formation, not a vast and far-off star system. For Roberts and many other astronomers photography had also replaced drawing as the technology of choice for depictions of nebulae and for mapping the heavens.

The chief find of the observers who had used the Leviathan of Parsonstown in the middle of the nineteenth century was that some nebulae exhibit a spiral shape. By the end of the century, however, astronomers had identified fewer than one hundred such spiral nebulae. In the late 1890s, the American astrophysicist James E. Keeler, with the aid of a 36-inch reflector at the Lick Observatory atop Mount Hamilton in California, decided to re-examine photographically all of the nebulae the Leviathan's observers had reckoned or suspected were spirals. When Keeler counted the number of spirals visible in a limited area of the sky, and then extrapolated to the number of spirals for the entire sky, he estimated the total at hundreds of thousands, making them by far the most numerous sort of nebula. Keeler had elevated the nature of the spirals into a major question for astronomers.

Nor did it seem feasible to link the spirals to the nebular hypothesis, at least in the form originally proposed by Laplace. As the nineteenth century drew to a close, astronomers increasingly reckoned Laplace's version of the nebular hypothesis as too flawed for it to offer a credible account of the development of the Solar System. One alternative was the so-called Chamberlin-Moulton or planetesimal hypothesis (termed by Moulton 'the spiral nebula hypothesis'). First advanced in full in 1905 by the leading geologist T. C. Chamberlin and his astronomer colleague at the University of Chicago, F. R. Moulton, they argued that when two stars pass close to one another the tidal pulls between them may cause matter to be ejected and scattered along spiral arms. Planetary bodies are formed as the result of the relentless accretion of the ejected particles, the planetesimals. No object on Keeler's photographs, they reckoned, matched the sort of nebula described by Laplace in which planets had condensed out of rings of material. The plates did, however, exhibit numerous nebulae with condensations in spirals' arms that appeared poised to evolve into planets.

Redshifts?

In 1898, H. G. Wells described a Martian invasion of London and its surrounds in his novel *War of the Worlds*. So 'vain is man', wrote Wells, 'and so blinded by his vanity, that no writer, up to the very end of the nineteenth century, expressed any idea that intelligent life might have developed there far, or indeed at all, beyond its earthly level'. But one astronomer who did was Percival Lowell, the wealthy founder of the Lowell Observatory at Flagstaff in Arizona and the main figure in a wide ranging and often acrimonious debate in the decades around 1900 on whether or not intelligent life existed on Mars.

Lowell had been deeply impressed by the Mars observations of the Italian astronomer Giovanni Schiaparelli. In 1877, Schiaparelli detected an intricate system of fine straight lines—lines that soon became generally termed 'canals'—that criss-crossed over the Martian surface. A little later, Schiaparelli even observed some lines to become double, where previously there had been only one line.

Some observers were sceptical about and others hostile towards the claims of canals. Schiaparelli, however, was usually very cautious in interpreting his observations, but not always—and in one article he wrote of a sophisticated system of canals and dikes operated by the Martian engineers. Others were less restrained. For Lowell, the Martians had confronted the slow advance of desert conditions over much of their planet and had developed the means to transport water from the ice capped polar regions to the planet's arid areas. Vegetation grew alongside the canals (and it was the tracks of vegetation that were supposed to be visible from the Earth rather than the canals themselves). Lowell's enthusiastic and tireless advocacy of the canals as Martian-built features won him attention and notoriety as the issue was fought over in scientific journals as well as books, newspapers, popular magazines, and in lecture halls.

But Lowell was interested in more than Mars. These broader interests led him to ask one of his assistants, Vesto Melvin Slipher, to examine the spectra of the spiral nebulae for clues to the origin of the Solar System. By late 1912 Slipher had four photographic plates on which he could distinguish not only lines in the spectrum of the Andromeda Nebula, but he could also tell that they had been shifted from their usual positions.

If a light source is in motion with respect to an observer, then the wavelengths of its spectral lines will move from the values they would have had in the absence of relative motion. The amount the lines shift, the Doppler shift, reveals the line-of-sight, or radial, velocity of the source. Assuming that the shifts of the lines were Doppler shifts, Slipher measured the Nebula's speed of approach as three hundred kilometres per second, the highest speed recorded to that time for an astronomical body, and a find so astounding that some astronomers did not believe it. By 1914, however, Slipher, despite the great difficulty of these measurements, was able to announce the radial velocities of fifteen spirals. A few of his results had been confirmed at other observatories. The sceptics had been quieted.

Most of the spirals measured by Slipher were receding, that is, the spectral lines were shifted towards the red end of the spectrum or redshifted. The fastest spiral was travelling—again assuming the spectral shifts were Doppler shifts—at around eleven hundred kilometres per second, several hundred kilometres per second faster than the fastest stars. If the spirals were protosolar systems, then, astronomers expected, they should move with about the same speeds as the stars. Slipher's results therefore came as a shock.

In 1912, Slipher had assumed the Andromeda Nebula was a solar system in formation. By 1917, however, prompted by the growing number of high redshifts he had measured, Slipher had joined those who reckoned the spiral nebulae might well be island universes after all. The point was that the spirals were apparently moving much too rapidly for them to be gravitationally bound to our own Milky Way Galaxy and so must be separate from it.

The Big Galaxy

As the spiral nebulae were scrutinized ever more closely, other astronomers mapped our own stellar system, the Galaxy, through a variety of observational and theoretical techniques. The researches on the nature and size of the Galaxy were also tightly related to the debates on the existence of external galaxies. Galaxies (if such there were) would necessarily lie beyond our own Milky Way system, so the distances to the possible galaxies as well as the extent of our Galaxy would be involved in settling the question.

Our immersion in our own galactic system, astronomers knew, made it very hard to picture the Galaxy as a whole. The astronomer usually recognized in the early twentieth century as having the best answers to this problem was J. C. Kapteyn, professor of astronomy at the University of Groningen. William Herschel had sought to map the Galaxy with the aid of stellar statistics. Kapteyn was also a 'statistical astronomer', but his approaches were much more fine grained and involved the collection of far more data, and so were far more reasonable to his colleagues than Herschel's 'star gaging' had been.

Opposite: V. M. Slipher examining his spectrograph at the Lowell Observatory.

Kapteyn's great goal was to solve what was called the 'Sidereal Problem', that is, to determine the distribution of the stars in our own stellar system and thereby elucidate the architecture of our Galaxy. Kapteyn was also one of a new sort of astronomer cum astrophysicist. Mathematically adept, he had a deep concern for observational evidence and the limitations and weaknesses of that evidence, both positional and astrophysical. Tackling the Sidereal Problem would demand, he reckoned, massive amounts of data. Kapteyn offered, however, 'the unique figure of an astronomer without a telescope'. He therefore relied on other astronomers to collect data while he concentrated on its analysis.

Kapteyn's statistical astronomy led him in 1908 to argue that the Sun is fairly centrally placed in our own galactic system (an opinion very much in line with other astronomers at this period) and that the limits of the stellar system are reached only at distances of about 30,000 light years, that is, it has a diameter of around 60,000 light years (something of an increase on previously accepted sizes). These conclusions were challenged in 1918 by an astronomer at the Mount Wilson Observatory, Harlow Shapley.

Shapley was puzzled by the distribution of the seventy or so known globular clusters, giant balls each of around one million stars that were nearly all concentrated into one part of the sky. What, however, if these globular clusters surround the Galaxy? From the galactic centre the globular clusters would be spread evenly around the sky. For Shapley, the peculiar distribution of the globular clusters follows then from the fact that our own Sun is not close to the galactic centre as Kapteyn and other astronomers long believed but is displaced many tens of thousands of light years from that centre. We view the globular clusters from such an off-centre position that they appear to crowd into one part of the sky. From the distances to the globular clusters Shapley reckoned our own stellar system has a diameter of around 300,000 light years, a startling five-fold increase on even the size proposed by Kapteyn, hence the name 'Big Galaxy'.

Shapley also came to believe that the spiral nebulae could not be systems comparable to the Big Galaxy and might be bodies driven off by the Galaxy. For him, by the late 1910s, there was only one visible galaxy, our own Milky Way Galaxy.

Other Galaxies?

The Lick Observatory astronomer H. D. Curtis took a very different position from Shapley on the spirals. In 1910, Curtis had taken charge of the Observatory's programme of nebular photography, and was also one of a group that can be termed the 'Lick School', astronomers at the Lick Observatory who while often sceptical of new developments in astronomy (such as Shapley's Big Galaxy) were advocates of external galaxies.

In the late 1910s, Curtis was also able to offer new evidence for galaxies in the form of novae in spirals. A very bright nova that flared in 1885 in the Andromeda Nebula had been widely viewed as impossible to square with the Nebula as an external galaxy. But the detection of further novae in spirals, starting with discoveries in 1917 by Curtis himself and George Ritchey at the Mount Wilson Observatory, pointed, Curtis reckoned, to the 1885 nova as anomalous and indicated distances to the spirals that put them well outside our own galactic system.

Curtis's energetic championing of island universes won him an invitation to participate in the so-called 'Great Debate' at the National Academy of Sciences in Washington, DC, in 1920 on 'The Scale of the Universe'. Curtis's debate opponent was Harlow Shapley. Curtis advocated external galaxies and criticized Shapley's Big Galaxy, as he was not persuaded of the worth of the distance determinations Shapley had exploited. Shapely defended his Big Galaxy, whose fate he thought was critical in any case for the spirals as external galaxies, but pitched his remarks at a much less advanced level than Curtis. Curtis and Shapley did however get to grips with each others' arguments in the published and much more technical versions of their talks.

What was needed to settle matters was a means of measuring distances to the spirals that everyone could agree was accurate. Such a method was forthcoming just three years after the Great Debate in the shape of Edwin Hubble's observations of stars in the Andromeda Nebula.

Hubble began graduate work at the University of Chicago's Yerkes Observatory in 1914, secured his PhD in 1917, but then enlisted in the US army to go and fight in Europe. Astronomy suffered greatly in the First World War and as a result of the social, political, and economic convulsions it spawned. Scientific communications were often cut; many astronomers devoted their efforts to war service and in so doing some of them, including the brilliant German astronomer Karl Schwarzschild, lost their lives. Hubble, however, never saw battle, and after his release from the Army in 1919 he joined the staff of the Mount Wilson Observatory where he had access to the most powerful telescope in the world, a 100-inch reflector.

In the early 1920s, Hubble followed up earlier studies by Curtis and others when he took numerous photographs of the Andromeda Nebula with the aim of detecting novae, the better to calculate the Nebula's distance. In October 1923, Hubble noticed on one of his plates what he took at first to be a nova and he marked it as such. But when he started to plot the brightness of the 'nova' over time, he found that it changed its light output in a regular and periodic manner, declining slowly and then brightening rapidly. The star's varying brightness was characteristic not of a nova, but of a type of variable star known as a 'Cepheid'. Hubble knew that by measuring the time between the peaks of the Cepheid's shifting brightness, there was a straightforward means of determining its distance (the crucially important relationship between the brightness of a Cepheid and the time from one peak in brightness to the next had been discovered in 1908 by the Harvard astronomer Henrietta Swan Leavitt). Hubble's answer came out at around 900,000 light years. This was hugely significant because, even if Shapley's diameter for the Galaxy of 300,000 light years was correct (an estimate itself founded on distance determinations using Cepheids), it clearly placed the Andromeda Nebula far outside of the Milky Way. The Andromeda Nebula must therefore be a distant star system.

Hubble swiftly found more Cepheids. Within a year or so, Hubble had collected enough evidence from the Cepheids and other methods of computing the Nebula's

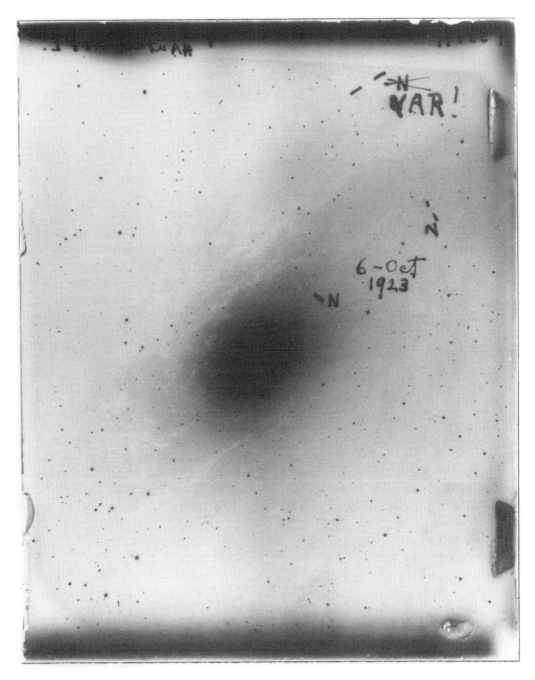

Photographic plate showing the location of the first Cepheid variable star that Hubble found in the Andromeda Nebula. Hubble initially believed he had observed a nova, but then changed his mind, crossed out `N' for nova and added `VAR' for variable.

distance to convince almost all astronomers that the outer regions of the Andromeda Nebula truly consist of clouds of stars and that the Nebula is indeed an external galaxy. The debate over the existence of visible galaxies was practically over. Many astronomers were involved in the 'discovery' of external galaxies—Slipher, Curtis, and others—but Hubble provided the evidence that clinched the argument and so a complex process has, misleadingly, often been collapsed down to Hubble being awarded all the credit.

What was soon accepted as persuasive proof that the Sun is eccentrically placed in our own Galaxy came in 1927 from studies by the Dutch astronomer J. H. Oort of a systematic rotation of the Galaxy. These pointed to the stellar system rotating around a distant centre of the Galaxy that lay in the same direction claimed by Shapley as the centre of the system of globular clusters, and so for him the centre of the Galaxy. A few years later, astronomers came to accept the existence of interstellar absorption—that is, the absorption of starlight by material spread throughout interstellar space. The effect of this material had been to block starlight and so make stars appear dimmer than they really were; Shapley had therefore overestimated distances. The size of Shapley's model was now cut to what seemed to many to be a more credible diameter of about 100,000 light years. The result was that by the mid-1930s, the notion of a Big Galaxy in which the Sun is placed eccentrically had become commonplace, although it was not as big as Shapley had claimed in the late 1910s.

The Expanding Universe

In December 1859, U. J. J. Le Verrier visited an obscure French doctor in the village of Oregères-en-Beauce. Several months earlier Edmond Modeste Lescarbault had reported to Le Verrier that he had observed with his small telescope an object he interpreted as a planet transiting the Sun. Lescarbault knew that Le Verrier had sought to explain anomalies in the motion of Mercury in terms of an unseen planet between Mercury and the Sun, much as Le Verrier had explained anomalies in the motions of Uranus in terms of a planet beyond Uranus (an explanation that led of course to the discovery of Neptune). Le Verrier's questioning of Lescarbault convinced him of both the observer's trustworthiness and the observation's veracity and so he announced the discovery of a new planet, Vulcan.

In the mathematical analyses that led to both Neptune and Vulcan, Le Verrier had exploited Newton's law of universal gravitation. But after Lescarbault's initial sighting, Vulcan stubbornly refused to reveal itself to other astronomers and the 'discovery' slid into obscurity. Einstein would show over fifty years later that the anomalies in Mercury's motions required a new theory of gravitation for their explanation. That same theory of gravitation would also be central to one of the major discoveries of twentieth-century cosmology: the expansion of the universe.

In 1917, Albert Einstein explored the cosmological consequences of his very recently developed and revolutionary theory of gravitation, general relativity. Here he speculated on the very size and nature of the universe and in so doing proposed an immense conceptual shift from an infinite Newtonian universe to a 'finite' universe where there were no boundaries. Einstein reckoned the universe must be static and, when his equations pointed to a non-static universe he introduced a term, the cosmological constant, in order to ensure its static nature.

Einstein was the director of the Kaiser Wilhelm Institute for Physics and a professor at the Humboldt University of Berlin in 1917 and so suffered at this period from the breakdown in scientific communications caused by the First World War. He was however in contact with an astronomer in neutral Holland, Willem de Sitter, who became thoroughly versed in general relativity; indeed de Sitter soon secured another solution to Einstein's equations different from that advanced by Einstein himself. De Sitter also predicted that there should be a 'redshift-distance relation' for objects at great distances. That is, the more remote an object, the larger the redshift should be of its spectral lines. As de Sitter supposed the spiral nebulae to be island universes (and so the most distant objects in the universe), he expected them to exhibit such a relation if his model approximated the properties of the universe. The exact form of the relation however was a matter of controversy. Moreover, the redshifts de Sitter predicted for remote objects were not due solely to Doppler shifts, but were, in de Sitter's opinion, largely a consequence of the properties of the 'four-dimensional space-time' that underpinned general relativity.

General relativity, with its exploitation of a 'four-dimensional space-time' and unfamiliar and highly abstract mathematics, seemed baffling to many working astronomers. However, when Einstein calculated the motions of Mercury using his new theory he claimed a very close agreement between theory and observation without needing to invoke, as Le Verrier had, an unseen planet. There was an important further test of the theory in 1919 connected with observations of the possible deflection of starlight passing by the Sun during a solar eclipse.

They 'really proved the bending of light around the Sun', wrote an excited Einstein to his mother when the results came in. Almost overnight Einstein became an international celebrity and a scientific 'superstar'. The cosmological consequences of general relativity had to be taken seriously, even by sceptical astronomers.

In 1922 and 1924, the Russian mathematician and meteorologist Alexander Friedmann found and analysed solutions to Einstein's equations of general relativity that were non-static, among which were solutions for an expanding universe. But Friedmann did not have the opportunity to develop his ideas, as he died in 1925 of typhoid fever at the age of 37.

The first person to join theory and observation in a way that would come to be widely seen as physically meaningful within the general framework of the expanding universe was a Belgian abbé and professor at the University of Louvain, Georges Lemaître. In 1927, Lemaître, who at the time was unaware of Friedmann's earlier studies, published what would later be recognized as a seminal paper on the expanding universe. But for a brief time, Lemaître's researches, like Friedmann's, drew no interest.

By 1927 there had also been, following de Sitter's paper of 1917, a decade's worth of speculation about a redshift-distance relation for very distant objects. Prompted by predictions of such a redshift-distance relationship, a few astronomers, most notably the Swedish astronomer Knut Lundmark and the German astronomer C. A. Wirtz, tried to determine observationally the form (if any) of the relationship by investigating the speeds and distances of spiral nebulae and globular clusters. But for the majority of astronomers the form of the relationship, if such there was, was still an open question.

It is against this background that we need to see the studies of a redshift-distance relation by Edwin Hubble and his Mount Wilson colleague Milton Humason. Hubble's first scientific paper on such a relation was published in 1929. Here he employed his own estimates of the distances of the galaxies together with radial velocities, nearly all of them obtained prior to 1923 by Slipher.

A much longer paper, co-authored with Humason, followed in 1931, and this included a number of redshifts measured by him. Humason measured one galaxy to be moving at nearly 20,000 kilometres per second. This was a staggering result when one remembers the consternation caused by Slipher's measurement of eighteen years earlier that the Andromeda Nebula was travelling at 300 kilometres per second. With this paper Hubble and Humason effectively brought to an end the debate over the existence of a redshift-distance relationship with the answer that there was indeed such a relation and that, at least in the first approximation, it was linear. That is, on the average, if a galaxy's distance is doubled, its redshift is doubled; if its distance is tripled, its redshift is tripled, and so on. The great majority of astronomers also interpreted the redshifts as velocities and so for them it was a velocity-distance relation (the relation would later be known as Hubble's Law). A. S. Edington, an influential British astronomer, and other theorists swiftly meshed Lemaître's calculations with Hubble and Humason's observational researches on the redshift-distance relation in the context of the expanding universe.

From the 1930s on, however, a small number of astronomers have felt uneasy interpreting the redshifts as Doppler shifts alone, and so have doubted the usual explanation of the expanding universe. Most notable among them was Hubble himself. He was always careful in print to avoid definitely identifying the redshifts as velocities and wrote of the velocities measured by Slipher and then Humason as 'apparent velocities'. Hubble's position on the redshifts makes the usual attribution of credit to him in astronomy textbooks for the discovery of the expanding universe especially misplaced.

For the astronomers who did accept an expanding universe there were new issues to confront. In particular, what began the expansion? Lemaître had initially conceived of the expansion as building very gradually from matter spread throughout the whole of space. But in 1931 he proposed a much more radical idea: that the start of the expansion marked the beginning of the entire universe. For Lemaître, the universe had begun as a 'primeval atom'. This highly unstable atom divided into smaller and smaller atoms by a kind of super-radioactive process. The 'last two thousand million years are slow evolution', Lemaître contended; they 'are ashes and smoke of bright but very rapid fireworks'. This was the first version of what later became known as

'Big Bang' cosmology and thereby offered an account of the creation of the universe. Many other cosmologists disliked such a notion and viewed it and the implied finite age of the universe with suspicion. How could a singular event at the very beginning of time be explained by the timeless laws of physics?

The expanding universe also led to a clash of timescales. If one ran the present expansion back in time and took no account of a possible speeding up or slowing down of the expansion, the time since Lemaître's 'very rapid fireworks' was only about 2×10^9 (that is, 2,000,000,000) years. This meant that the ages of many of the stars, indeed of the Earth at about three billion years, were longer than the age of the universe! To avoid this problem de Sitter, for example, suggested that the beginning of the universe and the start of its expansion could not be identified as the same event. The galaxies had lasted for longer than 2×10^9 years, but they had approached to a minimum separation before starting their expansion, so in fact the universe was far older than 2×10^9 years.

As well as the debates on the start of the expansion and origin of the universe, there was also renewed interest in its ultimate fate. The idea of the 'heat death' of the universe had been introduced in 1854 by Hermann von Helmholtz as a consequence of what we would now call the Second Law of Thermodynamics. Such a final state would be one of complete disorganization and the universe would be composed of a uniform, featureless mass at a constant temperature. In this state, evolution would have ended. In the late nineteenth century, what we might call the thermodynamic cosmos was a matter of very considerable religious and philosophical as well as scientific debate. Relatively few astronomers however were active participants and those that were tended to side with those who argued for cyclical cosmology and so a universe that could undergo renewal rather than an inevitable death. In the late 1920s, there was renewed interest in the heat death that reached a broad audience through popular writings and articles.

Astronomy Transformed

Cosmology before the Second World War was very largely the province of individual researchers be they observational or theoretical astronomers. Matters were different after the war as teams increasingly came to the fore in astronomy, as they did in many other areas of science. Astronomers were used to employing very large and expensive telescopes. The great 200-inch Hale telescope perched on top of Palomar Mountain in California was completed in 1948 and was the most powerful telescope ever built, the fruit of philanthropic support from the Rockefeller Foundation. Astronomers, however, used it in traditional ways, as individuals or as small groups. In contrast, the rapidly burgeoning field of radio astronomy was dominated by Big Science methods, including the use of interdisciplinary teams.

Before the Second World War, astronomers focused almost entirely on observing the universe in the wavelengths of visible light, but wartime developments greatly speeded the opening up of new regions of the electromagnetic spectrum to astronomical observation. One of the leading practitioners of radio astronomy after the war was the British scientist, Bernard Lovell. Trained as a physicist before the war, he spent the war years developing various kinds of radar techniques and learnt not only technical skills but also expertise in 'science politics' and a strong sense of what was needed to fund large-scale scientific enterprises. In the middle of the nineteenth century, astrophysics had come into being as a result of new findings to do with spectroscopy being applied to the study of the heavens. Now, after the Second World War, there was not just an 'invasion' of astronomy by new physical and chemical methods and instruments (many of which were made possible by advances in electronics during the war) but by physicists, such as Lovell.

Soon after war's end, Lovell conceived the idea of building a very large radio telescope, a steerable 'dish' to collect radio waves from astronomical bodies. But the wavelengths of radio waves are many times larger than the wavelengths of visible light. For a radio telescope to have comparable resolving power to an optical telescope, it must be many times larger than its optical equivalent. Lovell's goal was a dish some 76 metres in diameter, a size that posed enormous design demands. Lovell also expected its cost to be far higher than could be afforded by his university alone, the University of Manchester in the north of England. What became known as the Mark I radio telescope meant very extensive government funding, as well as the involvement of teams of scientists and engineers, and so the sort of science that became known as Big Science.

After many trials for its builders, the Mark I was completed in late 1957. Within a short time, radio astronomers were employing the Mark I to address a major puzzle: are the majority of sources of radio waves stars within our own Galaxy or are the sources really extragalactic objects? In the early 1950s, at the time of the first extensive surveys of the sky at radio wavelengths, radio astronomers had generally agreed that most of the sources were 'radio stars' spread fairly evenly over the sky and located inside our Galaxy. Could these radio stars also be identified with known, visible stars? In 1952, Walter Baade and Rudolf Minkowski used the 200-inch telescope on Palomar Mountain to hunt for visible objects that might be linked to powerful radio sources. One source seemed to be a remnant of a gigantic stellar explosion, a supernova, but the other source, Cygnus A, appeared to be in the location of a galaxy with an odd or peculiar appearance. Optical astronomers had estimated the galaxy to be about a thousand million light years distant. If the identification were correct, and Cygnus A really so far away, then it was producing prodigious quantities of energy in radio wavelengths.

Linking radio sources to optical objects, however, proved to be a difficult and slow process. The major problem was accurately pinning down the positions of the radio sources. Even a 76-metre dish, with radio waves of, say, 1 metre in wavelength, produces a resolving power of about one degree, twenty times worse than the naked eye, and a value far too large to be of much help in sorting out which optical objects corresponded to which radio sources. To overcome this handicap, radio astronomy groups at Sydney, Australia, and the University of Cambridge in England began to

Mark I Radio Telescope at Jodrell Bank, near Manchester in England.

develop 'interferometers', that is, instruments in which the radiation from a radio source is split into two or more parts. Even the Mark I radio telescope at Jodrell Bank was often used as an interferometer. By combining its observations with those made from a radio telescope some kilometres away, the resolving power of the two radio sources could be much increased over that of the Mark I working alone. Using interferometers, radio astronomers secured high resolution data that convinced them that most of the radio sources are in fact extragalactic. At first, extragalactic sources such as Cygnus A were widely supposed to be galaxies colliding into one another. Later, astronomers interpreted Cygnus A in terms of a 'double radio source', in which there are two 'hot spots' of radio emission and Cygnus A became one of the best examples of a sort of extragalactic radio source that astronomers decided was very common. Results from radio astronomy were also crucial for the hard-fought debates over the nature and origin of the universe.

A Steady State Universe?

In 1950, the astronomer Fred Hoyle delivered a series of radio talks on 'The Nature of the Universe' for the British Broadcasting Corporation. In the talk on the expanding universe he argued that:

[T]he assumption is that the Universe started its life a finite time ago in a single huge explosion. On this supposition the present expansion is a relic of the violence of this explosion. This big bang idea seemed to me to be unsatisfactory even before detailed examination showed that it leads to serious difficulties.

The name 'Big Bang' stuck, indeed, Hoyle spent a good part of the rest of his life (he died in 2001) attacking the Big Bang theory and advocating a so-called Steady State universe.

Hoyle, like his colleagues at Cambridge University, Hermann Bondi and Tommy Gold, was uneasy not only with the notion of a start to the universe, but also the conflict between the estimates of the ages of the Earth and the stars and the age of the universe. These problems disappeared in the 'Steady State theory', first proposed in 1948 by Bondi and Gold, and in a somewhat different form in the same year by Hoyle. All three agreed on a never-ending and unchanging universe (at least when viewed on a large enough scale) in which matter is created continuously throughout time and space rather than in one singular event at the start of the universe in a Big Bang. In the Steady State theory, the creation of matter in effect drives the expansion as, overall, the density remains constant despite the creation of new matter. The universe, then, really is in a kind of steady state.

But by the mid-1950s, astronomers had decided that Hubble's methods of measuring distances to galaxies were undermined by serious errors, such as mistaking clouds of glowing gas for very bright stars. Now it seemed that the earlier distances to distant galaxies had to be doubled, and this in turn meant that the rate of expansion of the universe was slower than believed before. The age of the universe had, as a result, been seriously underestimated, and the age of the universe was no longer shorter than the ages of the Earth and stars, removing one of the justifications for the Steady State theory. Further, what quickly became widely accepted as convincing evidence against the Steady State theory came in 1965 in the form of the 'cosmic microwave background'. In 1964, two physicists at the Bell Telephone Laboratories in New Jersey on the east coast of the United States began a series of experiments with an unusual radio antenna. Arno Penzias and Robert Wilson used the antenna to investigate 'noise' in satellite communications. They wanted to track down and root out all the sources of noise. At first, it seemed to Penzias and Wilson that their antenna might be producing more electrical noise than they anticipated. After careful tests and cleaning of the antenna (including the removal of two roosting pigeons and their droppings), they concluded that whatever they did one source of noise could not be eliminated and that it was arriving with equal intensity from all directions in space. What was its source?

The answer Penzias and Wilson received shortly from physicists and astronomers at nearby Princeton University was that the cosmic microwave background radiation could be interpreted as the remains of an extremely early period in the history of the universe, in fact as a kind of relic of a 'Big Bang' origin to the universe. Although incredibly hot at the very start of the expansion, so the argument went, the expansion had led to the radiation's initially rapid and then gradual cooling to a temperature of about 3 kelvins, that is, three degrees Celsius above absolute zero. George Gamow, Ralph Alpher, and Robert Herman had earlier predicted the existence of such cold radiation as evidence of a very early and hot period in the history of the universe, but it had taken Penzias and Wilson, unaware of these predictions, to find it. The 3 K radiation, generally interpreted as direct evidence of the Big Bang itself, was widely viewed as a hammer blow against the Steady State theory.

Astronomy from Space

The radio and microwave regions of the electromagnetic spectrum were not the only ones opened up to astronomical observation after the Second World War. The development of astronomy from high in the atmosphere using balloons and above the Earth's atmosphere by the use of rockets and satellites made available even more regions, as well as offering the prospect to optical astronomers of securing images and spectra of astronomical objects far sharper than those to be achieved with ground-based instruments.

Before the Second World War, astronomers could only dream of placing telescopes above the obscuring layers of the Earth's atmosphere. The available rocket technology was so crude that, at best, such telescopes lay far in the future. During the war the push to build weapons, however, sped the development of rocket technology. The biggest technological advances were made in Nazi Germany. With the provision of large amounts of the state's resources and with the forced labour of many thousands of concentration camp inmates, the Germans built the V-2 guided missile. The V-2s were terrible if inaccurate weapons, but they underlined that rockets capable of carrying astronomical instruments to space had been constructed.

Intercontinental ballistic missiles armed with nuclear warheads defined the Cold War. There were also major military-funded projects in the late 1940s and 1950s to better understand the upper atmosphere, the medium through which such missiles would travel. Scientists became intimately involved in this effort and often flew scientific instruments high into or above the atmosphere aboard rockets. A group at the Naval Research Laboratory near Washington, DC, for example, flew instruments in V-2 rockets to observe the Sun in ultraviolet wavelengths that, due to the shielding effect of the atmosphere, do not reach the ground.

The launch of the Soviet Union's *Sputnik I* in October 1957, the first human-built object to be sent into orbit, elevated space into a very public theatre of the Cold War and space astronomy became one weapon in the battle for scientific and technological prestige between the United States and the Soviet Union. Through a combination of scientific, technical, and political factors, space astronomy attracted a level of support

Launch of a 'RGV-G-4 Bumper' rocket (a combination of a V-2 rocket and a Corporal sounding rocket) at Cape Canaveral in 1950.

undreamt of even a few years before. Space astronomy 'both literally and metaphorically, took off'.

In the United States, it was NASA, the National Aeronautics and Space Administration, that managed the programmes to study astronomical objects from space. In the 1960s and 1970s, NASA missions gave a major boost to Solar System studies, with, for example, the flight of *Mariner* 2 that flew by Venus. In addition to the fly-bys, both the US and Soviet Union set out to land spacecraft on the planets with the first success being that of the Soviet *Venera* 7 on Venus in 1970. In mid-1976, two US spacecraft landed on Mars. Each *Viking* spacecraft comprised a 'lander' and an 'orbiter', the lander to transport a range of experiments to the Martian surface and the orbiter to scrutinize the planet from high above the surface. The landers hunted for signs of life, but the general conclusion of the *Viking* scientists was that they had not found such evidence, at least not in the immediate vicinity of the two landers. During the 1960s, the United States and the Soviet Union performed the major space science missions. This state of affairs began to change in the 1970s. By the following decade, the European Space Agency ran a very significant space science programme and Japan too was building and launching space missions. Hence when in 1985 and 1986, Halley's Comet made one of its regular returns to the inner regions of the Solar System, it was met by a small armada of craft from the European Space Agency, the Soviet Union, and Japan, but not the United States.

Astronomical instruments carried aboard spacecraft have made some sub-disciplines possible that would otherwise be impossible from the Earth's surface. X-ray astronomy is one example. X-rays from astronomical objects are absorbed by the Earth's atmosphere and so to pursue X-ray astronomy, instruments must be sent into space.

The key development in the early history of X-ray astronomy came in 1970 with the launch of *Uhuru*, the first satellite devoted to X-ray astronomy (*uhuru* is the Swahili word for freedom and was chosen because the satellite blasted off from close to the coast of Kenya). With instruments aboard rockets, X-ray astronomers were restricted to a few minutes of observing time before the rocket arched back into the atmosphere under the pull of gravity. With *Uhuru*, this crippling limitation went away. In 1963, Riccardo Giacconi, the leader of a research group at American Science and Engineering, a company based near Boston, had conceived of a plan for such an X-ray satellite (he would share the Nobel Prize in Physics in 2002 for his for pioneering contributions to X-ray astronomy). The group was already very experienced in launching satellites with X-ray instruments due to its work in 1961 and 1962 for the US Department of Defense in measuring the bursts of nuclear weapons at high altitudes, weapons which also emitted X-rays. Hence, as often in space astronomy, an ostensibly scientific project, *Uhuru*, owed much to a conflation of scientific and military interests.

Perhaps the major discoveries made with the aid of *Uhuru* were the rapid variations in the intensity of a few of the sources. How were these changes to be explained? One source, 'Cygnus X-1', drew particular attention and theoretical astronomers interpreted the X-rays from Cygnus X-1 as the result of emissions from hot gas that had been heated up by being dragged towards a very compact object. Some astronomers proposed the compact object was a 'black hole', a star that had ended its evolution with a density so great that not even light, which travels at about 300,000 kilometres per second, can escape from its surface. As physicists Robert Oppenheimer and Harlan Snyder argued in a classic paper in 1939, once a star had become a black hole it 'tends to close itself off from any communication with a distant observer; only its gravitation field exists'.

Two Soviet physicists, Yakov Zeldovich and Igor Novikov, suggested in the mid-1960s that black holes might be detected as X-ray sources. Our Sun is at a temperature of about 6,000 kelvins and so radiates most of its energy in visible wavelengths. The hotter an object, the shorter the wavelengths of the emitted radiation. Astronomers had long established that stars at temperatures of 50,000 kelvins, for example, emit most of their energy in the ultraviolet region of the spectrum. Zeldovich and Novikov proposed that as matter is pulled towards a black hole, it accelerates and this increased energy of motion is converted into heat energy by friction. Very close to the black hole, the temperatures are so high that, rather than visible light, the matter emits X-rays and gamma rays.

Was Cygnus X-1 really powered by a black hole? For many astronomers the evidence was ambiguous. From the early 1970s on, black holes nevertheless became a central topic for astronomers, and speculations were rife that they were perhaps not only the end point in the evolution of very massive stars, but also present in the hearts of galaxies, as well as the highly energetic quasars—objects which radiated far more energy than normal galaxies, but which filled a much smaller volume—that had been discovered in the early 1960s. In fact, the finds by astronomers of such highly energetic objects as quasars, as well as speculations on black holes and the intense outpourings of energy from radio galaxies over time undermined the widely held view of a universe of galaxies sedately cartwheeling through space that had been common since the 1930s. The universe, it seemed by the late 1970s, was a far more violent place than astronomers had imagined earlier.

During the 1970s and 1980s, there were also growing interchanges between astrophysics and high energy physics, and another 'invasion' of astronomy by physicists. Among the most striking examples of links between the two disciplines are the underground 'observatories' in mines and tunnels. These observatories have been designed and built not to observe, say, visible light or radio waves, but neutrinos. Some students of the Sun, for example, had made calculations of the number of neutrinos expected to be released by thermonuclear reactions deep in the Sun's interior. One means astronomers and astrophysicists pursued to test these predictions, and the theories on which they were based, was to count the neutrinos released.

When the first of these underground observatories was built in the 1960s, neutrinos had been studied extensively by physicists at particle accelerators and reactors, and their ability to penetrate prodigious quantities of dense matter was well known. Detecting these ghostly particles therefore presented many problems. First, the observatory designers wanted to shield out cosmic rays that might interfere with the neutrino measurements. To do this, the neutrino observatories were sited a mile or so underground, a distance to which the neutrinos can reach, but not the cosmic rays. But how to 'catch' the neutrinos? The solution adopted at the Homestake Gold Mine in South Dakota in the United States in the 1960s was to fill a huge tank with 100,000 gallons of cleaning fluid. The chlorine in the fluid can absorb neutrinos. When a neutrino interacts with a chlorine atom, the chlorine is converted into radioactive argon. The amount of radioactive argon released in a particular period can be measured, and the number of neutrinos counted. Such experiments found that the number of neutrinos from the Sun was less than half of the number expected, and this anomaly between observation and theory presented a major anomaly. However, in the early 2000s, new measurements of neutrinos from more advanced observatories than the Homestake Mine, most notably the Sudbury Neutrino Observatory in Canada, plus new theories of neutrinos resolved the problem and the predicted numbers of neutrinos were brought into much better agreement with the observed numbers.

Telescopes into Space

In April 1990, the space shuttle *Columbia* rose from its launch pad at Cape Kennedy on a pillar of smoke and flame. In its cargo bay was the Hubble Space Telescope, a reflecting telescope with a primary mirror of 2.4 metres in diameter and a joint effort of NASA and the European Space Agency. Built at a cost of over $2 billion, it is the most expensive telescope ever constructed. Roughly the size of a school bus, it was also designed to be a highly precise instrument. But soon after it was deposited in orbit by NASA's space shuttle, astronomers and engineers found that its primary mirror had been manufactured to an exact but incorrect shape, with the edges of the mirror a fraction too flat at the edges. The result was that the Hubble's focus was not as good as planned and for very dim objects, such as distant clusters of galaxies, this was a crippling problem. In late 1993, a repair mission was launched to the telescope in which space shuttle astronauts switched some instruments as well as installed a device to correct for the primary mirror's wrong shape.

This high-stakes mission proved to be a great success and made Hubble as capable as originally planned, though it did mean the loss of one of the scientific instruments.

The Hubble Space Telescope.

Indeed, over time, it became a more powerful telescope because during later space shuttle missions more scientifically and technically capable scientific instruments replaced older, less capable ones. By 2013, over $20 billion had been spent on building Hubble and maintaining it in orbit as well as in analysing and archiving the data it had returned to Earth to do with a host of astronomical problems. By any standards, Hubble is science on the biggest scale.

Observations with Hubble were also very important in confirming the observations that led to the rapid acceptance of the idea of the accelerating universe, bizarre as it the notion seemed initially, as it meant that the velocity of a receding galaxy would speed

'Ultra Deep' view of the universe secured by the Hubble Space Telescope. The great majority of the spots of light on this image are distant galaxies.

up with time rather than slow down as everyone had expected. Saul Perlmutter, Adam G. Riess, Brian P. Schmidt, and their research groups had observed in 1998 that very distant supernovae—as supernovae are extremely bright they can be detected to extreme distances—are systematically fainter than expected by comparison with more nearby supernovae and so at larger distances than anticipated. The universe is not only accelerating in this picture but will expand forever at an ever increasing rate driven on by what has been termed 'dark energy'. This remarkable discovery won the 2011 Nobel Prize for Physics for Perlmutter, Reiss, and Schmidt.

Conclusions

By the second decade of the twenty-first century, astronomy was being pursued by more astronomers with more sophisticated apparatus than ever before. National governments and private foundations had never provided so much support to the science. In 1800, very few astronomers had wanted to follow William Herschel in probing far beyond the Solar System. By the early twenty-first century, astronomers spoke confidently about the history and very creation of the universe, and drew on not just information from optical wavelengths to test and interpret their theories, but also data from other regions of the electromagnetic spectrum, including the gamma ray, X-ray, ultraviolet, infrared, and radio regions, as well as information secured from spacecraft from a range of places beyond the Earth. Astronomy had been remade in scientific, technical, and social terms.

11 Theoretical Visions

MATTHEW STANLEY

THE word 'theory' comes from the Greek term *theoria* ($\theta\epsilon\omega\rho\iota\alpha$), which was used to describe the witnessing of a spectacle in an exotic land. Theory, in this sense, was seeing novel sights, learning things beyond everyday life. So it is in physics. Physicists create systems of concepts and equations to provide a new sense of coherence, meaning, and unity to an otherwise confusing world. Newton showed how the combination of gravity and a handful of simple laws could explain the movements of all bodies, from cannonballs to comets—at least in principle. Theories like his revealed the universal principles hidden behind the bedlam of observation. Theory, as a way of organizing reality, argues for an unseen world governed by bizarre rules. It says we must move beyond our direct experience into a realm of concepts and ideas.

Of course, concepts and ideas are often wrong. History reveals a central tension in theoretical physics: how can we gain the benefits of exploration through theory while making sure the theory is reliable? Skepticism of the value of theory is often traced back to Francis Bacon's warning about the 'anticipation of nature'. He argued that human reason was weak, and profoundly shaped by individual prejudices, the restraints of language, and the dogmas of philosophy. These idols, as he called them, might make it impossible to use hypotheses to correctly understand the natural world. He cautioned that many philosophers created their theories solely out of their imagination, like a spider that 'spins webs out of his own substance'. The usual object lesson for this warning was Descartes, who filled the cosmos with vortices without any evidence that they existed.

Newton's achievements in the seventeenth century persuasively showed that theory was of some use in physics' quest for universal laws. But how could one uncover these laws of nature without falling victim to Bacon's idols? How could one make theory useful and not dangerous?

Restraining Speculation

These concerns were still very real when physics in the modern sense emerged near the beginning of the nineteenth century. It was at this time that laboratory methods became widespread. As physics began exploring the new world of electricity and magnetism, problems arose. The forces manipulated by experimenters were mysterious and invisible. There were deep concerns that novel phenomena such as electromagnetism

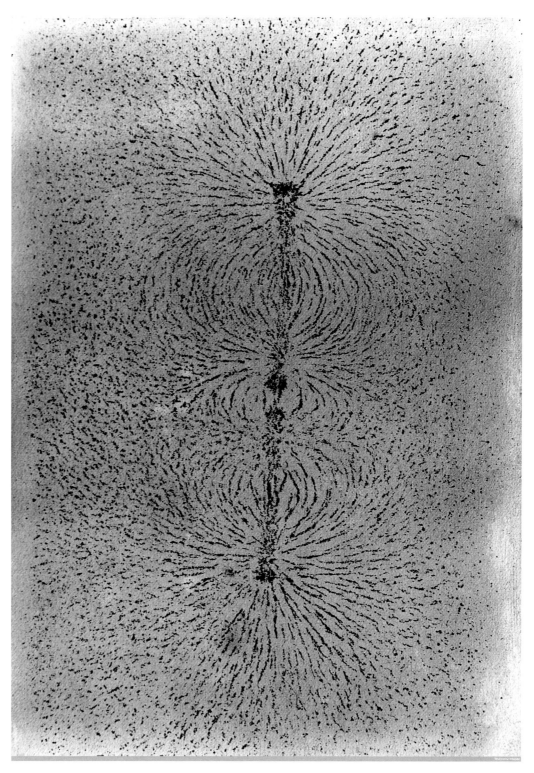

Michael Faraday's 'lines of force' rendered visible.

would allow for unchecked speculation. When investigators tried to make sense of this strange new world, what was to prevent them from simply inventing absurd explanations and projecting their fantasies onto nature?

In Britain, these concerns often led to theory being regarded with suspicion. Consider Michael Faraday, the self-taught pioneer of electrical and magnetic studies. He was particularly cautious about the twin dangers of abstraction without physical insight and the unjustified invention of new hypothetical entities whenever convenient. At the same time, he realized that his laboratory work required new ways of talking about electricity and magnetism—everyday physical intuition had clearly reached its limits. He developed his famous 'lines of force' as way out of this puzzle. The lines were used to show where electromagnetic effects were most intense, and how they were directed. These lines provided a way to think about invisible forces without any concrete hypothesis about their nature. They also allowed him to apply familiar physical categories (such as tension and density) to new situations. His drawings were a way to approach physics humbly, without any assumption that he understood the specifics of what was happening. They helped move investigations forward while restraining the human tendency to speculate and invent.

Another strategy for preventing speculation was the use of mathematics. Mathematics with its rigid rules and demands for consistency, could be an important check on imagination. Advocates for this approach pointed to the amazing achievements of the French mathematician Pierre Simon de Laplace. His *Traité de mécanique céleste* was published in several volumes in the first three decades of the nineteenth century. It solved the puzzle that had stumped Newton—how could the Solar System remain stable over long periods of time? The messy gravitational interactions among the planets had seemed to create an impossible situation. Laplace attacked the problem with staggering mathematical insight and new analytical tools. He successfully demonstrated that we need not worry about Jupiter's pull hurling the Earth out of its neat orbit. His approach was elegant. Take Newton's laws of motion and gravity (which everyone accepted as true) and apply them with unending deductive rigour. There was no need to propose new forces or subtle fluids. Eventually, Laplace suggested, all the phenomena of the world could be brought under mathematical description.

The success of French mathematical theory was inspirational across Europe. Cambridge University, the venerable home of Newton, resisted these new methods. Eventually a group of young men interested in physics imported these continental approaches. They called themselves the Analytical Society, and around 1812 began promoting the importance of modern mathematics in science. It included such future luminaries as the astronomer John Herschel, the mathematician and Cambridge leader William Whewell, computing pioneer Charles Babbage, and the Astronomer Royal George Airy. They were passionate about the new mathematics for its efficiency, and its potential for introducing innovation into an increasingly stale natural philosophy. They would eventually have a powerful influence on the credibility of theory in physics.

Herschel helped shape the debate over the proper role of theory in his widely read *Preliminary Discourse on the Study of Natural Philosophy*. He saw himself as following in Bacon's footsteps. Proper induction was empirical and rejected unjustified speculation. He wrote that natural philosophers should only invoke 'causes recognized as having a real existence in nature, and not being mere hypotheses or figments of the mind'. New phenomena should be explained in terms of already-understood principles, as Laplace did.

However, this group also acknowledged the necessity to move beyond mere observation. Whewell wrote of the need to formulate *laws* that connected individual facts. Natural laws were regularities of behaviour and activity, such as Newton's laws of motion. These could not be seen directly, and could only be recognized through mental reasoning. He argued that this meant discoveries in science were not accidental. Rather, they happened when someone was properly prepared for examining nature. This preparation usually took the form of intense study of natural laws. Recognition of these laws then allowed one to see patterns and truths that were otherwise hidden. His favourite examples were Kepler's laws. The minute observations of Tycho Brahe were useless until Kepler brought the new concept of the ellipse into play, whereupon the true movements of the planets were revealed. Pre-existing ideas (such as an ellipse), then, could help move science forward.

Many of Whewell's contemporaries were unhappy with this contention. This was dangerously close to Bacon's warnings about the anticipation of nature. What if a concept used for a theory was misleading, or simply wrong? The results would then be incorrect as well. Whewell's chief defence of the reliability of theory was theological. He wrote that God created both the natural world and the concepts available to the human mind. And since God wanted humans to learn about His creation, we could be confident that there was a guaranteed compatibility between our concepts and the world. This implied that mathematics, for instance, was innate to the human mind (a position not always popular with his contemporaries). Despite the controversy around his ideas, Whewell's position on the value of theory became extremely widespread.

This was largely due to his important position as the Master of Trinity College at Cambridge. He was a critical part of the revival of natural philosophy there, and his views became standard within a generation. The core of training in science at Cambridge was the Mathematics Tripos, an examination which produced generations of theoretical physicists. The Tripos was both mentally and physically rigorous and gave rise to a distinctive mathematical culture at Cambridge. Along with specific analytical techniques and problem-solving approaches, students there were inculcated with a deep reverence for the power of mathematics for exploring the natural world. The effects of this Cambridge culture were crucial to the development of theoretical physics over the next century. For example, the truths of nature were expected to be mathematical in form. Contemplation of natural laws and theoretical schemes could be useful for discovery. And perhaps most important was the confidence that the human mind could indeed understand the world rationally.

Industrial Theory

This confidence in our ability to understand and control the world was an integral part of the rapid leaps of industrialization in the nineteenth century. We think of the industrial age as an era of massive iron ships, ferocious steam engines, and roaring railroads. But it also gave rise to a host of theories about the intangible, the tiny, and the ephemeral. The growing ubiquity of steam power and its effects provided not only new phenomena for physicists to study, but also new ways of conceptualizing nature itself. Even as these engines drove factories to greater and greater profits, they raised increasingly difficult philosophical questions. It seemed that they destroyed coal, became hot, and created mechanical effects. But where did the heat go? Where did the mechanical effect come from? Could existing knowledge explain this, or was some new approach needed? Few people thought these questions could be solved through purely mental considerations. The physicists who developed theory to investigate these mysteries were almost always skilled in experiment as well. Theory was rarely expected to stand on its own.

Sadi Carnot, a French engineer, helped to begin this process. He wanted to understand the operation of all engines through some general principle, rather than studying each mechanism on its own. He imagined a subtle substance called *caloric* that produced the work of engines. Caloric was not consumed like fuel, but rather only moved from higher temperatures to lower temperatures. This movement produced mechanical effects, much as falling water could drive a water wheel. His work was largely ignored until it was rediscovered by the young Scottish physicist William Thomson (who later became Lord Kelvin).

Thomson grew up in Glasgow in the core of industrial Britain. He came of age surrounded by engines, ships, and factories. It is not an accident that the basic measurement of the effect of engines—'work'—is at heart an economic category. For people like Thomson, a deeper understanding of the physics of heat was inseparable from the application of that physics for real economic and social benefits. Eventually, the steam engine would become Thomson's model for the universe as a whole.

Like Faraday, Thomson was skeptical of attempts to explain phenomena through speculation about new unseen substances (such as Carnot's caloric). He was trained in the Scottish 'Common Sense' school of philosophy. This school warned that theories needed to be grounded in our everyday experience of the world. Guided by this philosophy, Thomson proposed that the conversion of heat into motion should be explained through a dynamical theory of heat. In this theory, heat was simply our crude observation of the motion of microscopic bodies. This motion could sometimes be manipulated into directed work (as in an engine), but sometimes could be lost (as in waste heat from that engine). Thomson developed his theory over several years to great success. It was very useful for understanding general behaviours of heat such as conduction. It also provided significant insight into specific engineering issues such as efficiency. A Manchester brewer, James Joule, also contributed to these ideas. Joule had noticed that beer batches increased in temperature while stirring. He conducted further experiments of amazing accuracy to determine a precise numerical relationship between a certain amount of motion and a certain amount of heat. While far from a confirmation, the ability to say that a given amount of heat equalled a precise amount of motion was a great support for this theory.

Thomson was, however, concerned about the foundations of his ideas. On the microscopic scale, what exactly was moving? There had been long-standing speculation that all matter was made up of tiny particles, variously called atoms or molecules at this time. This, however, was exactly the sort of reasoning of which Thomson was suspicious. No one had seen, or could see, an atom. They were therefore dangerously unreliable as an explanation. They could easily be figments of the human imagination. And yet, they were undeniably useful as a hypothesis. Other physicists, such as the German Rudolph Clausius, were less cautious than Thomson. In 1857 Clausius took this hypothesis more seriously and worked out some of its mathematical consequences. His goal was to see if the aggregate properties of hypothetical moving molecules could be correlated with any of the observed properties of gases.

Clausius's efforts were extended in 1860 by James Clerk Maxwell. Maxwell was, like Thomson, a Scottish graduate of the Cambridge Tripos system. He was both mathematically skilled and familiar with the industrial and experimental roots of these questions. He brought new mathematical techniques to the problem. Inspired by the statistical tools being developed to understand human society, he added randomness to the models of atomic motion. Another uncertainty had now joined the basic ignorance of what these atoms were—statistics meant that theorists had to give up on ever knowing the exact motion of an individual atom. They could only know their motions on average. For Maxwell this was an acceptable price to pay. The new statistical techniques allowed theories that explained many properties of gases, including diffusion and the well-known relationships among temperature, pressure, and volume. This kinetic theory was enormously successful, so long as one was willing to pay the conceptual price: do atoms really exist? Some investigators remained sceptical. The Austrian physicist and philosopher Ernst Mach famously confronted believers in atoms with the devastating question: have you ever seen one? Maxwell, on the other hand, suggested that the success of a theory based on the hypothesis of atoms was itself strong evidence for their existence, despite their invisibility.

The kinetic theory of heat was a great victory for understanding the connections between heat and motion raised by the steam engine. This was increasingly seen, however, as only one case of a broader series of transformations. Voltaic batteries had shown that chemicals could be changed into electricity. New motors showed that electricity could become motion. Even the humble candle demonstrated some fundamental links among heat, light, and chemistry. Natural philosophers across Europe were grappling simultaneously with these questions. Many came to a version of the idea that there was something *conserved* in all these processes. Despite manifest differences, perhaps there was an unchanging entity underlying all these. Joule's mechanical equivalent of heat suggested that there could be a similar relation for all of these different phenomena. A Berlin doctor with interests in physics, Hermann von Helmholtz, said that what they all had in common was the ability to do work. And the *amount* of work that could be done did not change from form to form, only the way it appeared. The capacity for work within a given amount of coal was transformed into the equivalent amount of mechanical effect. This force could neither be created nor destroyed, only transformed. Helmholtz even proposed the radical idea that this principle should apply to living things as well. Humans and animals were clearly able to do work, which meant that they too were analogues to the steam engine—they simply happened to be fuelled by food rather than coal. Life thus became a special case of physics.

British natural philosophers began describing this principle in its more modern form of the *conservation of energy*. Thomson and his colleagues pondered the possibility that energy was the most fundamental principle in the universe. If so, then all natural phenomena should be examined through the lens of this theory. For every process, there were now essential questions to ask: where does the energy come from? Where does it go? How does the transformation occur? The conservation of energy also suggested deep truths about the universe. It now seemed that there must be a fixed amount of energy in existence, which could never be increased or decreased. All human power would have to come from the transformation of this storehouse from one form to another. Even the Sun, it was realized, must eventually burn through its (still mysterious) fuel source and cool to darkness. Many physicists interpreted this in religious terms: only God could create or destroy energy, and this new law of nature was a clear demarcation between human and divine action. Other physicists understood the principle in terms of Romantic philosophy: a deep, hidden unity to the world that showed the interconnection of all things. Despite many ways of interpreting the conservation of energy, it quickly became one of the most powerful tools for theorizing about the natural world in the nineteenth century. Thomson and his friend Peter Guthrie Tait even took on the project of translating Newton's laws into energetic terms. They wanted Newton's great achievements to be seen as, again, just one more instance of a grand unifying principle.

While physicists widely embraced the conservation of energy as true, it was clear to many that it alone could not be a complete theory of the universe. Anyone who had sweated alongside a working steam engine knew that while heat was changed into motion, some of that heat was lost into the environment. That is, the process was not perfectly efficient. Some heat was always wasted during the conversion process. Further, the process could not always be reversed with perfect balance. If one converted the output motion of a steam engine into heat, the end result would be less heat than had been originally put into the engine. This might seem like a simple problem of engineering efficiency. But Thomson generalized it into a universal law: every energy conversion process will waste some energy. The energy was not lost—that would violate the conservation of energy. Rather, it dissipated into a less useful form (such as waste heat). If this was indeed a universal principle, it implied a constant depletion of useful energy in the universe. Even more, it suggested both an end of the world as we

NO MORE WATER.

The imagined future of the world as dominated by entropy.

know it (when all useful energy was dissipated) and a beginning (a point in the past when no energy had yet been dissipated). This was a completely unsurprising result for deeply religious figures such as Maxwell and Thomson, as it seemed to harmonize with the biblical narrative of Genesis to Revelation which they already accepted as true.

Thomson framed these insights as the laws of thermodynamics: the first law postulated the conservation of energy; the second law, the continuous dissipation of useful energy. The second law raised some interesting questions for the dynamical theory of heat. What was the microscopic correlate to this dissipation? Clausius proposed the concept of entropy. This was a measure of the disorder of the atomic motions which were observed as heat. The second law could then be rephrased as predicting a constant increase in the entropy (or disorder) of a system over time. However, Maxwell and Ludwig Boltzmann reminded their contemporaries that their theoretical knowledge of atomic motions was only statistical. There would always be random fluctuations and inherent uncertainties in any statistical theory. Therefore, if the second law was understood in terms of atomic motions, then it too would only be *statistically* true. This was a profound departure from earlier senses of natural laws, which were understood to be *always* true. Physicists now had to grapple with the idea of a law that was only *generally* true. It was certainly accurate enough for general use, but raised profoundly troubling issues of what it meant to say that a theory was true at all. Maxwell even came up with a thought experiment of a microscopic 'demon' that could circumvent the second law without any use of energy. Maxwell's demon has since become a constant target for theorists hoping to avoid the disturbing implications of a natural world that runs on statistics.

Not all theories at the time were so unsettling. Physicists could take refuge in a theory based on more familiar principles: that of the ether. This was a proposed continuous, all-pervasive substance that provided a medium through which light travelled. The theory that light was a wave was well-supported by experimental evidence, but seemed to require a substrate through which the wave could be conducted. Thus, some kind of ether must exist. It needed to have unusual properties. It had to be extremely rigid due to the amazing speed of light, had to be subtle enough to pass through transparent matter without trouble, and also fill the spaces between the stars without retarding the planets. Faraday, for instance, was skeptical that humans could accurately conceptualize such a strange entity. Most physicists in the nineteenth century, though, accepted that something like the ether must exist, even if it was not well understood.

Thomson became persuaded that there must an electrical ether as well. Maxwell took this idea further and was able to develop a mathematical formulation that encompassed Faraday's lines of force within the hypothesis of a real substance that propagated electrical effects. Maxwell's new mathematical tools allowed him to articulate many of the properties of this ether. He even proposed a mechanical model in which electrical effects were explained as deformations, stresses, and rotations of microscopic wheels and elastic spheres. It is unclear whether Maxwell thought this model was physically real. Another possibility was that it was only heuristic: the theory was perhaps only a way to conceptualize phenomena and direct new research, without proposing any actual new entities or structures. A theory can often be read in a range of ways, from wholly realistic to wholly imaginary.

As Maxwell developed his model further he increasingly discarded its 'physical' elements. He instead relied almost completely on the mathematical relationships suggested by his original model and constrained by the physics of energy. He came to accept that the actual internal workings of the ether were inaccessible to physics. His theory could only deal with the systematic relationships between the observable phenomena. He proposed a metaphor to explain the situation. Imagine, he said, a bell ringer who pulls ropes to accomplish his tasks. Even if he was forbidden to climb into the belfry and see how the ropes connected to the bells and each other, he could still, through diligent study, discern the patterns of rope movements and bell sounds. Such,

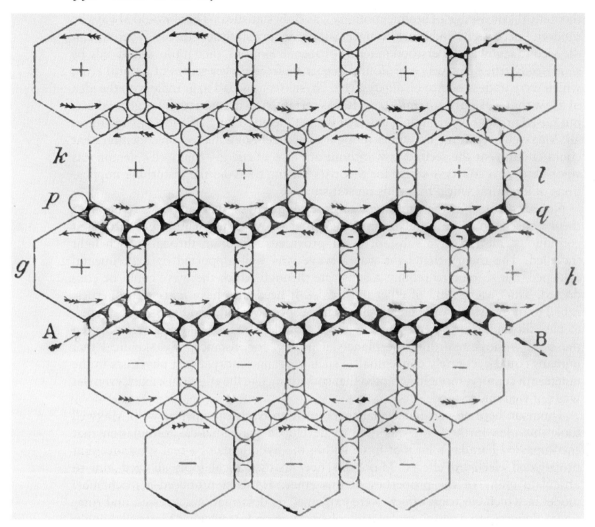

Maxwell's 'idle wheel' model of the ether as a mechanical system.

he said, was the state of our knowledge of the invisible world of the ether. It surely existed, though we might never truly understand it. Thomson rejected this strategy completely, calling it nihilistic. He argued that theories needed to stay intimately close to clearly visualizable mechanical processes. That, he said, was the only way to prevent unreliable speculation. Maxwell's willingness to trust mathematics and increasingly abstract principles seemed a dangerous road.

However, Maxwell's theory quickly chalked up persuasive victories. It not only articulated precisely well-known relationships between electricity and magnetism, it also predicted an entirely new phenomenon: the electromagnetic wave. Maxwell found that this hypothetical wave would travel at exactly the speed of light, and he leapt to the dramatic conclusion that light *was* an electromagnetic wave. His electromagnetic theory was extremely difficult to understand and use, but its success in unifying so many different phenomena was enormously appealing. Sadly, his untimely death meant he never saw his predicted electromagnetic wave created and measured by Heinrich Hertz in 1888. His ideas were also widely used by figures such as Oliver Heaviside in the rapid development of telegraph technology in the final decades of the century. Their utility for practical purposes made the ether seem ever more real: surely the victories of a theory indicated the truth of its basic hypothesis? The British wariness of the overreach of theory had been significantly moderated by the successes of the atomic hypothesis, the sophistication of ether physics, and the amazing reach and simplicity of energy theory. The rigour of mathematics and vigorous experimental checks of new predictions gave new confidence to the reliability of theory.

The Power of Thought

Not all physicists shared the skepticism that shaped the British approach to theory in physics. German scientists, for instance, had a different sense of what could be known solely through theoretical means. Their intellectual and institutional conditions made theory a more appealing way to investigate the natural world. Theory gained more of a life of its own.

The German university system was well funded and organized, and by the 1870s stressed the importance of original research. Physicists were encouraged strongly to produce new scholarship and publications. For much of the nineteenth century, this meant experimental investigations. However, a rapidly increasing need for physics instructors quickly outstripped the supply of experimental apparatus. This led to the novel development of professorships of theoretical physics: professors needed to teach classes, and were expected to produce research, but were not provided with any resources to develop an experimental laboratory. Theory thus became a route to professional development. These were originally junior positions intended to provide relief to more senior experimentalists, but the success of these theorists rapidly increased their standing in intellectual life. By the 1890s, theoretical physicists could be found in charge of entire German science institutes. This institutional support, combined with a legacy of superb theorists such as Helmholtz and Carl Friedrich Gauss, meant that by the end of the century theory was a recognizable and autonomous part of physics. Helmholtz famously declared that he wanted to gain an 'intellectual mastery over nature'. This meant understanding the laws of nature—theory. Material mastery of nature had to wait for a proper theoretical picture.

Physics was taught as an inseparable part of the overall German educational ideal of *Bildung*, the core humanistic knowledge that was thought to provide proper development of the intellect. This meant that students interested in physics became intimately familiar with philosophy as well as mathematics. Consider, for example, the young Albert Einstein. While he was still struggling to find a career in the sciences he made time to read Ernst Mach, John Stuart Mill, and David Hume. In tackling theoretical problems he found epistemology to be as useful a tool as mathematics.

This approach can be clearly seen in Einstein's famous 1905 papers on relativity, quantum physics, and atomic theory. Written while he was working as a patent clerk in Switzerland, they show both his idiosyncrasies and the firm philosophical roots of German theory. One paper, 'On the Electrodynamics of Moving Bodies', addressed a problem that physicists had been struggling with for some time. What were the precise interactions among matter, light, and ether for moving bodies? Most theorists had treated this as a problem of calculation and cleverness, of slightly modifying and tweaking existing theories. But Einstein did not begin his paper with a review of relevant experiments or a novel analytic technique. Rather, he proposed a simple thought experiment. Imagine, he said, an induction coil and a magnet moving past each other. If one thinks of the coil as being at rest and the magnet moving, electromagnetic theory provided one explanation for the resulting current. If one instead thinks of the magnet at rest, a different explanation was provided. The explanations were different even while the observed phenomena were identical. Einstein said this 'asymmetry' indicated a deep problem with the state of electromagnetic theory. New fundamentals were needed.

Einstein proposed two basic postulates as the basis of what would become his theory of relativity. First: the laws of physics were the same in all inertial reference frames (i.e. for observers who are not accelerating). Second: the speed of light was the same in all inertial reference frames. The first postulate was seemingly unproblematic: it was a formal statement of the symmetry he found lacking in his initial thought experiment. Einstein was simply insisting that all observers, no matter where they were or how they were moving, needed to agree on the laws of physics. There were no privileged observers.

In itself, this caused no problems. But then Einstein asked pointed epistemological questions. Taking a cue from Mach's rigorous positivism, he contended that our basic philosophical categories of space and time had no meaning beyond the way we *measured* those categories. Einstein developed further thought experiments with moving clocks, lightning strikes, and trains moving near the speed of light that showed rather surprising results. In order for all observers to agree on the laws of physics, they would sometimes disagree about their measurements. The length of measuring rods and the elapsed time between two events could only be established in relative terms. Universality of laws seemed to demand malleability of measurements.

Einstein's ideas were difficult to test at the time. They did help resolve some mysterious experimental results such as the Michelson-Morley interferometer's failure to detect the Earth's motion through the ether. However, relativity emerged from Einstein's rigorous epistemological approach to the fundamentals of physics, not out of any desire to resolve tiny experimental discrepancies. He relied largely on thought experiments, not laboratory work. This approach only made sense to someone with great confidence in the ability of the human mind to rigorously and correctly interrogate nature on its own. A thought experiment assumes that a theorist has reliable access to the principles on which the world worked. Similarly, the appeal of relativity was in the simplicity and elegance of its premises, not in its adherence to experiment. Its promise was in its potential to reframe the very nature of physics. Einstein dismissed the ether, the central theoretical construct of the previous decades, as 'superfluous'. In a relativistic world, an absolute frame of reference such as the ether was no longer needed.

Special relativity showed Einstein's insistence that theories be based on broad, universal principles (much as the laws of thermodynamics). As he tried to expand his theory to include gravity, he developed two new principles: that the laws of physics must be the same for all observers whatsoever (an expanded version of his previous postulate), and the equivalence principle (that observers cannot distinguish between the local effects of acceleration and gravity). Once he had finally achieved a stable university job, Einstein attempted to continue the Machian approach that worked for special relativity. This meant looking only at observable quantities and refusing to talk about what 'space' and 'time' were beyond their measurement. However, at the same time, the German mathematician Hermann Minkowski had proposed an alternative way to think about Einstein's ideas. Minkowski rejected Machian restrictions and said that relativity in fact taught us a great deal about the true nature of reality. He argued that Einstein had revealed that space and time were merely 'shadows' of a deeper four-dimensional world of 'space-time'. Einstein was originally resistant to accepting such a radical idea. But he was soon forced to admit that Minkowski's reworking of his equations were extremely helpful for solving the increasingly intricate puzzles he was encountering. As he used the concept of space-time, he became more comfortable with thinking of it as real.

Armed with the idea of a continuum of space-time, Einstein found he could solve many of the problems of his new 'general relativity' by thinking about that continuum being bent and warped by massive bodies. The mathematics of this turned out to be extremely complicated, and he soon regretted skipping so many classes as a student. With the help of friends he eventually mastered the non-Euclidean geometry that described his new four-dimensional world. In 1915 he finally presented the full field equations of general relativity. Einstein's goal of a completely uniform physics had led us to a universe where triangles could have 270 degrees, clocks ran slow near planets, and twins rapidly aged apart. The mathematical and conceptual strangeness of the theory was a serious obstacle to its acceptance.

Fortunately Einstein found some clearly testable observable consequences that could support his bizarre claims. One showed that a long-standing anomaly in the orbit of Mercury was actually the result of relativistic effects. Another, the gravitational redshift, was too difficult for the technology of the day. The third, the deflection of starlight as it passed the Sun, became the only real opportunity for a decisive confirmation. Unfortunately, it could only be seen at a total solar eclipse. Einstein presented his equations at the height of the First World War, when globe-spanning scientific expeditions were rather difficult. The prediction was finally confirmed in 1919 soon after the armistice by the British astrophysicist A. S. Eddington. Eddington, a Quaker objector to the war, saw a kindred spirit in the pacifist Einstein. He used the dramatic

A visual metaphor for how massive bodies deform four-dimensional space-time in Einstein's theory of general relativity.

timing of the war's end and the international character of the expedition to use relativity as a tool to restore the scientific internationalism ruptured by the war. Einstein's universal physics became a bridge to global science.

Revolutionary Theory

Relativity's origin in a war-torn world fits with our general sense of it as a theory that brought great violence to the elementary categories of our experience. Relativity is almost always described as a revolutionary theory—one that upset everything prior and changed everything going forward. But Einstein did not intend to break with the traditions of physics. He saw the theory's emphasis on universal, objective laws as preserving the best of science. Relativity, despite its strangeness, retained a world of uniform, continuous phenomena that obeyed clear laws of causation. Not all theoretical innovations of the time were so kind to the basic assumptions of physics. Quantum theory would come to question almost everything that physicists thought they understood about nature. This split was eventually articulated as classical (Newtonian mechanics, Maxwellian electromagnetism) versus modern physics (relativity and quantum mechanics).

Unlike relativity's genesis in profound principles, quantum physics emerged from a series of messy, practical attempts to match equations to empirical observations. One of the great experimental discoveries of the nineteenth century was the realization that gases emit and absorb light at startlingly precise wavelengths (i.e. specific colours) called spectral lines. This was quickly used to discern the chemical composition of both materials in the laboratory and celestial bodies such as the Sun. Despite continuous effort, however, no one was able to develop a satisfactory theory of how these spectral lines came to be. By the turn of the twentieth century physicists widely assumed that the answer lay in the structure of the atom itself. The new phenomena of radioactivity and cathode rays suggested a complicated atomic constitution and gave tools for exploring that arrangement. Initial models mixed electrons and protons together (the so-called 'plum pudding' atom). Ernest Rutherford's experiments made it clear that the protons were clustered at the centre, with the electrons somehow scattered around the perimeter. This was very frustrating to theorists. Electromagnetic theory predicted that the electrons in such a system should quickly radiate away their energy and crash into the nucleus. Their understanding of the atom did not even allow for stability, much less explaining spectral lines.

The log-jam was broken by an iconoclastic young Dane working in England, Niels Bohr. Beginning in 1911 he took a radical new approach to atomic structure. He ignored the theoretical problems involved and proposed new, apparently arbitrary rules for atomic models. Electrons orbited the nucleus at set distances known as 'stationary states'. Further, electrons only moved between these states. When they shifted, radiation was absorbed or emitted in discrete quantities ('quanta') equal to the difference in energy between the states. Bohr's model built on earlier work by Max Planck and Einstein that suggested light could be thought of as existing in particle-like chunks of energy (though neither initially took the idea as physically real). The model seemed preposterous, except for one huge benefit. The transitions between these hypothetical stationary states were exactly equal to the spectral lines emitted by hydrogen. This was a striking achievement. The match between Bohr's model and observation was remarkable. No one else had come close to explaining why spectra looked the way they did. This was at the cost of raising many new questions: why were the stationary states where they were? What made an electron jump from one to another? Wasn't light a wave, not a particle?

Nonetheless, this obviously incomplete theory was a major new inroad into the atom. It was therefore worth pursuing. Bohr was able to model the hydrogen atom quite well, but helium proved intractable. As he worked to develop his model further he proposed what came to be called the correspondence principle. This was a way of reconciling the odd behaviour of atoms and electrons with ordinary experience. The principle declared that any equation that applied on atomic scales should approach more familiar physical laws when extrapolated to large numbers of particles. This provided some guidance. More progress was made by developing purely mathematical, problem-oriented modifications to the theory. Einstein and Arnold Sommerfeld in Munich made serious contributions of this sort, providing methods for calculation without any understanding of the physical processes involved. By the early 1920s it had become possible to model the structure and spectra of quite large atoms. The theory worked, but it was far from satisfying.

Bohr was remarkably willing to grapple with the strange philosophical issues raised by this work. His institute in Copenhagen became a place of lively, unrestrained discussion about the fundamental principles governing the microscopic world. A great number of young, talented theoretical physicists came through the institute to be trained in quantum theory. There, they debated with Bohr and each other, developing new techniques and often discarding well-established ideals. Max Born suggested that the problems of the quantum world meant that all the concepts of physics might need to be reconstructed from the ground up.

Werner Heisenberg, a young student of Bohr's, brashly rejected all attempts to understand what was physically happening within atoms. He wanted to develop a quantum theory that used only observable quantities—that is, things that could be directly measured in the laboratory (at least in principle). The frequency of light emitted by an atom was acceptable; the orbital path of an electron was not. The theory abandoned visualization, causality, or any possibility of describing what was 'really happening'. His mathematical techniques were highly effective, but his insistence on the complete abandonment of physical intuition was unsettling to many. Heisenberg's hyper-positivist theory culminated in the uncertainty principle in 1927. He originally articulated the principle operationally: due to the way any measurement changed the system it was measuring, it would be impossible to ever know certain quantities exactly at the same time (e.g. the position and velocity of an electron). Bohr was unhappy with this presentation and suggested that the uncertainty principle actually revealed a far more fundamental principle about reality. Uncertainty in measurement was not the result of human error or limitation, it was an actual physical limit to what could be known. The essence of phenomena would be forever inaccessible to science.

One of Bohr's colleagues, Erwin Schrödinger, refused to accept Heisenberg's highly abstract approach. Instead, he tried to develop a quantum mechanics that retained some level of physical meaning. He was particularly interested in recovering visualizability, a tool that he saw as essential to any theory. In late 1925 he adopted an idea proposed earlier by the French aristocrat Louis de Broglie. Reversing the analogy that light behaves like a particle, de Broglie suggested that matter might behave like a wave as well. This explained why stationary states existed in particular places—they were simply where an electron's wave 'wrapped' evenly around an atom. Schrödinger was able to build on this wave idea to perform the same calculations as Heisenberg (he showed that their two systems were calculationally equivalent). It had the great advantage of using wave equations already familiar to generations of physicists, unlike Heisenberg's strange matrix mathematics. The unsettling instantaneous jumps from orbit to orbit were now replaced with a graceful fading away or emergence of a wave. Waves were easily visualized and manipulated for problem solving. It seemed, perhaps, that some level of sense could be brought to the quantum world. Unfortunately the wave equation introduced new mysteries. What exactly did the wave represent? The original sense of a 'smeared out' electron was soon found to be unsatisfactory. Born suggested that the waves were a representation of the probability of finding the electron in a given position or state. While this interpretation worked very well for calculation, not everyone was pleased to have a theory based completely on probabilistic reasoning (as we shall soon see).

Quantum theorists found themselves in a strange situation. They now had two conceptually incompatible theories—Heisenberg's and Schrödinger's—each of which was equally effective. Variations on this dualistic theme appeared frequently at the time. For example, light sometimes behaved like a particle, sometimes like a wave. Electrons did the same. The options seemed irreconcilable. Heisenberg and Schrödinger built their theories on completely different visions of what a theory could explain and what humans could know. Bohr tried to resolve these tensions with his principle of complementarity. This principle, which Bohr struggled to articulate clearly, suggested that there was no logical inconsistency in using incompatible models so long as they applied only to different situations. Since light only displayed wave-like and particle-like behaviour in completely different experimental set-ups, the models were complementary rather than contradictory. Profoundly dissonant theories could be embraced, rather than reconciled. Complementarity came out of Bohr's work in physics, but it quickly grew into a general philosophical principle which he applied to psychology and culture. When he was knighted in 1947 he even placed the principle on his coat of arms, represented through the Chinese principle of dynamic opposites (*yin* and *yang*).

A powerful consequence of these ideas was the implication that the act of measurement could affect the physical world itself. Setting up particular arrangements of detectors actually changed the state of what was being measured. If Bohr was right, an electron had no precise location until someone measured it. This apparent 'measurement dependence' of the microscopic world questioned basic assumptions about the objectivity and stability of the physical world. And the results of any such measurements were inherently probabilistic. They could be predicted very precisely en masse, but any individual event was unpredictable even in principle. There could be no strict causality in subatomic processes. The fundamental events that made up our world were completely unlike our ordinary experience. This general framework came to be known as the Copenhagen interpretation, after the site of Bohr's institute where so many of these ideas were hammered out.

Einstein refused to accept this vision of reality. He was deeply committed to the ideal of an objective physical world independent of human activity and adhering to strict causality. He regularly confronted Bohr and other Copenhagen adherents with the absurd implications of their ideas. Was the Moon really not there if no one was looking at it? Did God throw dice when designing the universe? Schrödinger agreed with Einstein that there was something clearly amiss with quantum theory. Imagine,

CONTRARIA SUNT COMPLEMENTA

Einstein and Bohr in the middle of one of their famous debates.

he said, a sealed box with a chunk of uranium, a cat, and a radiation-triggered poison vial. Radioactivity was a quantum process, and therefore inherently probabilistic. Thus the release of the poison, and the life or death of the cat, was represented by a quantum wave function. The equations of quantum mechanics suggested that until the box was opened (i.e. the existence of the cat was 'measured'), the cat was both alive and dead. Schrödinger posited this thought experiment as a clear demonstration of the absurdities of the current state of quantum theory.

The debates between Einstein and Bohr about the proper foundations for theoretical physics became legendary. Einstein would propose an apparently iron-clad thought experiment to circumvent or contradict complementarity or uncertainty; Bohr would come back the next day with a winning response. Einstein continued to insist that, despite the success of quantum mechanics, physical reality could not be inherently probabilistic and subjective. He argued that those theories must be 'incomplete'. He called for physicists to continue searching for theories that gave a deterministic, objective world that existed independent of observers. The 1935 paper known as 'EPR' for the initials of its authors (Einstein, Podolsky, and Rosen) laid

Opposite: Niels Bohr's coat of arms, prominently representing his concept of complementarity with Asian symbolism.

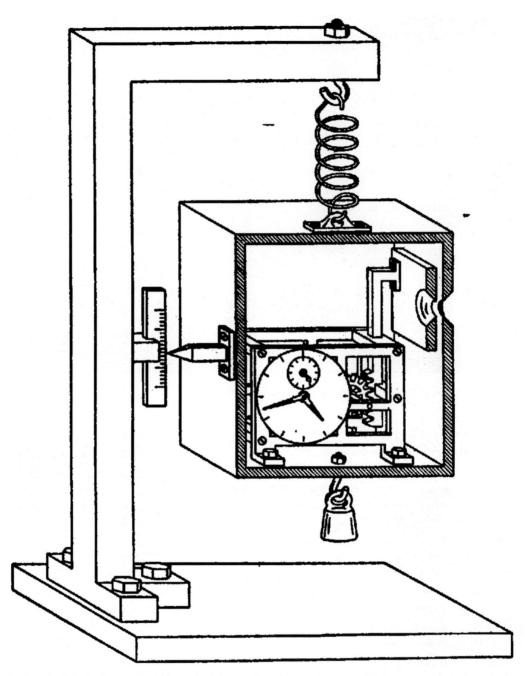

An imagined device from one of Einstein's elaborate thought experiments attacking quantum physics.

Opposite: Werner Heisenberg on one of the mountaineering trips that incubated quantum mechanics.

out the case that quantum mechanics was incomplete. Perhaps some deeper structure or hidden variables ultimately determined the seeming chaos of quantum phenomena. Ironically, this paper eventually gave rise to further theoretical developments (Bell's theorem) and a measurement indicating that there were no hidden variables to be found. Quantum theory, despite its strangeness, was an accurate description of reality.

Such grappling with the philosophical consequences of physics theories was a typical feature of quantum theory as practised and taught in Europe between the World Wars. Philosophy went hand in hand with calculation. Understanding the *meaning* of theories was essential. To the generation who came of age at this time, it seemed that physics advanced by great leaps of insight and bold new theories. Problems were to be solved by wholly novel approaches, not incremental shifts. For some time Bohr thought the best way to understand the mysterious phenomenon of beta decay was to discard the conservation of energy, the most beloved of physical principles. Every step forward brought new crises. Paul Dirac's addition of relativity to quantum theory brought the amazing prediction of anti-matter, but also seemed to make a full quantum electrodynamic theory of the electron impossible. His book ended with an appeal for yet another revolution in physics to fix things.

The generation of Heisenberg was a generation of revolutionaries. It was also a generation of theorists. The great work of physics could be done through heated conversations in ski chalets or during mountain hikes. Experiments could provide new puzzles to solve, or restrict speculation, but theory was something that could work independently. This can be seen concretely in the enormous contemporary expansion of theoretical publications—the percentage of physics papers focusing on theory—doubled in the 1920s. Unlike the era of Maxwell, a career built entirely on theory was now possible and respectable. Einstein was the mythic model for this, ending his career working alone in his study in Princeton. He mused that the keeper of a lighthouse would be the perfect occupation for a physicist: solitude with one's own thoughts and the laws of nature was all that was needed.

Theory Goes to War

Einstein's vision of the aloof theorist isolated from the distractions of the ordinary world was struck a great blow in the middle of the twentieth century. The years leading up to the Second World War, and the war itself, dramatically changed the nature of theoretical physics. The methods, goals, and geography of theory all underwent startling changes.

Some of these changes were shaped by politics. The rise of fascism chased Einstein and many other theoretical physicists of Jewish descent from their homes. The Nazis targeted theoretical physics in particular as being incompatible with their racial theories. Abstraction and speculation were seen as inherently Jewish and therefore un-Aryan. An exodus of European theorists scattered across the world, with many landing in the United States.

Opposite: Einstein enjoying the solitude of his office in Princeton.

Einstein settled in Princeton's Institute for Advanced Study, Hans Bethe at Cornell, Enrico Fermi at the University of Chicago. They had to adjust to a very different culture of theory in America. American theory was pragmatic and focused on calculations that could be checked by experiment. Philosophical pondering of the meaning of theory was strongly discouraged. Many of the leaders of theoretical physics in the country had been trained in Europe, including J. Robert Oppenheimer. So they were familiar with the philosophical debates over modern physics, but those were distinctly secondary to the *use* of theory. For example, the equations associated with the uncertainty principle were widely used, though very little effort was spent debating their deeper significance. Theoretical physics there had been strongly shaped by the American tradition of pragmatic philosophy (such as the work of Percy Bridgman).

These attitudes were dramatically accelerated as the United States entered the Second World War and theorists were recruited for the war effort. Many refugee physicists were passionate about defeating the Axis powers and threw themselves wholeheartedly into military projects. One project in particular was intimately tied to theory: the atomic bomb. Nuclear fission had been discovered in 1939, but the practical exploitation of nuclear energy remained speculative. This changed when Leo Szilard, a refugee theorist from Hungary, realized the possibility that a chain reaction could release enormous amounts of subatomic energy. Such a chain reaction had never been observed—it was purely theoretical. Nonetheless, the mere possibility of a bomb using that process spurred Szilard, Einstein, and others to press the United States government to develop it before the Nazis did. The Germans retained a great deal of talent in physics, including Heisenberg, and the threat felt very real.

This was the beginning of the Manhattan Project. Eugene Wigner argued that at this point, an atomic bomb existed only in the minds of scientists. Therefore, theory must lead. Oppenheimer, a theorist, was put in charge of the design of the bomb. Los Alamos became packed with European theorists looking to defeat their former governments. These refugees had to quickly learn the American style of theory and become comfortable working closely with experimentalists and engineers. Wartime pressure allowed no time for philosophizing. The theorists needed to figure out how to build a bomb as fast as possible. Don't worry about what a nuclear cross section actually represents—just get the calculation done so the construction can begin.

The bomb, of course, worked. The sudden surrender of Japan in August 1945 led to the creation of a new narrative in physics: theory won the war. And there was certainly an element of truth in that. Bethe's theory group at Los Alamos was the crucial centre with which everyone else interacted. It was impossible to imagine an implosion-style bomb without theory. Even the massive industrial edifices of Oak Ridge and Hanford that actually produced the fissionable material were built and organized with the guidance of the theorists. The essential industrial infrastructure of the Manhattan Project was largely obscured, and the bomb was painted as a product of theory. Oppenheimer became the model of what a theorist could contribute—the war was won by the mind of a physicist. This narrative had an unexpected side effect in the realm of Cold War nuclear secrecy. This was the idea of a 'secret formula',

Theoretical physicists at Los Alamos working on the first atomic bomb. J. Robert Oppenheimer can be seen in the centre.

a short equation or diagram that, when scrawled on a piece of paper, would enable an enemy to build their own atomic weapons. Theoretical secrets seemed to be portable in a way that engineering secrets were not. Theoretical physicists thus became both essential and suspicious (after all, many of them had foreign accents).

This sense that physics won the war changed everything about the discipline. The sudden value of physics for national security led to massive funding increases from the federal government. Even more, physicists were now a military asset. The government accordingly became interested in vastly increasing their number. The need for more numbers meant a need for increased speed of training. Physicists who had learned their discipline in small seminars, or over beer in Bohr's office, now had to teach enormous lecture classes. The teaching of theory increasingly shifted away from deep discussions or philosophical inquiry into the nature of measurement. Such things were impossible with the new influx of students. Instead, straightforward calculation and problem sets became the order of the day. This was a distinctly American shift in the teaching of theory. In West Germany and France, quantum theory continued to be taught in the framework of philosophical questions. But in the post-war world, the centre of gravity of theoretical physics had shifted to the United States. The European physicists who fled there decided to stay. Further, the massive growth in funding and personnel in America gave physicists there a huge advantage over their compatriots overseas. American theory increasingly became seen as the norm.

The shift in practice of post-war theory can be seen clearly in the story of quantum electrodynamics (QED). QED had the goal of a full theoretical understanding of quantum interactions between light and electrons. Attempts at this had been made since the 1930s, but were always disrupted by certain mathematical terms going to infinity and destroying the theory. Further efforts tried to re-invent the theory from the foundations, hoping for a new quantum revolution. This changed when a new generation of physicists tackled the problem. This young group came of age during the war, and their approach to theory was that of the Manhattan Project and the equally massive radar project: don't worry about a whole new theory, just figure out the problem and get calculating. Richard Feynman and Julian Schwinger were exemplars of these new techniques. By 1948 they both (along with Sin-Itiro Tomonaga in Japan) were able to 'renormalize' QED and allow for finite calculations. Freeman Dyson brought these different versions together. This renormalized QED was highly conservative. It was not a new theory per se—it had no new principles or grand visions—it was just the old theory in an improved form. The wartime generation was only concerned with tweaking the theory to the point of utility. Philosophical reconceptualizations were unnecessary. The new motto: 'shut up and calculate'.

This emphasis on the speed and efficiency of calculation was linked to the rise of novel theoretical techniques. Consider the now-ubiquitous Feynman diagrams. These simple line-drawings were originally presented as mnemonic aides to help with increasingly complicated quantum field theory calculations. They helped streamline fantastically difficult mathematics. There was debate, however, over exactly what they were. Were they actually pictures of physical processes? Did they have meaning unto themselves, or were they merely formalisms? Were they simply convenient, without any truth behind them?

These problems were not unique to Feynman diagrams. The post-war search for more efficient calculation produced other visual techniques, such as Penrose diagrams in general relativity. Electronic computers also became much more widespread. This often encouraged theorists to tailor their research to numerical, rather than analytical, solutions. That is, solutions to specific problems instead of general laws. Some theorists wondered if these results could provide insight or understanding the way classical theories did. Or had they become so focused on problem solving that meaning had been lost? Computer simulation further aggravated these issues. Were patterns seen in a simulation truly laws of nature? Or had theory devolved to just organizing numerical results or suggesting the next round of experimental trials?

Richard Feynman and Sin-Itiro Tomonaga, two of the pioneers of quantum electrodynamics. One of Feynman's eponymous diagrams can be seen on the board behind him.

The requirements of post-war 'Big Science' reinforced this kind of theory. At the gigantic particle physics laboratories theorists were essential. No particle collider could be constructed without theoretical guidance. But at the same time, theorists had to work intimately with experimenters and engineers. The techniques and organizations that worked so well during the Second World War had been preserved into the Cold War. Theory was just one specialized part of vastly complicated physics factories like CERN. Specialization meant it was increasingly rare that any individual physicist could be an expert in both theory and experiment. Fermi's skill in both became the stuff of a legendary past. At the same time, specialization meant that theory had less and less life of its own. The tangled connections with experiment made it difficult for a theorist to ponder philosophical implications on their own terms. The

Einstein model of the lone theorist sitting at a desk pondering the truths of the universe seemed less and less accessible.

Unified Theory

As much as Big Science fortified the pragmatic, institutional approach to theory, it also helped spur a resurgence of one of the great classical goals of theory: unification. Many theorists in the first half of the twentieth century saw themselves as following Maxwell's footsteps in unifying the laws of physics. Unifying electromagnetism and gravity was the great goal. The problem was that the former was governed by quantum mechanics, the latter by general relativity: two profoundly incompatible theories with completely different foundations and formalisms. Einstein, Eddington, and Schrödinger all spent the last years of their life dedicated to this project. This goal fell by the wayside with the general decline of philosophically oriented physics. Unification became re-established as an important goal in the late twentieth century. Achievements toward unification boosted its role as a core concept of physics, and helped re-establish the independence of theory.

The impetus for unification emerged from the new government-funded particle physics laboratories of the 1950s. The accelerators there began producing more and more bizarre subatomic particles. This 'particle zoo' was messy and confusing for theorists. The instinct of theorists was to try to bring order to the chaos through classification—were there families or other natural groupings among the particles? And was there any way to talk about *why* the particles had the properties they did? By 1961, Murray Gell-Mann at Caltech and Yuval Ne'eman from Israel proposed a system of symmetry to impose order on the zoo. Particles were grouped by properties into particular sets. These sets had various possible symmetries that were used to identify them (Gell-Mann named his SU(3) symmetry the 'Eightfold Way' after the Buddhist moral code). Sometimes these symmetry groups had gaps—that is, there was no particle occupying an expected spot in the set. However, these gaps were not seen as failures of the theory. Rather, they were seen as predictions: the theory as a whole was so successful that physicists thought there *must* be a particle occupying that spot. And those particles were often found when experimentalists were sent to check specifically for them (such as the omega minus hyperon in 1964).

SU(3) clearly worked. The deeper question was, *why* did it work? Was there a basic principle that explained this symmetry? Gell-Mann interpreted the pattern as stemming from a composite structure of strongly interacting particles—that is, hadrons (particles like protons and neutrons) were made up of smaller entities. In naming these smaller particles, Gell-Mann intentionally avoided anything physically descriptive. He called them 'quarks', a nonsense word borrowed from *Finnegans Wake*. George Zweig suggested calling them 'aces'. These choices were intended to remind physicists that no one understood the nature of these particles. The different 'flavours' of quarks reinforced this intentional ignorance: up, down, and strange (later joined by top, bottom, and charm). The names were chosen to be fanciful to prevent any temptation to apply everyday physical intuition to them. Gell-Mann wanted to avoid all philosophical discussion about quarks. He particularly evaded the question of whether quarks were real or merely a formalism. They worked for calculation, and that was what was necessary. This tension was aggravated by experimentalists' inability to detect individual quarks. Theorists eventually came to embrace this odd situation. They declared that it was *impossible* for quarks to exist outside the particles they composed. Whatever quarks were, they were now by definition unobservable and incomprehensible. The post-war approach of avoiding philosophical discussion had been raised to a physical principle.

Symmetry became more and more important as a guideline for theoretical physics. It was often interpreted in terms of *invariance*. This described a physical property or quantity that did not change in certain specified situations. For example, in relativity, the laws of physics were invariant even for observers in arbitrary states of motion. This was a goal of theory that goes back at least to the conservation of energy (and perhaps to the pre-Socratic philosophers): what are the properties of the universe that are constant? Some invariants were proposed to help explain various quantum phenomena, including those of charge, parity, and time-reversal. T. D. Lee and C. N. Yang in 1956 questioned the universal applicability of these, and their work led to experiments demonstrating that parity was not invariant in weak nuclear interactions. But the search for invariants continued. Modern theorists hope that the *combination* of charge, parity, and time (CPT) is invariant.

Invariants were valuable for theorists because they provided a basic framework for understanding and simplifying otherwise complicated situations. Many theorists in the 1960s and 1970s, including Sheldon Glashow in America and the Pakistani Abdus Salam, drove important developments based on gauge invariance. Steven Weinberg joined Salam in showing that symmetry considerations could allow both electromagnetic and weak nuclear forces to be explained through one set of particles. This 'electro-weak' theory proposed the existence of exotic particles that were detected at CERN and Fermilab over 1973–1974. Theorists were excited by this demonstration of the power of symmetry-based quantum field theories. They provided a compelling new vision of the fundamental unity of the universe. Salam and Weinberg had shown that at high energies, two apparently separate forces were actually one. Perhaps, then, at even higher energies all the fundamental forces (electromagnetic, weak nuclear, strong nuclear, and gravity) could be unified.

This framework set an agenda for theorists: find new kinds of symmetry that would meld all forces in a 'grand unified theory' (GUT). It helped set the agenda for experimental physics as well. More and more powerful accelerators would be required to test the increasingly esoteric empirical predictions of these tentative GUTs. The difficulty of setting up these experiments drove some particle physicists (e.g. Alan Guth) to cosmology, where they could ponder the ultimate particle collider—the Big Bang. In this way, the ontological hierarchy of particle physics (increasing energies revealed a deeper unity) also became a temporal hierarchy (closer to the beginning of the universe, perhaps all the forces were unified). Attempts to

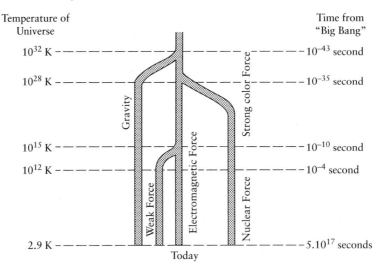

Temperature of Universe

10^{32} K

10^{28} K

10^{15} K

10^{12} K

2.9 K

Today

Time from "Big Bang"

10^{-43} second

10^{-35} second

10^{-10} second

10^{-4} second

5.10^{17} seconds

Gravity

Weak Force

Electromagnetic Force

Strong color Force

Nuclear Force

A conceptual diagram illustrating how a hypothetical unified force at the time of the Big Bang split into the fragmented forces we observe today.

understand the smallest constituents of matter had come full-circle to explain the universe as a whole.

One branch of the search for a unified theory gained increasing independence during the 1980s. 'String theories' tried to derive the properties of all known particles from hypothetical one-dimensional 'strings'. The different vibrational modes of these strings would hopefully explain the full spectrum of elementary particles. Some models were successful in doing this—as long as several extra tangled dimensions were allowed. One of the appealing features of string theory was that it could be developed almost entirely through considerations of mathematical elegance and conceptual clarity. This harkened back not only to Einstein's approach to physics, but also to his vision of the independent theorist. String theory's ability to give mathematical explanations for what was already known was impressive. But this was soured by the field's inability to predict previously unknown phenomena that could be tested. This damaged the close working relationship between theory and experiment that had become standard post-war. It also questioned some deep-seated assumptions about the nature of physics, and raised the possibility that string theory had no connection to reality. Sheldon Glashow spoke for many critics when he asserted that string theorists who never left their offices 'simply are not doing physics'.

Nonetheless, as the twentieth century ended, it was theorists who were the public face of physics. Stephen Hawking, the wheelchair-bound expert in general relativity, became a larger-than-life representative of the power of the human intellect to understand the universe. The Columbia physicist Brian Greene enthralled audiences with the beauty of string theory. Michio Kaku said that physicists' goal should be a 'theory of everything' that could fit on a T-shirt. Many theorists of this generation felt they were a hair's-breadth away from completely understanding the fundamental forces of the universe. Hawking wondered as early as 1980: 'Is the end in sight for theoretical physics?' Could there be a point at which there was no more to understand? Could theorists make themselves obsolete? Surely not—even if there was a theory that completely encompassed the four basic forces, there would still be plenty of puzzles in condensed matter physics and a dozen other sub-fields. But the trope of the end of physics is a persuasive and long-standing one. Feynman lamented it. William Thomson proclaimed it in the 1890s. Even Newton's contemporaries expressed disappointment that there was nothing left to discover. These statements are sometimes read as arrogant. What they really express is confidence that science is on the right track, that the present mysteries are solvable, and that nature is comprehensible. Every generation thinks that the final puzzle pieces are right around the corner.

The Nature of Theory

The proponents of theories of everything think of that goal as the key of what it means to do physics at all. But not everyone agrees that theory is the essential core of physics. The question of the proper importance of theory has been a central question in the way philosophers try to describe science. The logical positivists did their best to articulate a theory-free approach. They thought that theory, if unavoidable, should at least follow the footsteps of experiment. Karl Popper posited that the role of theory was to provide claims that can be tested. To him, a theory could never be proven, only not disproven yet. Thomas Kuhn argued nearly the opposite—that theory so constrained and shaped a scientist's experience of the world that it was almost impossible to disprove one. His notion of a paradigm suggested that theory always came first. Without it, no science could be done.

Philosophers will never agree on the correct use of theory. We can see historically that theory has filled an amazing range of roles—follower, interpreter, guide, aspiration. Bacon's suspicion of spiders remains with us in the critics of string theory. But few physicists today would accept similar criticism of the atomic hypothesis. And what should we make of the ether? It was a theory absolutely necessary to the development of electromagnetism, though now completely dismissed. Probably the most important lesson is this: theories are tools designed to solve problems. They never exist on their own. They are not discovered in a flash of insight. Like any tool, they are the outcome of difficult struggles over long periods of time. They are always incomplete, and always changing. And they are absolutely essential. Without them we can never move beyond our own experience. Invisible electric fields, subtle quantum waves, or even the universal processes of energy would be forever forbidden to those refusing theory. As the Greeks warned, theory is part of a journey to realms both frightful and fruitful, where the world is not as we thought.

12 Communicating Science

CHARLOTTE SLEIGH

A lion is mauling a buffalo. Red blood stains his mouth; his powerful claws rip at the buffalo's hide. It looks as though natural selection has scored a fateful blow at the buffalo's expense. But walk around this fighting pair and you will see, literally, another side to the story, for the buffalo's sharp horn also pierces the lion's shoulder. Each swipe, each movement, made by the king of beasts threatens to rend his own flesh against that dreadful point. He may be fatally gored at any moment. The animals are poised in the deadliest of dramas.

This momentary scene has been frozen in time for over one hundred years at the Powell Cotton Museum, in the East Kent seaside resort of Birchington. It looks like a piece of natural history on display—and of course it is—but what *exactly* does it communicate to the viewer? On the simplest level, it shows what animals may fight one another in a certain part of Africa. It may be an actual recreation of a fight between these two particular beasts, or a representative display of their species' usual fighting stances and techniques. On a more theoretical level, it seems to illustrate nature 'red in tooth and claw'—Darwin's 'struggle for existence'. There is, however, yet another side to the story, for which we have to step back from the case and view a torn and faded linen suit in another gallery.

The lion in question, carefully stuffed and mounted by the Piccadilly taxidermist Rowland Ward, nearly killed Percy Powell Cotton, its collector and the founder of the museum in 1896. Powell Cotton, a retired army major, had thought the creature shot dead and approached too close—whereupon it raised itself up for counter-attack. The lion swiped at Powell Cotton's shoulder—the point at which the prepared lion itself now receives a wound from the buffalo. Powell Cotton's suit, with the rip mark clearly visible, is on display at the museum's entrance. In a pleasing twist on the old trope of the Bible in the breast-pocket that protects the soldier from gunshot, Powell Cotton was—so the story goes—saved by a rolled up copy of *Punch* which he was carrying tucked inside his jacket. A final point of interest is that the African buffalo, *Bos brachyceros cottoni*, is one of many species to have received Powell Cotton's name in honour of its discovery by that trigger-happy naturalist.

Opposite: An African lion and buffalo fight it out in this taxidermized presentation of nature. Its autobiographical connection to its maker additionally tells a story about the nature of scientific knowledge.

499

Now the story has added dimensions in which we can read the buffalo as Powell Cotton himself. It becomes a story about adventure and bravery; perhaps also, as *Punch* takes on the lion, a story about Powell Cotton's self-presentation as unconventional naturalist, more interested in a good chuckle than in the serious learning of the British scientific elite. Visitors to the exhibition—trippers and holiday-makers from London—would have been drawn into this dramatic presentation of nature, and, by implication, perhaps also into its reframing of science as gung-ho adventure.

The buffalo and the lion demonstrate that when science is presented to the public, it does not cleanly convey something about nature. Viewers of this exhibit were not just taught something about nature; they were not even just given a subjective view of nature; they were also told a story about what the nature of science was. They were drawn into a relationship both with nature as seen by the show's creators, and with the nature of knowledge-production, or science, as practised by those creators. As for the lion and the buffalo, so for everything else in this chapter: a version of 'science' itself is produced through each re-presentation of nature. As historian James Secord puts it, 'science [*is*] a form of communication'.

Early-Modern Roots

The scientific revolution involved many ways of communicating nature, generally amongst equals—communicators and viewers were fellow participants in natural history and natural philosophy. Genteel men, cultured and wealthy, admired one another's collections of natural curiosities. As they looked, they talked; their learned discussions produced knowledge about nature. Because audiences were made up of fellow collectors, there was no distinction between the making and the communication of knowledge. In the seventeenth century, experimental science was born. Again, its audiences were select. Controlled groups of witnesses—made up of savants and their patrons—attested to the value of the demonstrations that they watched, and their conclusions were disseminated through print media that were as well protected as their producers could manage against plagiarism, piracy, and vexatious reinterpretation.

The eighteenth century saw a major expansion in the trading and consuming classes. Science was displayed in front of larger and less select groups: the same people who attended other events in the London season, or who attended street fairs and lower-class entertainments. It also progressed beyond the capital, notably into the emerging industrial areas of Britain. In some respects (as Larry Stewart has argued), the eighteenth century can be seen as the heyday of public science: the era in which the usefulness of discovery was taken for granted, and in which the public rehearsal of experiment (begun the previous century) put the seal on accepted methods of science. During this period, chemistry—including electricity—was the favoured science for exploration through display. There were still demonstrations for extremely select audiences of the rich and influential, such as at the Royal Society. But there were also industrial speculators watching displays of Newtonian science in the hope of making canny investments. Others just turned up for a bit of a laugh. Electrified monks, ladies, and children made for a good show. The newly discovered 'airs' could stimulate political or erotic effects amongst their experimental subjects and, in a short step (dangerously short for their critics), their audiences.

The taming of nature: the Royal Institution

By the late eighteenth century, science was becoming a mass commodity in its own right; an entertainment whose audience expected neither to make immediate financial investment nor to draw cultural cachet from what they saw. Its audiences did not participate; they watched. The seal on this new identity for science was arguably set by the foundation of the Royal Institution (RI) in 1799, although one might not perhaps infer it from its written aims of participation and improvement:

... for diffusing the knowledge, and facilitating the general introduction, of useful mechanical inventions and improvements; and for teaching, by courses of philosophical lectures and experiments, the application of science to the common purposes of life.

The collective motives of the Royal Institution's backers included national economic competitiveness and the betterment of the poor, as well as personal interest and investment. There was also, perhaps, a desire to tame or at least provide an alternative to the incendiary presentations of chemistry that were all too common. This last aim was personified in the hiring of Humphry Davy as lecturer and soon professor. Davy, a compelling stage presence, had hitherto been involved with the insalubrious 'factitious air' (nitrous oxide or laughing gas) but now modified his performances to suit an audience that Thomas Carlyle described in 1836 as 'all manner of fashionable people ... a kind of sublime *Mechanics' Institute* [of] the upper classes'. The intention to educate Britain's technical labourers had fallen by the wayside.

There was also a new development at the Royal Institution that had not been originally planned, namely the development of a serious research laboratory in its basement. Here, demonstrations could be prepared ready for display to the audiences upstairs. In a move reminiscent of the original Royal Society demonstrations, seats were installed for expert witnessing: not just shows, but knowledge was being produced backstage. Upstairs and downstairs, Davy mastered both the use of rhetoric and his instruments to show that he was in control of nature, an authority in chemistry.

The ultimate master of the Royal Institution's laboratory was Michael Faraday. Faraday's ascent to national celebrity from humble origins was the quintessential Victorian myth of self-betterment. He worked tirelessly on his self-presentation as well as on his mastery of the scientific apparatus that he used to demonstrate the phenomena of nature. Even more than Davy, he sought to convince his audiences that it was not his dexterity that they had come to watch, but rather *nature speaking through him*. Thus Faraday came to be celebrated very differently from Davy in terms of both style and content. In his heyday, Davy was a wildly popular figure—sexy would not be too strong a word—whose romantic science (he also dabbled in poetry) was all the rage. Once

Scientific Researches! — New Discoveries in PNEUMATICKS! — or — an Experimental Lecture on the Powers of Air

The cartoonist Gillray here casts doubt upon Davy's transition from dubious demonstrator to doyen of establishment science—though audiences were convinced.

Faraday was in vogue, Davy appeared, in hindsight, rather eighteenth century: too elevated, too blowy, with a rather embarrassing account of science that leaned towards profit and entertainment. Sober Faraday was feted for his clarity, his focus.

Diversifying display: science in the hands of workers

The original intended audience for the Royal Institution, finding it closed to them in practice, had found other venues to patronize. When George Birkbeck expounded his vision for adult education at the Crown and Anchor Tavern on the Strand in December 1823, an astonishing two thousand people attended. Birkbeck's London Mechanics' Institution, as it became known, was one of a large number of such organizations scattered around the country. Often housed in buildings specially commissioned by their industrial sponsors, they comprised libraries and lectures where working men could learn the latest in science.

These libraries, of course, required contents, and the early nineteenth century saw an explosion in the number of titles intended for the working classes, facilitated by the cheap and fast production of the steam-driven printing press. Not only was this new technology a brilliant new medium so far as its mostly Whig producers were concerned; it was also a model of the new sciences and their place in the nation. Entrepreneurship, education, and national success went hand in hand; the new machines could both educate the masses and then provide profitable work for them.

The Society for the Diffusion of Useful Knowledge was one such Whig organization, founded by a lawyer and politician, Henry Brougham, and an established publisher, Charles Knight. The Society's aim was to 'impart useful information to all classes of the community', and to this end it published cheap magazines and cyclopaedias which contained everything from facts about electricity to household hints, from home remedies to religious guidance. Some commentators feared the effects of all this education upon the masses, while for others it did not go far enough. They feared, with some justification, that this method for the development of a skilful workforce was aimed in truth at cultivating docility.

Meanwhile there were also attempts to convey overtly conservative science to the masses. Most famous of these, perhaps, was the result of a bequest by the eighth Earl of Bridgewater, who in 1829 left £8000 to pay for a thousand copies of a book *On the Power, Wisdom, and Goodness of God, as manifested in the Creation*. Eight authors were appointed by the Royal Society, each of whom wrote one of the eight 'Bridgewater Treatises', as they became known. These varied in their scientific content and piety as well as their approachability by non-specialists. Apart from those copies donated to Mechanics' Institutes, rather few made their way into the hands of the poorer classes, since they were expensively priced—cheaper editions finally appeared in the 1850s. Different groups of readers gleaned very different things from their readings. Some found their theology confirmed, or their knowledge consolidated. Others read from a point of little knowledge, or on a peer-level. Some even combed the books' pages in the hopes of finding hidden or inadvertent vindication of radical, atheist politics.

Members of this last group were also inclined to consume printed matter about phrenology, the nineteenth-century science that promised knowledge for all based only on a fingertip feel of the skull, and a classless system of education and advancement based on individual capabilities. Most sensational of all, however, was the description of nature contained in an anonymously published book of 1844, *The Vestiges of the Natural History of Creation*. This ambitious book synthesized science to date in order to demonstrate how natural laws could account for the development of all aspects of nature from stars to species. Radicals loved it, and established gentlemen of science rushed to criticize it.

There was, therefore, a wide variety of motives for communicating nature amongst audiences and writers of the steam-press era, and what was intended by the author or publisher was not necessarily what was gleaned by the reader. The meanings of nature, and the nature of science, were negotiated in a complex web of politics, economics, social advancement, philosophy, and theology.

As periodicals and pamphlets gushed forth from the steam-press, the Society for the Encouragement of Arts, Manufactures, and Commerce was finding other initiatives to sponsor, similar to those originally intended for the Royal Institution. In 1828 it opened the Gallery of the National Repository, on the site of the present National Gallery. The Repository included models of industrial machinery and ships, and a functioning telegraph suitable for communication between the rooms of a house. Visitors were not well behaved, 'probing the exhibits down and leaping upon the models'. The exhibition moved to Leicester Square in 1833, but even in this more popular quarter it failed to provide sufficiently appealing entertainment to last very long.

More successful were the Adelaide Gallery and the Royal Polytechnic Institution, whose aims also resembled the original, technical vision for the RI. They succeeded in attracting a large number of mechanics and artisans, due in part to their emphasis on lecture and demonstration rather than simple display. The National Gallery of Practical Science, as the Adelaide was initially known, was founded in 1832 by the American entrepreneur Jacob Perkins. It garnered around half a million visitors over its lifetime, thanks to attractions that included a 21 m water tank containing model ships, a magnet capable of supporting 238 kg, a steam gun that discharged one thousand bullets in a minute, and an electrical eel 'in full life and vigour'.

The Adelaide Gallery competed for visitors with the Royal Polytechnic Institution (f. 1838), located in Regent Street. The Royal Polytechnic Institution traded on its own marine marvel, a diving bell in its Great Hall that was large enough to accommodate several people (paying an extra shilling over the entry price for the privilege) and which descended into a tank of water. Other notable features were a prototype spin-dryer and beef steaks cooked by optically focused rays. The Royal Polytechnic Institution also had lecture theatres, a laboratory, and a great deal of equipment and machinery. But despite these trappings of original science, the Royal Polytechnic Institution was frequently disparaged as a place of juvenile entertainment by critics such as the *Mechanic's Magazine*. Mogg's 1844 guide to London stuck up for its higher value, however, stating that visitors would be 'introduced to a knowledge of every mathematical, mechanical, and musical invention and improvement'.

The style of lecturing and demonstration at the Adelaide and Royal Polytechnic Institution was at odds with Faraday's at the Royal Institution, even though the topic of electromagnetism was common to all three institutions. The Adelaide's William Sturgeon emphasized his 'judicious arrangement of the phenomena', meaning that he was proud of his intricate experimental devices and wanted his audiences to be equally impressed at their design and his mastery of them. By contrast, Faraday was famously focused upon 'facts'—not what he had made, but what belonged to nature itself. This was more socially palatable in an age of political turbulence, as is illustrated by an electrical experience of the Duke of Wellington at the Adelaide Gallery. Contemporaries reported that he:

put his hands into the trough (containing water) used for electro-magnetic purposes, he was fastened there as completely as if his hands had been locked in a vice, and...the conqueror of Europe, was as helpless as an infant...

If such a paralysis were explained by the powers of nature—essentially the powers of God—then no social hierarchies were threatened. But if such a humbling were touted as a demonstration of the experimenter's powers—and Sturgeon shared similar humble origins to Faraday's—then it was a dangerously radical manoeuvre in the politically unstable 1830s. For Sturgeon to suggest, however circumspectly, that his cleverness resembled the cleverness of the Creator, was a claim too far. Demonstrators made a terrible error if they thought they could combine overtly impressive skills with an authoritative display of nature; Sturgeon was quietly excluded from respectable science and hence—until recently—from the history books. Today, historians contend that such venues as the Adelaide and the Royal Polytechnic Institution were worked by mechanics and instrument-makers in a serious attempt to appear before the public as men—if not gentlemen—of science.

This lack of cultural permission to combine original research with public display perhaps helps to explain the move towards overt entertainment in the latter part of the nineteenth century. Public institutions had to seek a different identity, as in this account of the Adelaide in 1849:

Then came a transition stage . . . at first stealthily brought about. The oxy-hydrogen light was slily [*sic*] applied to the comic magic-lantern; and laughing gas was made instead of carbonic acid. By degrees music stole in; then wizards; and lastly talented vocal foreigners from Ethiopia and the Pyrenees. Science was driven to her wit's [*sic*] end for a livelihood, but she still endeavoured to appear respectable. The names of the new attractions were covertly put into the bills, sneaking under the original engines and machines in smaller type. She was an exemplification of the old story of the decayed gentlewoman, who, driven to cry 'Muffins' for her existence, always hoped that nobody heard her.

A similar change in policy at the Royal Polytechnic Institution can be dated to 1854, when its management was taken over by John Henry Pepper. Under his direction there were musical entertainments, magic lantern shows, illusions, and electrical wonders. Above all, the Royal Polytechnic Institution was famed for its optical displays, and of these, the most celebrated was 'Pepper's ghost', a remarkable illusion for which a succession of dramatic scripts were rapidly commissioned by way of vehicle. As well as agreeable fright and puzzlement, the ghost's provocations included debate as to whether Pepper had gone too far in presenting entertainment at the expense of scientific improvement. For some visitors at least, a large part of the fun was figuring out how the tricks were done, which was no simple task and, arguably, a good training in scientific thinking. Indeed, many scientists confessed that Pepper's *Boy's Playbook of Science* was a significant factor in their scientific development.

Yet even amidst these increasingly effect-focused displays of science, historians have found active research that has had theoretical as well as practical consequences. The physics of light and sound—and in particular the visual registration of non-visual phenomena—were developed for public display. These went on to form the basis for cinematographic technologies, but at the time comprised genuine, new knowledge in their own right.

The famous optical illusion 'Pepper's ghost', first developed by Henry Dircks, was demonstrated to the British Association in 1858 and rapidly became a popular staple of the Victorian stage.

Science for the nation (1): the Great Exhibition and its successors

The Great Exhibition of 1851 successfully placed Britain at the centre of the world, and put science and the technical arts at the centre of Britain. It also embodied in the most remarkable form to date both the brilliance of the sciences and their uncertain nature. It highlighted questions about whether state, monarch, or capitalist was the legitimate patron of science; about the relationship between science and the arts; and about the exact nature of the improvement that was wrought upon the general public by their inspection of this spectacle.

The Great Exhibition was not the first show of its kind. Eleven expositions had already been held in Paris, and the Dublin National Exhibition of 1850 attracted a respectable three hundred thousand through its doors. The Royal Society of Arts's Manufacturers Exhibition of 1849 had welcomed one hundred thousand. But the show of 1851 was a behemoth; by the end of its 141 days, over six million people had come to view its collected items. Its scale was overwhelming, as contemporaries reported:

A glass-hall nearly three-quarters of a mile long and two hundred feet high! The Monument might stand in one of its recesses. The glass roof would cover twenty-five acres of ground. Of its contents,—its works of art, its displays of the natural and manufactured products of all countries, its gardens, and its museums of curiosity and science,—a large volume would be but briefest catalogue.

The Exhibition made Britain the global centre at which all colonies and indeed all other nations offered up their treasures. In some cases this was figurative, in others, literal; just seven years later, Britain assumed direct governance of India. William Whewell, avoiding allusion to the imperial implications of the Exhibition, found philosophical virtue in the fact that visitors could compare the state of the arts around the world contemporaneously—something that would be impossible if one travelled for real, since in the time of travelling situations would change. Whewell's delight at the panoramic potential of the Exhibition was a more high-flown version of a widely shared public pleasure. A great globe, 60 feet across, was created for the show—a panorama within a panorama—upon whose interior the whole Earth was projected. Unfortunately, the project was considered by the Exhibition organizers to be too overt a marketing device for its creator's map-making business. It was successfully mounted as a separate show in the gardens of Leicester Square instead.

Science and the useful arts were integral to the Great Exhibition; eleven of its original twenty-four overseers, the Royal Commissioners, were Fellows of the Royal Society. When the Crystal Palace was re-erected at Sydenham in 1854, it was claimed that:

In science,—geology, ethnology, zoology, and botany receive appropriate illustrations; the principle of which has been to combine scientific accuracy with popular effect, and in its ultimate development the directors are bold enough to look forward to the Crystal Palace of 1854 becoming an illustrated encyclopaedia of this great and varied universe, where every art and every science may find a place.

For others, the Exhibition's science was a monument to advances in the useful arts. The *Northern Tribune* pooh-poohed other newspapers' reporting of the Royal Family's attendance and instead urged its readers to rejoice with it at the real stars of the show:

Well done, industrious and daring capitalists! Every way well done! Does it not show us also what might be done if a whole nation was capitalist and contriver...?

And yet the show did rather poorly in terms of identifying successful technologies or sciences for future development, despite the prizes that it offered. For some Whigs and Liberals—those backers of publishing ventures in popular science—invention and manufacture was not the point. For them, the Exhibition was simply a vindication of free trade around the world. Depending on whether one was inclined towards royalty or trade, then, science might illustrate very different things, and its success might be variously judged.

The Eiduranion projected the movements of planets onto a screen, in this instance at the English Opera House at the Strand, London. Somewhat similar displays continued through the latter half of the twentieth century at the London Planetarium (1958–2006).

Whatever it was that it exhibited, the sheer stupendousness of the Great Exhibition meant that it had a long legacy in the public production and consumption of science. The very architecture of the Great Exhibition was itself part of the story of conspicuously brilliant Victorian civil engineering, a tradition of bridges, railways, and tunnels

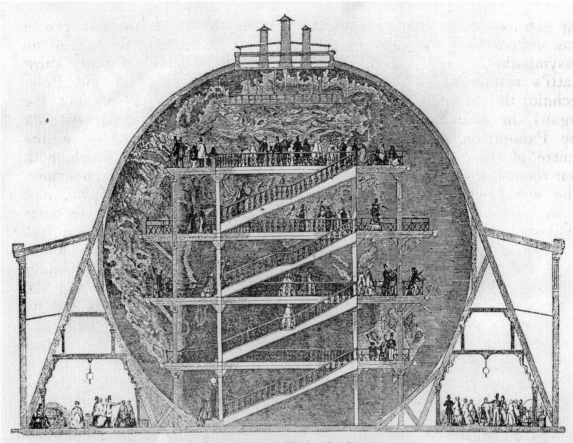

INTERIOR OF WYLDE'S GLOBE.

Too commercial for the Great Exhibition of 1851, Wyld's Globe dazzled its visitors with a view of the whole world in the more downmarket environs of Leicester Square.

that was hitched in the literature of Samuel Smiles and others to a narrative of self-improvement and national success. One of its specific major effects was to precipitate the creation of the diverse group of institutions in South Kensington—the museums, the Royal School of Mines, the Albert Hall, and so on. A total of ninety-six different institutions were proposed for the site, and although it is now best known for its museums, it was originally planned as a highly practical complex of research, training, and commerce, with the museums serving their purpose within this wider remit. The display of science was wedded to education, and education was not seen, primarily, as a business for universities.

The Great Exhibition also inspired direct emulation in science shows around the world, well into the twentieth century. During the 1870s four shows of select exhibits in London pulled in visitors in the high hundreds of thousands. There was a significant scientific strand in the London Festival of Empire of 1911 and in the London British

Empire Exhibition of 1924–1925. This latter focused in particular on chemistry, textiles, and engineering, and achieved attendance figures more than treble the apparently unbeatable Great Exhibition. The opening ceremonies of the 2012 London Olympics and Paralympics presented science in a recognizably similar mesh of patriotism, nationalism, and trade, with the added ingredients of self-deprecation and post-industrial embarrassment. The Olympics show featured, rather briefly, Tim Berners-Lee, developer of the Internet, whilst the Paralympics positively bloomed with science: Stephen Hawking, Higgs bosons, Isaac Newton, and umbrellas. Of this last, the programme notes explained:

The umbrella, or brolly, is a quintessentially British object that protects us from the rain. The steel-framed version was invented in 1852 by the British industrialist, Samuel Fox. It's a triumph of design and transformation... and full of comic potential.

Science for the nation (2): natural history

Natural-historical collections of exotic plants and animals, with their implications of voyage, trade, and conquest, had been a demonstration of monarchical or state power long before the Great Exhibition was dreamt up. The Jardin du Roi, which opened to the public in 1640, was a statement of the international reach of Louis XIII and most especially Louis XIV. The Zoological Society of London and its menagerie (1828) were no less emblems of colonial trade and dominion, albeit of a different age. The collection was largely the achievement of Sir Stanford Raffles, a tireless advancer of the British Empire in the East Indies. It was a respectable alternative to a vibrant culture of menageries—commerce and exhibition of exotic species and curios—that had lasted since the eighteenth century. Even many of the animals were royal—unwanted gifts donated by Prince Albert. The Linnaean Society cooperated with Raffles in the zoo's establishment, and the original remit of the gardens was to facilitate the study of animals (alive or dead) by specialists. But the real governors of the zoo were members of the nobility, colonialists, and upwardly mobile industrialists, and its governance followed the general nineteenth-century pattern of middle-class enfranchisement. For its first two decades, only members of the ZSL and their guests were allowed into the gardens. By 1846 financial difficulties forced the Society to open its gardens to the paying public.

For every animal that had been captured live around the world there were dozens that had been shot. Once taxidermized, these formed the core of collections around Britain. The British Museums Act of 1845 enabled civic museums to be established around the country, meaning that it was not just Londoners who could benefit from orderly displays of nature and empire. Increasingly, owners of stately homes opened their private exhibitions to members of the paying public, brought to them, as the century progressed, on an expanding rail network. The Powell Cotton Museum was one of these; rather better known is the Natural History Museum at Tring, originally the private museum of the second Baron Rothschild, built in 1889 and opened to the

This charming owl is one of a set of magic-lantern slides showing items of natural history. Other magic lantern displays utilized complex moving parts to show the movements of the planets and stars—a reinterpretation of the earlier Eiduranion (page 388).

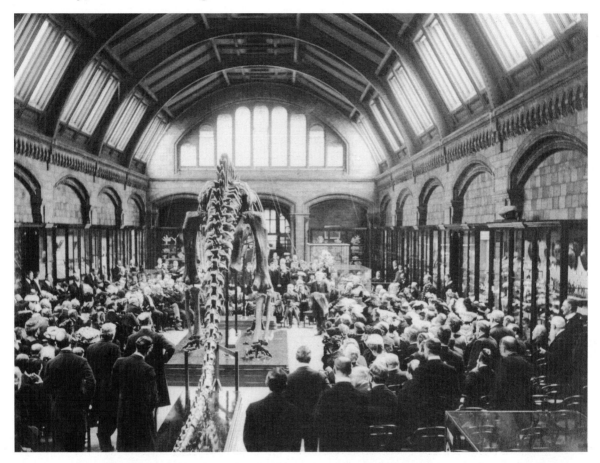

For many people, the great Hintze hall of the Natural History Museum seems designed to hold its famed Diplodocus skeleton, but in fact it has only inhabited that space since 1979. It was the cast of a real skeleton acquired by Scottish-born millionaire Andrew Carnegie. King Edward VII saw a sketch of the creature at Carnegie's Scottish castle home and remarked that it would look good in the Museum; it was installed in the Reptile Gallery in 1905. 'Dippy' is due to be removed in 2017.

public in 1892. Towards the end of the century, the great US museums of natural history, with their dramatic dioramas, were also established.

Although natural theological works continued throughout the nineteenth century (particularly for children), one does not get the sense that any strong theological message was conveyed by displays of nature in the same period. It was not exactly clear how the sight of an orchid or a kangaroo should uplift or improve the self. One commentator proposed a three-fold benefit: 'pleasing and innocent recreation' afforded by the sight of beauty; the 'various advantages' or practical benefits which may be derived from animals; and the incitement to admire divine wisdom and power. Later, he added that the zoological gardens were an 'adornment' to the city that should provoke admiration for the country. Not all visitors were obedient to such high-flown aims, however. Whilst mulling over his mechanism for evolution Charles

Darwin famously encountered an orangutan at the gardens, drinking from a saucer. 'Let man visit Ouranoutang in domestication, hear its expressive whine, see its intelligence when spoken to; as if it understands every word said', he mused, 'and then let him boast of his proud pre-eminence.' Queen Victoria limited her reaction to noting that she found the species 'disagreeably human'.

What one can say for certain is that the social discipline imposed upon visitors to civic and genteel spaces—whether metropolitan, provincial, or rural—imposed a tacit but powerful framework that contained their specimens: a framework of what it meant, socially and culturally, to participate in science. In general such spaces were established in wealthier metropolitan areas (or attracted wealth to them); they were healthful and civilized places away from the pressing growth of urbanism, attracting those who professed or aspired to gentility. The term 'zoological gardens' emphasizes the continuity of what we now think of as animal displays with the parkland prom-enades of Kew and elsewhere; originally, the experiential emphasis was at least as much on the external space. A nice illustration of this is given by the *Zoological Keepsake* of 1830, which describes the visits of an upper-class family to the Gardens (admitted thanks to a personal connection with one of the Zoological Fellows). After the mother has expounded at length on how she would re-landscape the area to decrease crime and ugliness, her daughter remarks with unintentional mordancy: 'When Primrose Hill is enclosed [as per her plans], all…shall be called, not the Regent's Park but dear Mamma's!' Similarly, a visit to Kew's Royal Botanic Gardens was a more salubrious form of entertainment than a turn around the Vauxhall Pleasure Gardens. Elsewhere in Europe, the creation of rational garden spaces was often part of an intentional and planned programme of urban renewal.

Nineteenth-century visitors to Kew enjoyed the sense of seeing the world in mini-ature (a similar experience to that achieved in panoramas, and the Great Exhibition), although it is not clear that all would have felt themselves to be stakeholders in the dominion that its exhibits commemorated. For many visitors, instead, the rational pleasures of a trip to Kew both fed and were fed by botanizing at home. From quite early in the century, artisanal classes had been keen collectors and information-swappers concerning plants; by the mid-century, the middle classes—more likely to afford the Kew entrance fee—were caught up in a craze of fern collecting. Zoological displays were also paralleled by a mid-century craze of domestic imitation, in the form of marine aquaria. Thus the gentility of science performed at home and in public spaces was mutually reinforcing.

The great exception to the socially conservative message of biological nature was, of course, evolution. Like that dangerous wielding of electrical equipment by non-gentlemen, so the dissection of the body (human or animal) could be used to draw destabilizing conclusions, if it were done by the wrong kind of person or in the wrong kind of way. During the 1830s, comparative anatomists in Edinburgh and London—many of them young students—eagerly seized on the radical implications of French evolutionists as they contemplated the structures shared by animals and human beings. But their work was not publicly displayed.

The story of the split of the Natural History Museum from its parent, the British Museum, is in part a tale of the tension between competing models of public natural history, in an era when the transmutation of species was less politically charged. The core of the British Museum's natural history collection was material gathered by Sir Hans Sloane in the eighteenth century, some of it still on show today in its Enlightenment Gallery. Although new items were acquired, by the mid-nineteenth century, many had been sold, lost, damaged, and destroyed; the natural history collection was not considered an important feature of the Museum in comparison with it archaeological treasures. The arrival of Richard Owen as curator of natural history in 1856 changed all that. Owen was a comparative anatomist who in his work on fossils coined the word 'dinosaur'. He began serious lobbying for a new building, which was joined by other voices and happily coincided with the development of the Albert-approved project on the South Kensington site.

T. H. Huxley's advice for the Commissioners of the Manchester Natural History Society, given in 1868, set out a vision for a museum comprised of two spaces, one given over to visitors and the other to researchers. The former should 'illustrate all the most important truths of Natural History', without being 'so extensive as to weary and confuse ordinary visitors'. Meanwhile the visitors were not to get under the feet of the curator and scientific students. This was definitely a didactic vision for a museum, and although it was not (on the face of it) intended for the South Kensington building it is a good description of the 'iceberg' nature of the museum that was eventually constructed, with vast collections not on show but available to researchers. Plans were drawn up in 1864 but, unsurprisingly given the massive nature of the building, it was not opened until 1881.

What did visitors learn about nature by seeing the British Museum (Natural History)? Under Owen's regime, they did not learn about the transmutation of species, since Owen organized his specimens according to a four-fold typology of basic forms rather than to tell a story of gradual change. Even the relief sculptures that decorated the building were spatially sorted by living and extinct species. But what visitors did learn was that biology was an endeavour so worthy, so elevated, that it required the architecture of a cathedral to contain it. In this, at least, Owen and his enemy Huxley were agreed. Huxley lobbied hard for the Museum to cleave from the British Museum, and stand culturally in its own right. When Huxley arranged for Darwin's burial in Westminster Abbey—an exceptional honour for a man of science—Huxley's version of evolutionary biology was finally enshrined in an equivalent architectural and cultural space of its own. The directors of the Natural History Museum (separated from the British Museum in 1963) finally completed Huxley's mission in 2009. They removed Owen's statue from the Museum's entrance hall and replaced it with Darwin's, at first as a temporary measure for Darwin's bicentenary, and then as a permanent fixture. At the time of writing, Huxley's portrait remains, rather surprisingly, at the centre of the National Portrait Gallery's triptych of evolutionary heroes (the other two being Lyell and Darwin). However, the ape and anthropoid skulls that originally lay on the table beside him—casting an entirely different light on the traditional *memento mori* human skull that he holds in his hand—have been mysteriously over-painted, a discreet act of veil-drawing that perhaps indicates his cultural domination was not so complete as he might have wished.

Science in the marketplace of the long nineteenth century

Huxley's effective advocacy for science (and for himself) was rooted in part in the lecture theatre. He was a talented and compelling lecturer, but just one part of a large lecture circuit that inhabited various civic spaces around the country, catering to an insatiable public demand. Polite society, working men, women, children—all attended. The subjects varied greatly, from electricity to molluscs. The science was likewise diverse, and could be hitched to conservative views of God's world or to the most outrageous brand of anarchism.

Lecturing created a punishing schedule for those who made their living by it. Lecturers could give up to 120 talks in a season, and they seem to have been remarkably diligent in preparing fresh material for different audiences. Tours abroad were possible, but these were risky, placing lecturers practically and financially in the hands of fixers whom they had not met. There was also a trend amongst audiences towards greater visual demands upon their speakers. The natural historian, J. G. Wood, for example, practised extensively in order to produce vast drawings live on stage, made rapidly and fluidly in uninterrupted lines. They were met with huge applause and appreciation. (A noteworthy hangover from the Victorian lecture format was seen in 2006, with Al Gore's filmed lecture on anthropogenic climate change, *An Inconvenient Truth*. This sprang from a tour of lectures and was itself screened in small and local settings, as well as appearing on major release and DVD.)

Lecturers, of course, were very often also writers, and the market for written science continued to expand during the later part of the nineteenth century. There were many children's books, and many richly illustrated texts too. Natural history and insects in particular were favourite topics, both in Britain and around the world. Jean-Henri Fabre's intimate observations of Provençal mantises and caterpillars, for example, won him readers from France to Australia, via Britain and India. Lightman has compiled figures on the bestselling books of the century, and finds that Brewer's *Guide to the Scientific Knowledge of Things Familiar* (1847) had had a combined print run of 195,000 by 1892. Wood's *Common Objects of the Country* (1858) had reached 86,000 by 1889. These books were frequently 'guides to', 'stories of', or even 'parables of' nature. When it came to entomology, insects often figured as 'fairy' intermediaries that inducted children into the world of nature but also, subtly, made a plea for a more free-spirited and imaginative approach to the subject than the perceived rigid empiricism of the scientific naturalists. Arabella Buckland, meanwhile, proposed the whole of science as a 'fairy-land'. Some of these books were overtly theistic, while others cultivated a vague admiration for the Creator that may have been expressed for the sake of convention. Other authors radically re-wrote evolution

as a feminist, socialist, or anarchist process. The 1870s onwards also saw a bloom of biological fiction, whose most famous practitioners were H. G. Wells and Grant Allen. Journals for learned readers prospered during this period, mixing elite science, philosophy, and fiction.

But although there was a substantial marketplace for science writing there were divisions within it, with men such as Huxley attempting to define a higher echelon of expertise that was limited to men, and often to other categories too, such as the secular middle classes. Understanding late nineteenth-century science as a story of professionalization is an approach that has gone in and out of favour with historians, but there is little doubt that some kind of distinctions of legitimacy were in play, and that Huxley was a major force in establishing them.

The development of nature on film very much followed through from the Victorian traditions of lecturing and optical display. The first scientific film, *Cheese Mites* (1903) quickly became implicated in pre-existing markets for natural history; microscope kits were repackaged to include a sample of the newly famous creatures so that buyers could recreate their cinematic viewing experiences at home. Nature, in the sense of living things, was absolutely the favourite early topic for factual reels. The American cinematographic entrepreneur Charles Urban laid out the value of the medium for science in a manifesto of 1907:

The time has now arrived when the equipment of every hospital, scientific laboratory, technical institute, college and private and public school is as incomplete without its moving picture apparatus as it would be without its clinical instruments, test tubes, lathes, globes, or maps.

Urban seems to have been serious about this (and his arguments read convincingly) but the booklet mostly functioned as a rhetorical ballast for entertainment-based sales. (Perhaps the US context explains why science seemed like a novel application for Urban; in his memoirs he recalled his earliest encounters with Edison's kinetoscope as occasions for viewing such items as 'Loie Fuller's Skirt Dance' and 'The Kiss'.) Almost immediately—as one might expect given established traditions in the demonstration of optical wizardry—impressive special techniques were developed for the filmic display of nature, such as microscopic lenses, time-lapse work, and colour. Urban's list of topics already filmed in 1907 includes the circulation of protoplasm in an amoeba, the life of bees, the birth of a crystal, and beriberi in Borneo. On moving to Britain, Urban employed a talented cameraman, Frank Percy Smith, who later struck out on his own as maker of the *Secrets of Nature* films. As the century went on, a number of naturalists toured Britain with their films and accompanying lectures, a phenomenon that echoed the Victorian circuit and competed with contemporary music halls and similar entertainments.

Opposite: T. H. Huxley poses casually yet authoritatively beside the dramatic blackboard. The skilfully—yet apparently effortlessly—drawn skull reflects the extremely high artistic abilities exhibited by successful science lecturers of the Victorian age.

F. Percy Smith (1880–1945) pioneered magnification and time-lapse techniques in nature filming, demonstrated in reels such as *The Birth of a Flower* (1910), *The Strength and Agility of Insects* (1911), and *Magic Myxies* (1931).

The nineteenth century, according to Lorraine Daston and Peter Galison, saw an increasing demand for 'objectivity' in science, by which was meant a mechanical recording of phenomena, uncorrupted by human mediation. In this sense, film might have been seen as yet another triumph. It was particularly useful for animal ethologists, who could slow down and review their observations of animal behaviour—early adopters of the technology included Karl von Frisch and Konrad Lorenz. But natural history was already losing prestige as most biology went into laboratories, and film's associations with popular theatre did not help its cause. A cinematic audience was not, in general, an authoritative witness for the knowledge they aimed to produce.

The filming of animals, arguably, enhanced the public's desire to see animals in a 'natural' setting, although this 'nature' was itself inevitably a construct of films' *mises en scène*. By the 1930s, London Zoo began to face criticism from visitors that cinemas showed animals in a more exciting fashion. Partly in response to this, and partly in response to criticism of caging practices, it (along with other world zoos) developed the moat and terrace enclosure system in the early twentieth century. It also joined forces with animal defence organizations (who in other circumstances were their critics) to lobby for the restraint of the film industry in rigging violent encounters between animals.

The coming of mass media

During the development of film, radio also emerged as a major medium for the communication of knowledge about nature. Like film, it had clear roots in earlier media, particularly the lecture and the book—but without the visual richness which audiences had come to demand. As such, it provided a home for the discussion of physics, evolution, and other more theoretical species of science which film-makers had been disinclined to explore. From the time of its incorporation in 1927, the BBC relied heavily on the prestige of science to secure its standing as an elite public-service institution. Ralph Desmarais has estimated that the percentage of talks dedicated to science varied from 5 to 15 per cent during the 1930s and 1940s. The public scientific intellectuals created through radio (by 1930 it was estimated that half of all households had a set) had a reach beyond anything previously possible. Far and away the most prolific of these, Gerald Heard, was not a professional scientist—but of the remainder, almost all were academic scientists, such as Julian Huxley, James Jeans, and J. B. S. Haldane. This generation was able to promulgate its models of science (university based, socially obligated) for a wide audience, although by the end of the 1930s, the influence of left-wingers at the BBC was considerably diminished. Besides lectures, the BBC also included science in programmes such as the *Brains Trust*; until 1930, *Children's Hour* featured the Zoo Man 'Uncle Leslie [Mainland]', formerly of the *Daily Mail*, who used a 'wireless pram... like a three-masted schooner', a wheeled contraption weighing three-quarters of a ton and trailing wire behind it, for his on-site broadcasts.

These public intellectuals of science also operated in the media of film and the written word. The 1920s saw a burst of self-education titles written by left-wing scientists, a trend which lasted through the 1940s and was complemented by the institution of the expert-written Pelican imprint of Penguin (1937–1984). Science was given presence in newspapers by Britain's first dedicated journalists in the field, Peter Ritchie Calder and J. G. Crowther. Scientific authors, and others in their circles, also helped to diversify the kinds of science shown on film, going beyond its starting point of natural history. During the 1930s and 1940s, a substantial number of documentaries were made celebrating technology and the virtues of a technologically planned future—something that seemed all too necessary for audiences caught between the hangover of the Great Depression and a looming Second World War. Telephones, aeroplanes, new towns, scientific healthcare: all of these things were praised and promised. Some of these films were sponsored by businesses, such as the Shell petroleum company; the British government paid for others, initially through the Empire Marketing Board, and subsequently (from 1933) through a film unit based in the General Post Office (GPO). During the war responsibility was passed to the Ministry of Information, which became the Central Office of Information in 1946.

For a time after the Second World War it seemed that the filmic vision of a technological, planned future was broadly shared by politicians, industrialists, and even consumers and workers. Its high point was the 'defiant modernism' of the Festival of Britain (1951). Science was at the heart of this festival. Although it was in many ways a trade fair like those of the nineteenth century and the colonial era (indeed it took place at the centenary of the Great Exhibition), it was also an invitation for the general public to participate, through consumption of a modernist aesthetic. Polymers (including plastic) were the material of choice, atoms and molecules the favoured motifs of decoration. The public was primed, aesthetically and patriotically, for the famous photograph celebrating Crick and Watson as discoverers of the molecular structure of DNA in 1953. Technology at the Festival was not so much to be admired as something that companies might trade (though it was that too), but as something that could be *bought by* individual consumers—as vacuum cleaners, toys, wallpapers.

Just as the Great Exhibition was the catalyst for the creation of the Science Museum, so the Festival of Britain provoked a re-thinking and refreshing of its aims. Under the directorship of Frank Sherwood Taylor (1950–1956) it attempted 'to learn from shop displays, advertising, posters, films and strip cartoons'. Over the following three decades its annual visitor figures would rise from under one million to over five million, whilst it continued to wrestle (as it still does) with the key questions of whether its business was to portray historic or current science, and whether its audience was adults or children.

Young people were particularly keen consumers of science. Science fiction, in books and magazines, celebrated the culture and aesthetic of space-age invention; indeed, there is a good case for saying that its imagination actually helped to drive it. The generation that began writing and reading in the 1920s and 1930s came through the war and went on to work in engineering and design, creating the infrastructure and goods that epitomized 'modernity'. For the post-war generation of youngsters, and particularly in the US, the nationalistic dimension of science and its fictions was refocused through the fight against familiar Cold War bogeys. There was even a sub-genre of atomic comics that rewrote the Manhattan Project as an object lesson in 'teamwork' and nuclear science as the search for energy sources.

Opposite: This wallpaper design from the Festival of Britain visualizes the molecular structure of insulin. Britain was, at that time, host to a number of chemists trying to work out such structures (most famously, DNA). In 1958, the British biochemist Frederick Sanger was awarded a Nobel Prize in chemistry for his successful elucidation of insulin, amongst other proteins.

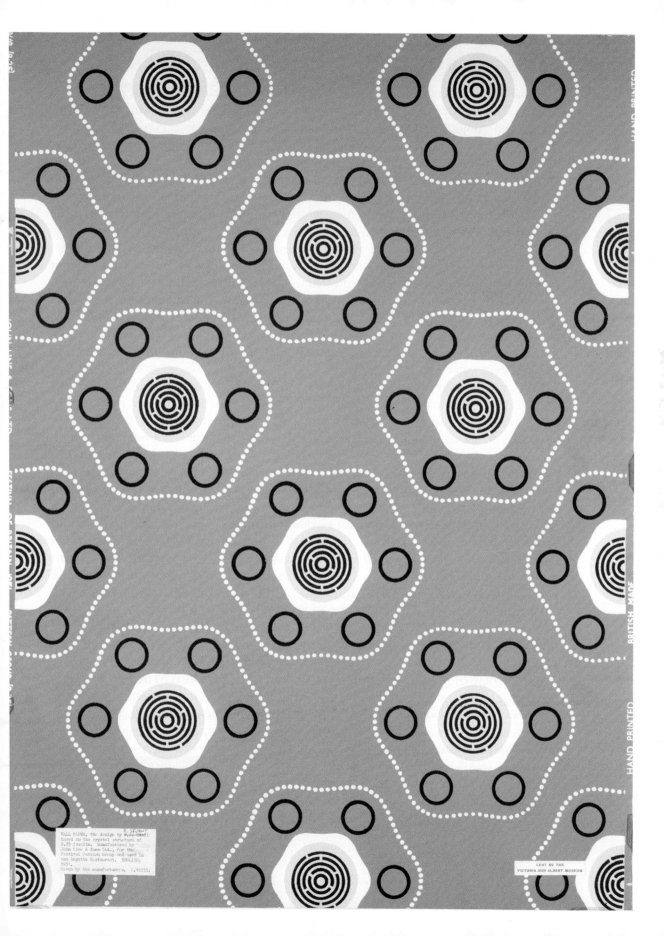

British participation in national science through consumption was provided by ERNIE, created by the ever-modern GPO as a display of 'computing' in 1957. Challenged to find a way to control investment, Chancellor Harold Macmillan came up with a scheme of government bonds which, rather than paying interest, would be randomly selected for cash prizes. A machine was developed by the GPO to perform these selections, built from recycled components of the wartime decryption machine Colossus—a practical choice that also had pleasing metaphoric overtones of the swords-into-ploughshares variety. ERNIE (Electronic Random Number Indicator Equipment) became a personified icon of mass culture and recipient of 'his' own fan-mail. His bonds were a popular gift for babies: a way of buying into their scientific, modern futures. ERNIE was not in fact a computer, working on physical rather than electronic properties, but its public presentation as the former was an essential part of its claim to be part of a truly modern system of citizenship. But this was not without its challenges; Emily Jane Roe has shown that there were tensions between the presentation of the machine as autonomously 'intelligent' and its answerability to human norms of fairness, and between the glamour of its presentation and its perceived scientificity.

Consumers could also buy televisions. Whereas entire streets gathered around single sets to watch the coronation of Elizabeth II in 1953, by the 1960s many households possessed one of their own. Accordingly, the long 1960s saw the birth of science on television, and, correspondingly, a major decline in non-fictional science cinema. The author of the guide to science at the Festival of Britain, Jacob Bronowski, would become an important public face for science on TV. Other scientific figures—both individually and institutionally—continued to challenge the BBC's autonomy in determining what kind of science to portray, as they had done during the war. But the BBC remained firm, actively discussing what its programming and principles should be, and sticking by them. '[A]s a foundation to our policy, we have firmly decided *that the broadcasting of science shall be in the hands of broadcasters.*' The BBC experimented with a wide range of topics, themes, and approaches during the 1950s and 1960s, with two of its longest-lasting series (in any genre) being established in the latter decade. The technophilic aspect of the Festival of Britain was given a lasting existence in the shape of *Tomorrow's World* (1965–2003), although by the 1980s the show had evolved a complicated mode of self-deprecation, with viewers relishing devices that were particularly unbelievable or, better yet, malfunctioned. A more serious display of nature (rather than technology) was provided by *Horizon* (1964–present). This abandoned the essentially lecture-based format of onscreen presenters for an invisible narrator and a selection of talking heads, and introduced a strong viewer-centred narrative as it took its audience on a journey of discovery and disclosure. Besides communicating science to adults, these programmes have arguably been even more significant in their communication of science to scientists-in-the making, enculturating young people into the tribe of science.

Televisual natural history co-evolved with zoo practices over the latter part of the twentieth century, coming to focus (in different ways) on conservation. London Zoo,

This must-see televisual moment of interspecies primate intimacy looks quite shocking today; touching is frowned upon for both its capacity for disease transmission, and for its power to disrupt social behaviours developed and maintained over many generations.

constrained by its original layout, was far behind its rivals around the world in terms of the scientific display of its collection, whether by classification or biogeography. However, it joined them in the latter part of the twentieth century in making conservation its asserted *raison d'être* and educational message. The standard information board on any given species now typically gives such information as size, location, number of offspring, and concludes with its human-posed threat. Adoption schemes, pioneered by London Zoo during the Second World War, now carry the moral kudos of personally saving a species.

David Attenborough's earliest TV programmes echoed an earlier era of zoos and were in fact made in conjunction with London Zoo, the eponymous patron of the first two series of *Zoo Quest* (1955–1961) and partial sponsors of the third. These programmes were remarkably Victorian in their narratives of discovery and capture, and in their inclusion of humans living in their biogeographical zones. In 1956 London Zoo transferred its allegiance to the commercial TV company Granada, producing the show *Zoo Time*. This, like the US predecessor *Zoo Parade* (1950–1957) confined itself to domestic footage.

Later in the century, as conservation came to the fore, it became normal to display nature on television as 'pristine', without human fingerprint. (This word was still used prominently in the promotional literature for Sebastião Salgado's photographic exhibition *Genesis* at the Natural History Museum in 2013). Thus, the most memorable moment of the seminal 1979 BBC series *Life on Earth*—arguably the climax of the whole series—came when Attenborough momentarily transgressed this boundary, physically touching a gorilla and speculating about the human condition and our place in the animal kingdom. Maintaining the distinction between humans and nature, the BBC's 2011 series *Frozen Planet* quarantined discussion of climate change to its final episode, which was moreover made optional to overseas broadcasters upon syndication. The exception to this untouched-by-human-hand presentation is the inclusion of a 'making-of' feature in most TV nature series and their accompanying DVDs, dating right back to *Life on Earth*. These extras perform the function of highlighting the skill of the programmes' makers, and in this sense at least they recall proudly self-conscious science shows right the way back to the mid-nineteenth century. On the other hand they also—through their incongruity and humour—reassert a distance between technology and nature.

It might seem as though the conservation impetus of zoo displays, predicated on human threats to biodiversity, is at odds with the pristine display of nature on television. However, they can be seen as two sides of the same coin. In both cases 'nature' is by definition pristine—the parts of the world that have nothing to do with humans. Humans are exterior to nature, meaning that they can either be removed from it (as in television) or pose rather superficial and soluble threats to it (zoos). Either way, this is not a promising model for the complex interactions of inorganic and organic, nature and culture, that form our world, or for the problems that face it.

In part through Attenborough's stint as controller and director of BBC TV channels, televisual science has become dominated by natural history. Around the world, the National Geographic and Animal Planet channels, and to a certain extent Discovery, carry such content. Other areas of televisual science are supplementary: mostly coverage of natural disasters and technology. Quality science programming for adults is mainly the preserve of public service channels (TV and radio).

Television has not seen off books about science; indeed glossy coffee-table volumes were and are produced to go alongside spectacular science series, particularly those concerning natural history. Books have, however, long since ceased to be a channel for the dissemination of new research or theories. The possible exception to this rule would be biology; *Sociobiology* and—to a lesser extent—*The Selfish Gene* were considered as general announcements of new perspectives. Stephen Hawking's *Brief History of Time* was a publishing sensation, although notoriously it was more often left lying around homes for visitors to notice than actually read. Science books have, like science TV, become chic again in the twenty-first century, regularly gracing the best-seller lists and with a wider and larger array of titles than has been seen for several decades—maybe even since the nineteenth century.

Science 2.0

Notwithstanding the continuing presence of science on TV, the mass medium par excellence of the late twentieth century, scientists felt under attack. The 1950s and long 1960s were generally regarded as a honeymoon period for science, supported by government and public. But towards the end of this time, and into the 1970s and 1980s, scientists perceived their funding decreased and their cultural status diminished, with a 'brain drain' of talented scientists going abroad. Lacking media savvy, they turned their propagandizing efforts fairly ineffectually towards the government (as, for example, Neil Calver and Ralph Desmarais have explored). In 1985 the Royal Society-commissioned Bodmer Report concluded that an enhanced public understanding of science was necessary, for reasons ranging from the idealistic (science as cultural achievement) to the pragmatic (an educated workforce; discerning voters and consumers). The Committee on the Public Understanding of Science (COPUS), set up the following year as a joint initiative of the RI, the British Association for the Advancement of Science, and the Royal Society provided funding and rewards for communicational initiatives. It was bolstered by government funding from 1993, and from 1995 Research Councils were charged with enhancing the public understanding of science as a part of their remit. COPUS is not generally regarded as having done a good job, tending to conceive of the public as ignorant, assuming that the problem was a lack of factual knowledge (rather than an understanding of how science works), and that more factual knowledge would automatically and uncritically result in a higher regard for science.

A variety of contests about natural knowledge in the 1990s, implicating scientists, authorities and consumers, helped to focus the issues at stake. In northern Europe, there were questions about the long-term safety implications of the Chernobyl explosion (1986); across Europe but especially in the UK there was a debate about the human transmissibility of Bovine Spongiform Encephalopathy; and around the world there were rapidly growing possibilities for genetic manipulation. In the US, the Smithsonian exhibition *Science in American Life* (1994), caused the American Physical Society amongst others to complain of 'profound dismay' over the show's 'portrayal of science that trivialize[d] its accomplishments and exaggerate[d] any negative consequences'. The London Science Museum had its subsidy for free visits removed; in the space of three years from 1987, annual numbers plummeted from regularly over three million to a little over a million. In tandem with this scepticism of the 1990s, a new communicational technology platform was developing that would characterize the next century: the Internet. The initial Internet medium was the web, which came into general use around the mid-nineties, at just the same time as these debates. To the champions of a certain kind of science, it seemed like a perfect storm. Homeopathy, Feng shui, and self-medication ruled; forums and discussion boards for conspiracy theories and bogus information bloomed. The web, where all opinions were (in its early days) given an even platform, seemed to instantiate the cultural crisis of science.

The medical profession quickly came to accommodate these alternative information sources, setting up authoritative medical guides of their own (whether from
commercial or state-based healthcare providers) and encouraging or at least tolerating
online patient communities that performed the soft end of medical care (sharing
experiences and encouragement). But the public's ability to participate in, or even
read, frontline research remained completely out of the question. The sorry case of
Andrew Wakefield, with his unproven and epidemiologically dangerous ideas about
immunization, was used ruthlessly over the following fifteen years as a cudgel with
which to beat non-expert participation in science. Commercial science organizations,
notably pharmaceutical and petrochemical companies, have stayed under the radar.
Although there have been critiques of their products and business practices, their
research remains largely invisible. The UK science journalist Ben Goldacre has fought
a long-running campaign to bring them within the idealistic norms of science, most
notably to oblige Big Pharma to publish the results of neutral or negative drug trials,
rather than selecting only positive ones for publication. Meanwhile, large non-profit
research institutions (Universities, CERN, NASA, etc.) have been obliged to beef up
their PR departments in order to dominate media streams that might otherwise be
flooded with others' accounts of their work—or worse yet, not mention them at all.
Despite complaints about media 'misrepresentation', many newspaper stories about
the latest discoveries are lifted virtually word for word from their press releases.

We are still in the early days of Internet media, and it remains an open question to
what extent their use will change the place of science in broader culture, and the
relationship of the general public with science. The first decade of the twenty-first
century has seen something of an embrace of science amongst a certain sector of the
young, fashionable, and educated, although it is too early to say whether this will be a
lasting trend. In the UK, popular comedians such as Bill Bailey and Dara O'Briain
make science a part of their business (the latter hosting his *Science Club* on BBC2). For
these broadly liberal comedians and their viewers, science functions as a cultural
bulwark against neo-conservatism (especially climate change denial) and fundamen-
talisms of both Christian and Muslim varieties. British and Australian 'Bright Clubs'
attract a similar audience ('bright' having been coined as an elective alternative for
'atheist') for comedy and science discussion—these somewhat resemble the grass-
roots (but non-comic) Café Scientifique movement, started in Leeds in 1998. Science
as decorative accomplishment has been advanced through its visibility in fine art since
the 1990s. The representations of nature thus produced have ranged from the frightening to the fond, from confrontational to elegant. The Wellcome Trust's massive
sponsorship of art/science endeavours (and latterly Google's through the Global
Science Gallery Network) has provided useful support and visibility for art/science,
but may be in danger of subordinating it to a celebratory science-communication
agenda. At the lighter end of the science spectrum, the Facebook group *I fucking love
science* reached almost ten million 'likes' in 2013, and scientists form the cast of the
(2007–) US sitcom *Big Bang Theory*—a genre in which fondness for, and identification
with, the central characters is judged as commercially essential. In general, the term

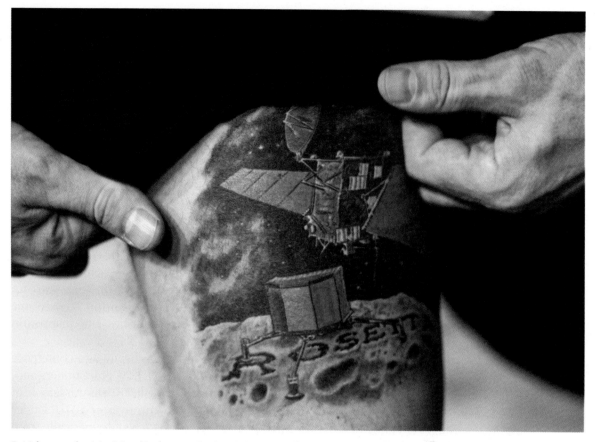

British astrophysicist Matt Taylor was Project Scientist on the Rosetta mission, landing the Philae probe on Comet 67P/Churyumov–Gerasimenko (the first time such a feat was achieved). His tattoos—of which this is one—were praised on social media, successfully keying into a trend of 'geek-chic'.

'geek' has been reclaimed as a badge of honour for these kinds of audiences. Originally connected with all things computer-ish, geekdom's fashionability has arisen in tandem with the consumer kudos of, above all, Apple devices. Science has ridden its coat-tails, with all kinds of T-shirts and mugs available to buy into membership of the geek tribe, and a thriving subculture of science/tech tattoos.

One much-discussed Internet-based phenomenon is the notion of 'citizen science', whereby ordinary people contribute their observations to large-scale research projects, or work on analysing large data sets such as in the classification project Galaxy Zoo. This can undoubtedly be fun and may serve as an introduction to science, but it might also seem to a historian little more than a fancy name for something that has been going on for centuries. Darwin, one might say, crowd-sourced his science from a network of correspondents sending him specimens and observations; this is simply the same thing facilitated through different media. One might also question sceptically the degree of participation in citizen science. Citizen scientists fulfil the tasks set of them, but it is hard to imagine the circumstances in which they might actually be allowed to critique the research agenda or set a new one. Mostly they provide cheap, or free, basic labour. Emergent open access schemes, making published research freely available, might enable this level of engagement in theory; it remains to be seen whether this will occur in practice.

Crowdfunding is currently touted as an important movement that can enable the public direction of research, whereby scientists raise the means for their research from public sources. The research only happens if the public votes with its wallets. This is much more talked about than it is actually practised, and the amounts of money involved are tiny compared to the full economic costings of serious research. The few success stories that can be highlighted serve to show how unusual the phenomenon is, requiring immense skill in manipulating new media.

Crowdfunding is often, and understandably, spoken of within the circles of tech-enthusiasts, including some prominent spokespeople who like to imagine that they personally embody the democratic mass of the Internet. In their discourse, crowd-funding elides with the practice of 'angel investment', whereby the wealthy Internet technology elite gamble their personal and companies' resources on tech start-up companies. These people are in an extraordinarily powerful position to advance their definitions of science and technology and their social identities; one institution that exemplifies them is TED.

TED stands for Technology, Entertainment, Design, and it was begun as a conference at which developers in those fields could showcase their ideas to fellow practitioners and, crucially, investors. Since it was bought by the self-made publisher Chris Anderson in 2001, TED has grown and grown, to a thirty million dollar turnover in 2010. Under Anderson's direction, topics covered by TED speakers have broadened dramatically from their original remit to encompass the sciences and arts. Perhaps TED's greatest triumph is the way that it manages to combine extraordinary exclusivity with a mythos of openness. Attendance at the main conference in 2013 cost a minimum of $7,500; becoming a TED patron cost $125,000 over five years. It is a networking opportunity of the most exclusive order. However, in 2006 TED began webcasting its talks free of charge, announcing its billionth view in November 2012. The public online accessibility of TED's talks enhances the warm glow of its patrons, basking in the knowledge that several million people have clicked to watch the clever and witty speaker that they met in person. The public, viewing online, can tag or select TED talks according one of a number of pre-determined labels—'persuasive, courageous, ingenious, fascinating, inspiring, beautiful, funny, inform-ative'—that necessarily affirm the brilliance of speakers and the taste of their patron-listeners. It is, in so many ways, an interesting reflection of such early nineteenth-century institutions as RI: industrialists practising a peculiar mixture of profit and philanthropy, linked by a genuine belief that all boats can be lifted by the new science. Like the purveyors of books and pamphlets from the steam-driven press, they believe that their media are not just media, but potent engines of scientific and social change. Likewise, both sets of investors treat science as tech-embedded and entrepreneurial, and, where not immediately applicable, as a cultured, decorative acquirement.

As the new tech-elite penetrate markets and cultures of emerging nations, they will take their vision of science with them. What does 'science' mean in the emerging nations of India, China, Nigeria, Brazil? Is it the late-Victorian ideal of disinterested knowledge, perpetuated through the twentieth century, or something closer to what the Society for the Encouragement of Arts, Manufactures, and Commerce imagined? The communication of 'nature', as opposed to nature–culture hybrids of technological prowess, may turn out to have been a fairly brief and localized historical phenomenon, lasting from the late nineteenth to the late twentieth century. The eighteenth-century version of science communication, a marketplace that did not trouble to distinguish between knowledge of nature and artful knowledge, may yet return. One notes with interest that the new science supplement of the British *Observer* newspaper is to be called *Observer Tech Monthly*. This title would have been hard to imagine fifty years ago, when technology derived from science, rather than, as here, 'tech' being the all-embracing term that includes knowledge of nature. But one could imagine that George Birkbeck, or Prince Albert, might have approved.

FURTHER READING

1. SCIENCE IN THE ANCIENT MEDITERRANEAN WORLD

Barton, T. *Ancient Astrology* (London: Routledge, 1994).

Bliquez, J., and R. Jackson. *Roman Surgical Instruments and Other Minor Objects in the National Archaeological Museum of Naples* (Mainz: von Zabern, 1994).

Clagett, M. *Ancient Egyptian Science: A Source Book*, 3 vols (Philadelphia: American Philosophical Society, 1989–1999).

Cuomo, S. *Ancient Mathematics* (London and New York: Routledge, 2007).

Evans, J. *The History and Practice of Ancient Astronomy* (New York: Oxford University Press, 1998).

Gillings, R. J. *Mathematics in the Time of the Pharaohs* (Cambridge, MA: Harvard University Press, 1972).

Hankinson, R. J., ed. *The Cambridge Companion to Galen* (Cambridge: Cambridge University Press, 2008).

Harley, J. B., and D. Woodward, eds. *The History of Cartography*, Vol. 1: *Cartography in Prehistoric, Ancient, and Medieval Europe and the Mediterranean* (Chicago and London: University of Chicago Press, 1987).

Jouanna, J. *Hippocrates*, trans. M. B. DeBevoise (Baltimore and London: Johns Hopkins University Press, 1999).

Kirk, J. S., and J. E. Raven, *The Presocratic Philosophers* (Cambridge: Cambridge University Press, 1960).

Landels, J. G. *Engineering in the Ancient World* (Berkeley: University of California Press, 1978).

Lindberg, D. C. *The Beginnings of Western Science: The European Scientific Tradition in Philosophical, Religious, and Institutional Context, Prehistory to A.D. 1450*, 2nd edn (Chicago: University of Chicago Press, 2007).

Lloyd, G. E. R. *Early Greek Science: Thales to Aristotle* (New York: Random House, 1970) and *Greek Science after Aristotle* (New York: Vintage, 2013).

Mattern, S. P. *Prince of Medicine: Galen in the Roman World* (Oxford and New York: Oxford University Press, 2013).

Neugebauer, O., and A. Sachs. *Mathematical Cuneiform Texts* (New Haven: American Oriental Society, 1945).

Riddle, J. M. *Dioscorides on Pharmacy and Medicine* (Austin: University of Texas Press, 1985).

Robson, E. *Mathematics in Ancient Iraq: A Social History* (Princeton: Princeton University Press, 2008).

Roller, D. W. *Eratosthenes' Geography* (Princeton and Oxford: Princeton University Press, 2010).

von Staden, H. *Herophilus: The Art of Medicine in Early Alexandria* (Cambridge: Cambridge University Press, 1989).

Some works of ancient science, philosophy of nature, and pseudo-science that are still engaging and accessible today (arranged roughly chronologically):

Hesiod. *Works and Days*, trans. Dorothea Wender in *Hesiod and Theognis* (Harmondsworth: Penguin, 1973).

Plato. *Timaeus*, trans. D. Lee in *Timaeus and Critias* (London: Penguin Classics, 1977).

Hippocrates. 'Nature of Man' and 'Aphorisms' trans. W. H. S. Jones in *Hippocrates*, Vol. 4 (Cambridge, MA and London: Cambridge University Press, 1931).

Aristotle. *On the Heavens*, trans. W. K. C. Guthrie (Cambridge, MA and London: Cambridge University Press, 1939).

Aratus. *Phaenomena*. The verse translation by A. Poochigian (Baltimore: Johns Hopkins University Press, 2010) is compact and enjoyable. The edition and translation by D. Kidd (Cambridge: Cambridge University Press, 1997) has a rich commentary.

Aristarchus of Samos. *On the Sizes and Distances of the Sun and Moon*, trans. T. L. Heath in *Aristarchus of Samos: The Ancient Copernicus* (Oxford: Oxford University Press, 1913) (for those who are not put off by geometry).

Geminos's Introduction to the Phenomena: *A Translation and Study of a Hellenistic Survey of Astronomy*, trans. J. Evans and J. L. Berggren (Princeton: Princeton University Press, 2006).

Lucretius. *The Nature of Things*, trans. A. E. Stallings (London and New York: Penguin, 2007).

Strabo. *Geography*, Books 1 and 2, ed. and trans. H. L. Jones (Cambridge, MA and London: Harvard University Press, 1959).

Pliny the Elder. *Natural History: A Selection*, trans. J. F. Healy (London: Penguin, 1991).

Ptolemy's Geography: An Annotated Translation of the Theoretical Chapters, trans. J. L. Berggren and A. Jones (Princeton and Oxford: Princeton University Press, 2000).

Ptolemy. *Tetrabiblos*, trans. F. E. Robbins (Cambridge, MA and London: Harvard University Press, 1940).

Martianus Capella. *The Marriage of Philology and Mercury*, trans. W. H. Stahl in *Martianus Capella and the Seven Liberal Arts*, Vol. 2 (New York: Columbia University Press, 1977).

2. SCIENCE IN ANCIENT CHINA

Cullen, C. *Astronomy and Mathematics in Ancient China: The Zhou bi suan jing* (Cambridge: Cambridge University Press, 1996).

Graham, A. C. *Yin-Yang and the Nature of Correlative Thinking* (Singapore: Institute of East Asian Philosophies, 1986).

Harper, D. *Early Chinese Medical Literature: The Mawangdui Medical Manuscripts* (London: Kegan Paul International, 1998).

Harper, D. 'Warring States Natural Philosophy and Occult Thought', in *The Cambridge History of Ancient China*, ed. Michael Loewe and Edward L. Shaughnessy (Cambridge: Cambridge University Press, 1999).

Kalinowski, M. *Cosmologie et divination dans la Chine ancienne: Le Compendium des cinq agents* (Paris: École Française d'Extrême-Orient, 1991).

Lloyd, G., and N. Sivin. *The Way and the Word: Science and Medicine in Early China and Greece* (New Haven: Yale University Press, 2002).

Needham, J. *Science and Civilisation in China*, Vol. 3: *Mathematics and the Sciences of the Heavens and the Earth* (Cambridge: Cambridge University Press, 1959).

Shaughnessy, E. *I Ching: The Classic of Changes* (New York: Ballantine Books, 1996).

Unschuld, P. *Huang Di nei jing su wen: Nature, Knowledge, Imagery in an Ancient Chinese Medical Text* (Berkeley: University of California Press, 2003).

Unschuld, P., and H. Tessenow. *Huang Di nei jing su wen: An Annotated Translation of Huang di's Inner Classic—Basic Questions* (Berkeley: University of California Press, 2011).

3. SCIENCE IN THE MEDIEVAL CHRISTIAN AND ISLAMIC WORLDS

Brentjes, Sonja, with Robert G. Morrison. 'Sciences in Islamic Societies', in *The New Cambridge History of Islam*, Vol. 4: *Islamic Cultures and Societies to the End of the Eighteenth Century*, ed. Robert Irwin (Cambridge: Cambridge University Press, 2010), 564–639.

Burnett, Charles. *Arabic into Latin in the Middle Ages: The Translators and Their Intellectual and Social Context* (Farnham, Surrey, UK: Ashgate/Variorum, 2009).

Caroti, Stefano, and Pierre Souffrin, eds. *La Nouvelle physique du XIVe siècle* (Florence: Olschki, 1997).

Courtenay, William J. *Changing Approaches to Fourteenth-Century Thought*, The Étienne Gilson Series 29 (Toronto: Pontifical Institute of Mediaeval Studies, 2007).

de Ridder-Symoens, Hilda, ed. *A History of the University in Europe*, Vol. 1: *Universities in the Middle Ages* (Cambridge: Cambridge University Press, 1992).

Edson, Evelyn, and E. Savage-Smith. *Medieval Views of the Cosmos: Picturing the Universe in the Christian and Islamic Middle Ages* (Oxford: Bodleian Library, 2004).

Endress, G. 'Die wissenschaftliche Literatur', in *Grundriss der arabischen Philologie*, ed. H. Gätje, Band 2: *Literaturwissenschaft* (Wiesbaden: Verlag Reichert, 1987), S. 400–506.

Gutas, Dimitri. *Greek Thought, Arabic Culture: The Graeco-Arabic Translation Movement in Baghdad and Early 'Abbasid Society* (London: Routledge, 1998).

Kaye, Joel. *Economy and Nature in the Fourteenth Century: Money, Market Exchange, and the Emergence of Scientific Thought* (Cambridge and New York: Cambridge University Press, 1998).

Lawn, Brian. *The Rise and Decline of the Scholastic 'Quaestio Disputata' with Special Emphasis on its Use in the Teaching of Medicine and Science* (Leiden: Brill, 1993).

McGinnis, Jon. 'Arabic and Islamic Natural Philosophy and Natural Science', in *Stanford Encyclopedia of Philosophy*, ed. Edward N. Zalta (Winter, 2013), <http://plato.stanford.edu/archives/win2013/entries/arabic-islamic-natural/>.

North, John. *Chaucer's Universe* (Oxford: Clarendon Press, 1988).

Siraisi, Nancy G. *Medieval and Early Renaissance Medicine* (Chicago: University of Chicago Press, 1990).

Sylla, Edith D., and Michael McVaugh, eds. *Texts and Contexts in Ancient and Medieval Science: Studies on the Occasion of John E. Murdoch's 70th Birthday* (Leiden: Brill, 1997).

Wisnovsky, Robert, Faith Wallis, Jamie Fumo, and Carlos Fraenkel, eds. *Vehicles of Transmission, Translation, and Transformation in Medieval Textual Culture* (Turnhout: Brepols, 2011).

4. SCIENCE IN THE PRE-MODERN EAST

Blue, G., P. Engelfried, and C. Jami. *Statecraft and Intellectual Renewal in Late Ming China: The Cross-Cultural Synthesis of Xu Guangqi (1562–1633)* (New York: Brill Academic Publishing, 2001).

Bray, F. 'Science, Technique, Technology: Passages between Matter and Knowledge in Imperial Chinese Agriculture'. *The British Journal for the History of Science* (41/3) 2008, 336.

Bray, F., and J. Needham. *Science and Civilisation in China*, Vol. 6: *Biology and Biological Technology* (Cambridge: Cambridge University Press, 1984).

Bray, F., G. Métailie, and V. Dorofeeva-Lichtmann. *Graphics and Texts in the Production of Technical Knowledge in China: The Warp and the Weft* (Leiden: Brill, 2007).

Chan Hok-lam. *Legitimation in Imperial China. Discussions under the Jurchen-Chin Dynasty (1115–1234)* (Seattle and London: University of Washington Press, 1984).

Elman, B. *From Philosophy to Philology: Intellectual and Social Aspects of Change in Late Imperial China* (Cambridge, MA: Harvard Univeristy Press, 1984).

Feng, J. *Chinese Architecture and Metaphor: Song Culture in the Yingzao Fashi Building Manual* (Honolulu: University of Hawai'i Press, 1993).

Furth, C., J. Zeitlin, and Hsiung Ping-chen. *Thinking with Cases: Specialist Knowledge in Chinese Cultural History* (Honolulu: University of Hawai'i Press, 2007).

Goldschmidt, A. M. *The Evolution of Chinese Medicine: Song Dynasty* (New York: Routledge, 2009).

Handlin Smith, J. *The Art of Doing Good: Charity in Late Ming China* (Berkeley: University of California Press, 2009).

Hanson, M. *Speaking of Epidemics in Chinese Medicine: Disease and the Geographic Imagination in Late Imperial China* (Milton Park, Abington, and New York: Routledge, 2011).

Ki Che Leung, A. 'Organized Medicine in Ming-Qing China: State and Private Medical Institutions in the Lower Yangzi Region'. *Late Imperial China* (8/1) 1987, 134–66.

Li Cho-ying. 'Contending Strategies: Collaboration among Local Specialists and Officials and Hydraulic Reform in the Late Fifteenth Century Lower Yangzi Delta'. *East Asian Sicence, Technology, Society (EASTS)* (2) 2009, 229–53.

Liu H. Picturing Yu. *Controlling the Flood: Technology, Ecology and Emperorship in Northern Song China*, in *Cultures of Knowledge: Technology in Chinese History*, ed. D. Schäfer (Leiden: Brill, 2012), 91–126.

Lloyd, G. E. R., and N. Sivin. *The Way and the Word: Science and Medicine in Early China and Greece* (New Haven: Yale University Press, 2002).

Needham, J. *Science and Civilization in China*, Vol. 3: *Mathematics and the Sciences of the Heavens and the Earth* (Cambridge: Cambridge University Press, 1954).

Needham, J., L. Wang, and J. D. de Solla Price. *Heavenly Clockwork: The Great Astronomical Clocks of Medieval China*, Antiquarian Horological Society (Cambridge: published in association with the Antiquarian Horological Society at the University Press, 1960).

Popplow, M. 'Models of Machines: A Missing Link between Early Modern Engineering and Mechanics', MPIWG-preprint (225) 2002.

Rossabi, M. *Khubilai Khan: His Life and Times* (Berkeley and Los Angeles: University of California Press, 1988).

Ruitenbeek, K. *Carpentry and Building in Late Imperial China: A Study of the Fifteenth-Century Carpenter's Manual Lu Ban Jing* (Leiden, New York, and Köln: Brill, Sinica Leidensia, 1993).

Savage-Smith, E. 'Attitudes Toward Dissection in Medieval Islam', *Journal of the History of Medicine and Allied Sciences* (50.1) 1995, 67–110.

Shinno, R. 'Medical Schools and the Temples of the Three Progenitors in Yuan China: A Case of Cross-Cultural Interactions', *Harvard Journal of Asiatic Studies* (67.1) 2007, 89–133.

Siebert, M. *Pulu: Abhandlungen und Auflistungen. Zur materiellen Kultur und Naturkunde im traditionellen China* (Wiesbaden: Harrasowitz, 2006).

Siebert, M. 'Making Technology History', in *Cultures of Knowledge: Technology in Chinese History*, ed. D. Schäfer (Leiden: Brill, 2012), 253–82.

Sivin, N. 'Science and Medicine in Imperial China—State of the Field', *The Journal of Asian Studies* (47.1) 1988, 43–4.

Sivin, N. *Science in Ancient China: Researchers and Reflections* (Brookfield, VT: Variorum, Ashgate Publishing, 1995).

Sun, X., and Y. Han. 'The Northern Song State's Financial Support for Astronomy', *East Asian Science, Technology, Medicine* (38) 2013.

Tucker, M. E., and John Berthrong. *Confucianism and Ecology: The Interrelation of Heaven, Earth, and Humans* (Cambridge, MA: Harvard University Press, 1998).

Tze-ki Hon. *The Yijing and Chinese Politics: Classical Commentary and Literati Activism in the Northern Song Period, 960–1127* (New York: State University of New York Press, 2004).

Van Dalen, B. 'Islamoc and Chinese Astronomy under the Mongols: A Little-Known Case of Transmission', in *From China to Paris: 2000 Years Transmission of Mathematical Ideas*, ed. Y. Dold-Samplonius, J. Dauben, M. Folkerts, and B. van Dalen (Stuttgart: Franz Steiner Verlag, 2002), 327–56.

Vogel, H. U., G. Dux, and M. Elvin. *Concepts of Nature: A Chinese-European Cross-Cultural Perspective, Conceptual History and Chinese Linguistics* (Leiden and Boston: Brill, 2010).

5. THE SCIENTIFIC REVOLUTION

Biagioli, Mario. *Galileo, Courtier: The Practice of Science in the Culture of Absolutism* (Chicago: University of Chicago Press, 1993).

Cohen, H. Floris. *The Scientific Revolution: A Historiographical Inquiry* (Chicago: University of Chicago Press, 1994).

Cohen, H. Floris. *How Modern Science Came into the World: Four Civilizations, One 17th-century Breakthrough* (Amsterdam: Amsterdam University Press, 2010).

Dear, Peter. *Discipline and Experience: The Mathematical Way in the Scientific Revolution* (Chicago: University of Chicago Press, 1995).

Fantoli, Annibale. *Galileo: For Copernicus and for the Church*, trans. G. V. Coyne (Vatican City State and Notre Dame: Vatican Observatory Publications/University of Notre Dame Press, 1994).

Frank, Robert G. *Harvey and the Oxford Physiologists: Scientific Ideas and Social Interaction* (Berkeley: University of California Press, 1980).

Funkenstein, Amos. *Theology and the Scientific Imagination from the Middle Ages to the Seventeenth Century* (Princeton: Princeton University Press, 1986).

Gaukroger, Stephen. *Descartes: An Intellectual Biography* (Oxford: Clarendon Press, 1995).

Grant, Edward. *Planets, Stars, and Orbs: The Medieval Cosmos, 1200–1687* (Cambridge: Cambridge University Press, 1994).

Hahn, Roger. *The Anatomy of a Scientific Institution: The Paris Academy of Sciences, 1666–1803* (Berkeley, University of California Press, 1971).

Harrison, Peter. *The Bible, Protestantism, and the Rise of Natural Science* (Cambridge: Cambridge University Press, 2001).

Harrison, Peter. *The Fall of Man and the Foundations of Science* (Cambridge: Cambridge University Press, 2009).

Henry, John. *The Scientific Revolution and the Origins of Modern Science*, 3rd edn (Basingstoke: Palgrave Macmillan, 2008).

Hunter, Michael. *Establishing the New Science: The Experience of the Early Royal Society* (Woodbridge, Suffolk: Boydell Press, 1989).

Moran, Bruce T. *Distilling Knowledge: Alchemy, Chemistry, and the Scientific Revolution* (Cambridge, MA: Harvard University Press, 2005).

Newman, William R. *Atoms and Alchemy: Chymistry and the Experimental Origins of the Scientific Revolution* (Chicago: University of Chicago Press, 2006).

Park, Katharine, and Lorraine Daston, eds. *The Cambridge History of Modern Science*, Vol. 3: *Early Modern Science* (Cambridge: Cambridge University Press, 2006).

Pumfrey, Stephen. *Latitude and the Magnetic Earth* (Cambridge: Icon Books, 2002).

Rossi, Paolo. *Francis Bacon: From Magic to Science* (London: Routledge and Kegan Paul, 1968).

Shapin, Steven, and Simon Schaffer. *Leviathan and the Air-Pump: Hobbes, Boyle, and the Experimental Life* (Princeton: Princeton University Press, 1985).

Smith, Pamela H. *The Body of the Artisan: Art and Experience in the Scientific Revolution* (Chicago: University of Chicago Press, 2004).

Webster, Charles. *The Great Instauration: Science, Medicine and Reform 1626–1660* (London: Duckworth, 1975).

Westfall, R. S. *Never at Rest: A Biography of Isaac Newton* (Cambridge: Cambridge University Press, 1980).

Westman, Robert S. *The Copernican Question: Prognostication, Skepticism, and Celestial Order* (Berkeley: University of California Press, 2011).

Wilson, Catherine. *The Invisible World: Early Modern Philosophy and the Invention of the Microscope* (Princeton: Princeton University Press, 1995).

6. ENLIGHTENMENT SCIENCE

Clark, William, Jan Golinski, and Simon Schaffer, eds. *The Sciences in Enlightened Europe* (Chicago and London: University of Chicago Press, 1999).

Darnton, Robert. *Mesmerism and the End of the Enlightenment in France* (Cambridge, MA: Harvard University Press, 1968).

Delbourgo, James. *A Most Amazing Scene of Wonders: Electricity and Enlightenment in Early America* (Cambridge, MA: Harvard University Press, 2006).

Delbourgo, James, and Nicholas Dew, eds. *Science and Empire in the Atlantic World* (New York and London: Routledge, 2008).

Egerton, Judy. *Wright of Derby* (London: Tate Gallery, 1990).

Fara, Patricia. *Newton: The Making of Genius* (Basingstoke: Macmillan, 2002).

Feingold, Mordechai. *The Newtonian Moment: Isaac Newton and the Making of Modern Culture* (New York and Oxford: The New York Public Library / Oxford University Press, 2004).

Fontenelle, Bernard le Bovier de. *Conversations on the Plurality of Worlds*, ed. Nina Rattner Gelbart, trans. H. A. Hargreaves (Berkeley, Los Angeles, and Oxford: University of California Press, 1990).

Gascoigne, John. *Joseph Banks and the English Enlightenment* (Cambridge: Cambridge University Press, 1994).

Golinski, Jan. *Science as Public Culture: Chemistry and Enlightenment in Britain, 1760–1820* (Cambridge: Cambridge University Press, 1992).

Heilbron, J. L. *Electricity in the 17th and 18th Centuries: A Study of Early Modern Physics* (Berkeley: University of California Press, 1979).

Jackson, Joe. *The World on Fire: A Heretic, an Aristocrat, and the Race to Discover Oxygen* (New York: Viking Penguin, 2005).

Koerner, Lisbet. *Linnaeus: Nature and Nation* (Cambridge, MA and London: Harvard University Press, 1999).

Pancaldi, Giuliano. *Volta: Science and Culture in the Age of Enlightenment* (Princeton: Princeton University Press, 2003).

Parrish, Susan Scott. *American Curiosity: Cultures of Natural History in the Colonial British Atlantic World* (Chapel Hill, NC: University of North Carolina Press for the Ohomundro Institute of Early American History and Culture, 2006).

Porter, Roy, ed. *The Cambridge History of Science*, Vol. 4: *Eighteenth-Century Science* (Cambridge: Cambridge University Press, 2003).

Porter, Roy, and Mikuláš Teich, eds. *The Enlightenment in National Context* (Cambridge: Cambridge University Press, 1981).

Riskin, Jessica. *Science in the Age of Sensibility: The Sentimental Empiricists of the French Enlightenment* (Chicago and London: University of Chicago Press, 2002).

Schiebinger, Londa. *The Mind Has No Sex? Women in the Origins of Modern Science* (Cambridge, MA and London: Harvard University Press, 1989).

Schofield, Robert E. *The Enlightenment of Joseph Priestley: A Study of His Life and Work from 1733 to 1773* (University Park, PA: The Pennsylvania State University Press, 1997).

Schofield, Robert E. *The Enlightened Joseph Priestley* (University Park, PA: The Pennsylvania State University Press, 2004).

Smith, Bernard. *European Vision and the South Pacific*, 2nd edn (New Haven and London: Yale University Press, 1985).

Spary, E. C. *Utopia's Garden: French Natural History from Old Regime to Revolution* (Chicago and London: University of Chicago Press, 2000).

Stewart, Larry. *The Rise of Public Science: Rhetoric, Technology, and Natural Philosophy in Newtonian Britain, 1660–1750* (Cambridge: Cambridge University Press, 1992).

Terrall, Mary. *The Man Who Flattened the Earth: Maupertuis and the Sciences in the Enlightenment* (Chicago and London: University of Chicago Press, 2002).

Withers, Charles W. J. *Placing the Enlightenment: Thinking Geographically about the Age of Reason* (Chicago: University of Chicago Press, 2007).

Woolf, Harry. *The Transits of Venus: A Study of Eighteenth-Century Science* (Princeton: Princeton University Press, 1959).

7. EXPERIMENTAL CULTURES

Alder, Ken. *Engineering the Revolution: Arms and the Enlightenment in France, 1763–1815* (Princeton: Princeton University Press, 1997).

Buchwald, Jed Z. *The Creation of Scientific Effects: Heinrich Hertz and Electric Waves* (Chicago: University of Chicago Press, 1994).

Buchwald, Jed Z., and Andrew Warwick, eds. *Histories of the Electron: The Birth of Microphysics* (Cambridge, MA: MIT Press, 2001).

Bud, R. F., and G. K. Roberts. *Science versus Practice: Chemistry in Victorian Britain* (Manchester: Manchester University Press, 1984).

Cahan, David. *An Institute for an Empire: The Physikalisch-Technische Reichsanstalt, 1871–1918* (Cambridge: Cambridge University Press, 1989).

Canales, Jimena. *A Tenth of a Second: A History* (Chicago: University of Chicago Press, 2009).

Fox, Robert. *The Savant and the State* (Baltimore: Johns Hopkins University Press, 2012).

Galison, Peter. *How Experiments End* (Chicago: University of Chicago Press, 1987).

Galison, Peter, and Bruce Hevly, eds. *Big Science: The Growth of Large-Scale Research* (Stanford: Stanford University Press, 1992).

Gooday, Graeme. *The Morals of Measurement: Accuracy, Irony and Trust in Late Victorian Electrical Practice* (Cambridge: Cambridge University Press, 2004).

Helmholtz, Hermann von. *Science and Culture: Popular and Philosophical Essays*, edited with an introduction by David Cahan (Chicago: University of Chicago Press, 1995).

Hoddeson, Lillian, Adrienne Kolb, and Catherine Westfall. *Fermilab: Physics, the Frontier and Megascience* (Chicago: University of Chicago Press, 2008).

Hughes, Jeff. *The Manhattan Project: Big Science and the Atomic Bomb* (London: Icon Books, 2002).

Hunt, Bruce. *The Maxwellians* (Ithaca, NY: Cornell University Press, 1991).

James, Frank, ed. *The Correspondence of Michael Faraday*, 6 vols (London: Institution of Electrical Engineers, 1991–2011).

Levitt, Theresa. *The Shadow of Enlightenment: Optical and Political Transparency in France, 1789–1848* (Oxford: Oxford University Press, 2009).

Morus, Iwan Rhys. *Frankenstein's Children: Electricity, Exhibition and Experiment in early Nineteenth-Century London* (Princeton: Princeton University Press, 1998).

Morus, Iwan Rhys. *When Physics Became King* (Chicago: University of Chicago Press, 2005).

Navarro, Jaume. *A History of the Electron* (Cambridge: Cambridge University Press, 2012).

Nye, Mary Jo. *Before Big Science: The Pursuit of Modern Chemistry and Physics, 1800–1940* (Cambridge, MA: Harvard University Press, 1999).

Smith, Crosbie. *The Science of Energy: A Cultural History of Energy Physics in Victorian Britain* (London: Athlone Press, 1998).

Traweek, Sharon. *Beamtimes and Lifetimes: The World of High Energy Physicists* (Cambridge, MA: Harvard University Press, 1989).

Tresch, John. *The Romantic Machine: Utopian Science and Technology after Napoleon* (Chicago: University of Chicago Press, 2012).

Wheaton, Bruce. *The Tiger and the Shark: The Empirical Roots of Wave-Particle Duality* (Cambridge: Cambridge University Press, 1983).

Wise, M. Norton, ed. *The Values of Precision* (Princeton: Princeton University Press, 1995).

8. EXPLORING NATURE

Allen, David. *The Naturalist in Britain: A Social History* (Princeton: Princeton University Press, 1976).

Ballard, Robert. *The Eternal Darkness: A Personal History of Deep-Sea Exploration* (Princeton: Princeton University Press, 2000).

Barber, Lynn. *The Heyday of Natural History, 1820–1870* (London: Jonathan Cape, 1980).

Bogerhoff-Mulder, Monique, and Wendy Logsdon. *I've Been Gone Far Too Long: Field Trip Fiascos and Expedition Disasters* (Muskegon, MI: RDR Books, 1996).

Driver, Felix. *Geography Militant: Cultures of Exploration and Empire* (Oxford: Blackwell Publishers, 2001).

Endersby, Jim. *Imperial Nature: Joseph Hooker and the Practices of Victorian Science* (Chicago: University of Chicago Press, 2008).

Goodall, Jane. *Through a Window: Thirty Years with the Chimpanzees of Gombe* (London: Weidenfeld and Nicholson, 1990).

Gupta, Akhil, and James Ferguson, eds. *Anthropological Locations: Boundaries and Grounds of a Field Science* (Berkeley: University of California Press, 1997).

Humboldt, Alexander. *Personal Narrative of a Journey to the Equinoctial Regions of the New Continent* (abridged) (London: Penguin, 1995).

Jardine, Nick, James Secord, and Emma Spary, eds. *Cultures of Natural History* (Cambridge: Cambridge University Press, 1996).

Lever, Christopher. *They Dined on Eland: The Story of the Acclimatisation Societies* (London: Quiller Press, 1992).

Lindsay, Debra. *Science in the Subarctic: Trappers, Traders and the Smithsonian Institution* (Washington: Smithsonian Institution Press, 1993).

Livingstone, David. *Putting Science in Its Place: Geographies of Scientific Knowledge* (Chicago: University of Chicago Press, 2003).

Lloyd, Christopher. *Mr Barrow of the Admiralty: A Life of Sir John Barrow, 1786–1848* (London: Collins, 1970).

Lyell, Charles. *Principles of Geology* (London: Penguin, 1997).

Klingle, Matthew. 'Plying Atomic Waters: Lauren Donaldson and the "Fern Lake Concept" of Fisheries Management', *Journal of the History of Biology* 31 (1998) 1–32.

Kohler, Robert. *All Creatures: Naturalists, Collectors and Biodiversity, 1850–1950* (Princeton: Princeton University Press, 2006).

Rees, Amanda. *The Infanticide Controversy: Primatology and the Art of Field Science* (Chicago: University of Chicago Press, 2009).

Rozwadowski, Helen. *Fathoming the Ocean: The Discovery and Exploration of the Deep Sea* (Cambridge, MA: Belknap Press, 2005).

Stafford, Robert. *Scientist of Empire: Sir Roderick Murchison, Scientific Exploration and Victorian Imperialism* (Cambridge: Cambridge University Press, 1989).

Van Riper, A. Bowdoin. *Men among the Mammoths: Victorian Science and the Discovery of Human Prehistory* (Chicago: University of Chicago Press, 1993).

Vetter, Jeremy. 'Cowboys, Scientists and Fossils: The Field Site and Local Collaboration in the American West', *Isis* (99) 2008, 273–303.

Vetter, Jeremy, ed. *Knowing Global Environments: New Historical Perspectives on the Field Sciences* (New Brunswick, NJ: Rutgers University Press, 2011).

9. THE MEANING OF LIFE

Allen, Garland E. *Life Science in the Twentieth Century* (New York: Willey, 1975).

Bowler, Peter J. *The Mendelian Revolution: The Emergence of Hereditarian Concepts in Modern Science and Society* (London and Baltimore: Athlone/Johns Hopkins University Press, 1989).

Bowler, Peter J. *Monkey Trials and Gorilla Sermons: Evolution and Christianity from Darwin to Intelligent Design* (Cambridge, MA: Harvard University Press, 2007).

Bowler, Peter J. *Evolution: The History of an Idea*, 25th anniversary edn (Berkeley: University of California Press, 2009).

Bowler, Peter J., and John V. Pickstone, eds. *The Cambridge History of Science*, Vol. 6: *The Modern Biological and Earth Sciences* (Cambridge: Cambridge University Press, 2009).

Browne, Janet. *Charles Darwin: Voyaging* (London: Jonathan Cape, 1995).

Browne, Janet. *Charles Darwin: The Power of Place* (London: Jonathan Cape, 2002).

Darwin, Charles. *On the Origin of Species by Means of Natural Selection: Or, the Preservation of Favoured Races in the Struggle for Life* (London: John Murray, 1859; facsimile reprint introduced by Ernst Mayr, Cambridge, MA: Harvard University Press, 1964).

Desmond, Adrian. *The Politics of Evolution: Morphology, Medicine and Reform in Radical London* (Chicago: University of Chicago Press, 1989).

Desmond, Adrian, and James R. Moore. *Darwin's Sacred Cause: How a Hatred of Slavery Shaped Darwin's Views on Human Evolution* (London: Allen Lane, 2009).

Gayon, Jean. *Darwinism's Struggle for Survival: Heredity and the Hypothesis of Natural Selection* (Cambridge: Cambridge University Press, 1998).

Henig, Robin Marantz. *A Monk and Two Peas: The Story of Gregor Mendel and the Discovery of Genetics* (London: Wiedenfeld and Nicolson, 2000).

Keller, Evelyn Fox. *The Century of the Gene* (Cambridge, MA: Harvard University Press, 2000).

Kevles, Daniel. *In the Name of Eugenics: Genetics and the Uses of Human Heredity* (New York: Knopf, 1985).

Kohn, David, ed. *The Darwinian Heritage* (Princeton: Princeton University Press, 1985).

Morus, Iwan Rhys. *Shocking Bodies: Life, Death and Electricity in Victorian England* (Stroud: History Press, 2011).

Olby, Robert C. *The Origins of Mendelism*. Rev. edn (Chicago: University of Chicago Press, 1985).

Richards, Robert J. *Darwin and the Emergence of Evolutionary Theories of Mind and Behavior* (Chicago: University of Chicago Press, 1987).

Secord, James A. *Victorian Sensation: The Extraordinary Publication, Reception, and Secret Authorship of 'Vestiges of the Natural History of Creation'* (Chicago: University of Chicago Press, 2000).

Smith, Roger. *The Fontana/Norton History of the Human Sciences* (London and New York: Fontana/W.W. Norton, 1997).

Stepan, Nancy. *The Idea of Race in Science: Great Britain, 1800–1960* (London: Macmillan, 1982).

Sulloway, Frank. *Freud: Biologist of the Mind: Beyond the Psychoanalytic Legend* (London: Burnett Books, 1979).

Watson, James D. *The Double Helix* (New York: Atheneum, 1968).

10. MAPPING THE UNIVERSE

Becker, Barbara. *Unravelling Starlight: William and Margaret Huggins and the Rise of the New Astronomy* (Cambridge: Cambridge University Press, 2011).

Crowe, Michael J. *The Extraterrestrial Life Debate, 1750–1900* (Cambridge: Cambridge University Press, 1986).

Crowe, Michael J. *Modern Theories of the Universe: Herschel to Hubble* (New York: Dover, 1994).

DeVorkin, David, and Robert W. Smith. *Hubble: Imaging Space and Time* (Washington, DC: National Geographic, 2011).

Dick, Steven J. *Discovery and Classification in Astronomy: Controversy and Consensus* (Cambridge: Cambridge University Press, 2013).

Gingerich, Owen, ed. *The General History of Astronomy*, Vol. 4: *Astrophysics and Twentieth Century Astronomy to 1950, Part A* (Cambridge: Cambridge University Press, 2010).

Gregory, Jane. *Fred Hoyle's Universe* (Oxford: Oxford University Press, 2005).

Herrmann, Dieter B. *The History of Astronomy from Herschel to Hertzsprung* (Cambridge: Cambridge University Press, 1984).

Hirsch, Richard F. *Glimpsing an Invisible Universe: The Emergence of X-Ray Astronomy* (Cambridge: Cambridge University Press, 1985).

Hoskin, Michael. *Discoverers of the Universe: William and Caroline Herschel* (Princeton: Princeton University Press, 2011).

Kragh, Helge. *Cosmology and Controversy: The Historical Development of Two Theories of the Universe* (Princeton: Princeton University Press, 1996).

Lankford, John, ed. *History of Astronomy: An Encyclopedia* (New York: Garland, 1997).

North, John. *Cosmos: An Illustrated History of Astronomy and Cosmology* (Chicago: University of Chicago Press, 2008).

Smith, Robert W. *The Expanding Universe: Astronomy's 'Great Debate', 1900–1931* (Cambridge: Cambridge University Press, 1982).

Tucker, Wallace, and Riccardo Giacconi. *The X-Ray Universe* (Cambridge, MA: Harvard University Press, 1987).

Wright, Helen. *Explorer of the Universe: A Biography of George Ellery Hale* (Woodbury, NY: American Institute of Physics, 1994).

11. THEORETICAL VISIONS

Bohr, Neils. *Atomic Physics and Human Knowledge* (Mineola, NY: Dover, 2010).

Galison, Peter. *Image and Logic* (Chicago: University of Chicago Press, 1997).

Galison, Peter. *Einstein Clock's and Poincare's Maps: Empires of Time* (New York: W.W. Norton, 2004).

Greene, Brian. *The Elegant Universe: Superstrings, Hidden Dimensions, and the Quest for the Ultimate Theory* (New York: W.W. Norton, 2010).

Heisenberg, Werner. *Physics and Philosophy*. Penguin Modern Classics (Harmondsworth: Penguin, 2000).

Herschel, John F. *Preliminary Discourse on the Study of Natural Philosophy* (London: Forgotten Books, 2012).

Holton, Gerald. *Thematic Origins of Scientific Thought: Kepler to Einstein* (Cambridge, MA: Harvard University Press, 1988).

Hughes, Jeff. *The Manhattan Project: Big Science and the Atom Bomb* (London: Icon, 2003).

Hunt, Bruce. *Pursuing Power and Light: Technology and Physics from James Watt to Albert Einstein* (Baltimore: Johns Hopkins University Press, 2010).

Kaiser, David. *How the Hippies Saved Physics: Science, Counterculture, and the Quantum Revival* (New York: W.W. Norton, 2012).

Kragh, Helge. *Quantum Generations: A History of Physics in the Twentieth Century* (Princeton: Princeton University Press, 2002).

McCormmach, Russell. *Night Thoughts of a Classical Physicist* (Cambridge, MA: Harvard University Press, 1991).

Morus, Iwan Rhys. *When Physics Became King* (Chicago: University of Chicago Press, 2005).

Smith, Crosbie. *The Science of Energy* (Cambridge: Cambridge University Press, 1998).

Smolin, Lee. *The Trouble with Physics: The Rise of String Theory, the Fall of a Science, and What Comes Next* (Boston and New York: Houghton Mifflin, 2007).

Snyder, Laura J. *The Philosophical Breakfast Club: Four Remarkable Friends who Transformed Science and Changed the World* (New York: Broadway Books, 2011).

Toulmin, Stephen, ed. *Physical Reality: Philosophical Essays on Twentieth Century Physics* (New York: Harper, 1970).

Warwick, Andrew. *Masters of Theory: Cambridge and the Rise of Mathematical Physics* (Chicago: University of Chicago Press, 2003).

Weinberg, Steven. *Dreams of a Final Theory: The Scientist's Search for the Ultimate Laws of Nature* (New York: Vintage, 1994).

12. COMMUNICATING SCIENCE

Allen, David Elliston. *The Naturalist in Britain: A Social History* (Princeton: Princeton University Press, 1976).

Allen, David Elliston. *Books and Naturalists* (London: HarperCollins UK, 2010).

Altick, Richard D. *The Shows of London* (Cambridge, MA: Harvard University Press, 1978).

Baratay, Eric, and Elisabeth Hardouin-Fugier. *Zoo: A History of Zoological Gardens in the West* (London: Reaktion Books, 2004).

Boon, Timothy. *Films of Fact: A History of Science in Documentary Films and Television* (London: Wallflower Press, 2008).

Bowler, Peter J. *Science for All: The Popularization of Science in Early Twentieth-Century Britain* (Chicago: University of Chicago Press, 2009).

Broks, Peter. *Media Science before the Great War* (London: MacMillan Press Ltd, 1996).

Early Popular Visual Culture, special issue: *Victorian Science and Visual Culture* (10.1) 2012.

Fyfe, Aileen. *Science and Salvation: Evangelical Popular Science Publishing in Victorian Britain* (Chicago: University of Chicago Press, 2004).

Fyfe, Aileen, and Bernard Lightman, eds. *Science in the Marketplace: Nineteenth-Century Sites and Experiences* (Chicago: University of Chicago Press, 2007).

Gates, Barbara T. *Kindred Nature: Victorian and Edwardian Women Embrace the Living World* (Chicago: University of Chicago Press, 1998).

Gates, Barbara T., and Ann B. Shteir, eds. *Natural Eloquence: Women Reinscribe Science* (Madison: University of Wisconsin Press, 1997).

Knight, David. *Public Understanding of Science: A History of Communicating Scientific Ideas* (London: Routledge, 2006).

LaFollette, Marcel Chotkowski. *Science on American Television: A History* (Chicago: University of Chicago Press, 2013).

Lightman, Bernard, ed. *Victorian Science in Context* (Chicago: University of Chicago Press, 2008).

Lightman, Bernard. *Victorian Popularizers of Science: Designing Nature for New Audiences* (Chicago: University of Chicago Press, 2009).

Mitman, Gregg. *Reel Nature: America's Romance with Wildlife on Film* (Seattle: University of Washington Press, 2013).

Morris, Peter John Turnbull. *Science for the Nation: Perspectives on the History of the Science Museum* (Basingstoke: Palgrave Macmillan, 2010).

Morus, Iwan Rhys. *Frankenstein's Children: Electricity, Exhibition, and Experiment in Early-Nineteenth-Century London* (Princeton: Princeton University Press, 1998).

Pyenson, Susan Sheets. *Cathedrals of Science: The Development of Colonial Natural History Museums during the Late Nineteenth Century* (Montreal: McGill-Queen's University Press, 1988).

Ritvo, Harriet. *The Animal Estate: The English and Other Creatures in the Victorian Age* (Cambridge, MA: Harvard University Press, 1989).

Schaffer, Simon. 'A Science Whose Business Is Bursting: Soap Bubbles as Commodities in Classical Physics', in *Things That Talk: Object Lessons from Art and Science*, ed. Lorraine J. Daston (New York: Zone Books, 2004), 147–94.

Secord, James A. *Victorian Sensation: The Extraordinary Publication, Reception, and Secret Authorship of Vestiges of the Natural History of Creation* (Chicago: University of Chicago Press, 2000).

Stewart, Larry. *The Rise of Public Science: Rhetoric, Technology, and Natural Philosophy in Newtonian Britain, 1660–1750* (Cambridge: Cambridge University Press, 1992).

White, Paul. *Thomas Huxley: Making the 'Man of Science'* (Cambridge: Cambridge University Press, 2003).

Willis, Martin. *Vision, Science and Literature, 1870–1920: Ocular Horizons* (London: Pickering and Chatto, 2011).

Yanni, Carla. *Nature's Museums: Victorian Sciences and the Architecture Of Display* (Princeton: Princeton Architectural Press, 2005).

PICTURE ACKNOWLEDGEMENTS

150	Science Museum/Science & Society Picture Library
152	Science Photo Library/ROYAL ASTRONOMICAL SOCIETY
154	Getty Images/Willem and Joan Blaeu
156	Getty Images/De Agostini Picture Library
160	Mary Evans/Iberfoto
164	Shutterstock/ Delpixel
166	Wellcome Library, London
170	Large oval dish moulded in relief with reptiles and fish (earthenware), Palissy, Bernard (1509–90)/Louvre-Lens, France/Bridgeman Images
172	Getty Images/Print Collector
176	Getty Images/DEA/G. NIMATALLAH
178	Scene of Witches (oil on canvas), Rosa, Salvator (1615–73)/Collection of Earl Spencer, Althorp, Northamptonshire, UK/Bridgeman Images
185	Science Museum/Science & Society Picture Library
186	The Orrery'. c.1766 (oil on canvas), Wright of Derby, Joseph (1734–97)/Derby Museum and Art Gallery, UK/Bridgeman Images
188	Madame du Chatelet-Lomont (oil on canvas), Tour, Maurice Quentin de la (1704–88) (after)/Private Collection/Bridgeman Images
190	Science Museum/Science & Society Picture Library
192	Wellcome Library, London
194	Photo, The Philadelphia Museum of Art/Art Resource/Scala, Florence
196	Sanders of Oxford
198	Science Photo Library/NATURAL HISTORY MUSEUM, LONDON
201	Omai, Joseph Banks and Dr Daniel Solander, c.1775 (oil on canvas), Parry, William (1742–91)/National Museum Wales/Bridgeman Images
203	Planisphere of the Earth from the Most Recent Astronomical Observations by James Cassini, copperplate printed in Amsterdam by Frans Halma, after 1696/De Agostini Picture Library/ Bridgeman Images
206	Marco Simola
208	Wellcome Library, London
216	Wellcome Library, London
217	Getty Images/Science & Society Picture Library
220	Author's collection
222	Glow Images
227	J. Goldiner, Berlin - http://www.zeno.org - Zenodot Verlagsgesellschaft mbH
228	University of Cambridge
229	Getty Images/Oxford Science Archive/Print Collector
232	Wellcome Library, London
235	National Library of Medicine (NLM)
237	Wellcome Library, London
242	Everett Collection Historical/Alamy Stock Photo
244	ZUMA Press, Inc./Alamy Stock Photo
247	culture-images/Lebrecht
249	Oxford University Museum of Natural History
251	Geographie der Pflanzen in den Tropen-Ländern, Louis Bouquet after Alexander von Humboldt, Schönberger and Turpin (1807)/Morn/Public Domain/Wikipedia Commons
255	© Royal Geographical Society

INDEX